The Rise of Birds

SANKAR CHATTERJEE

225 MILLION YEARS OF EVOLUTION

The Rise of Birds

THE JOHNS HOPKINS UNIVERSITY PRESS BALTIMORE AND LONDON

To Sibani, with affection and appreciation for her support

Printed in the United States of America on acid-free paper

06 05 04 03 02 01 00 99 98   5 4 3 2

The Johns Hopkins University Press
2715 North Charles Street
Baltimore, Maryland 21218
The Johns Hopkins Press Ltd., London

LIBRARY OF CONGRESS CATALOGING-IN-PUBLICATION DATA

Chatterjee, Sankar.
The rise of birds : 225 million years of evolution / Sankar Chatterjee.
p.   cm.
Includes bibliographical references (p.   ) and index.
ISBN 0-8018-5615-9 (alk. paper)
1. Birds—Evolution. 2. Birds, Fossil. 3. Birds—Flight. I. Title.
QL677.3.C48 1997
598.13'8—dc21   97-9474
CIP

A catalog record for this book is available from
the British Library.

Contents

Foreword

I am sure that May 21, 1986, will remain among the most memorable days of my career. On my way to Albuquerque, New Mexico, I had stopped in Lubbock, Texas, to study the Triassic archosaurs collected by Sankar Chatterjee from the Dockum beds of West Texas. I expected to see fossils of the large-headed carnivore *Postosuchus*, the armored herbivore *Desmatosuchus*, and the crocodile-like parasuchians. But that Wednesday afternoon, as I was in the lab taking notes on some mundane bit of basal archosaur skull, Sankar shyly interrupted and invited me to have a look at some intriguing new fossils. He related how they had been discovered with a jackhammer and initially identified as a baby theropod dinosaur, perhaps something like *Coelophysis*. He now thought that the fossils were, in fact, not those of a conventional dinosaur, but rather those of that peculiar sort of dinosaur we call bird—a Triassic bird more than 75 million years older than *Archaeopteryx*! Because I had been working on avian skull evolution, Sankar solicited my opinions of the fossils, and I agreed that many of the features were very birdlike. I left Lubbock feeling that there was a very good chance that those fossils, which he had informally dubbed *Protoavis texensis*, represented the remains of a Triassic bird.

Later that summer, the National Geographic Society, who funded the project, issued the press release that ignited the storm of controversy that persists to this day.

The present volume should only help fan the flames. Although it may be missed by the casual peruser (or hasty reviewer), this book is about much more than *Protoavis*, offering exciting and novel insights into the origin of avian flight and the evolution of numerous functional systems that stand largely independent of *Protoavis*. It is fair to say, on the other hand, that no work on birds by Sankar Chatterjee can be separated from *Protoavis*. It is probably worthwhile to devote some space to examining why this group of fossils has been so controversial. My intent here is to take a cold look not only at *Protoavis* but also at the criticisms leveled at it.

From the beginning, the fossils themselves have been contentious, and this is to be expected whenever a fossil with such important implications is discovered. No feather impressions were preserved, which is neither surprising given the geology of the locality nor unusual given that most avian fossils lack feather impressions. Nevertheless, it would be interesting to know how different the fossils' reception would have been had such impressions been preserved. Ironically—and this point is often overlooked—although the type of preservation precludes traces of feathers, the specimens can be (and have been) prepared free of the rock matrix and thus are available for study from all sides. It is actually because such spectacular fossils as *Archaeopteryx* preserve beautiful feathers that their bones will forever remain partially obscured, embedded in their stone slabs.

The fossils of *Protoavis* often are dismissed as being a chimera, that is, a mixture of several animals. Although Chatterjee has steadfastly maintained that the taphonomy of the site supports the association into single individuals, the more important issue is whether *any* of the bones are avian, because even if only *one* of the dozens of bones are actually avian, then there really is a Triassic bird. So the chimera complaint comes up empty.

It is sometimes mentioned that the fossils are too poorly preserved to be diagnostic. It is true that some of the specimens are not all that good and could be almost anything. Yet some of the pieces, such as the braincase, are well enough preserved to be diagnostic, and virtually all of Chatterjee's identifications are understandable. In sum, it is clear that the fossil material of *Protoavis* requires a fair amount of interpretation, and opinions may differ sharply as to what the fossils actually represent. Nevertheless, it is equally clear that the fossils cannot be summarily dismissed. They cannot just go away.

Yet, of course, it is not just the nature of the fossil specimens that has caused the controversy but also their supposed significance and implications. If *Protoavis* is indeed a bird, then it is by far the oldest bird, unseating *Archaeopteryx* from its venerable position. It is fair to suggest that some measure of the opposition to *Protoavis* relates to the challenge posed, as it were, to *Archaeopteryx. Archaeopteryx* has been such an important player in the debates on not just the origin of birds and of avian flight but also evolution in general that it has achieved an almost legendary, even iconic, status; thus, there will always be some resistance to any competing fossil. There is an interesting irony here, however, in that Chatterjee's phylogenetic hypothesis on the origin of birds does *nothing* to challenge either the position of *Archaeopteryx* as the most basal bird or the prevailing orthodoxy of the origin of birds from dromaeosaurid coelurosaurian theropods. According to Chatterjee's analysis, *Protoavis* is just another avian fossil that slots in above *Archaeopteryx* with no other major effects on tree topology. Thus, given that *Protoavis* really does not require a drastic overhaul of prevailing phylogenetic hypotheses, why all the fuss?

Although *Protoavis* does not affect the *topology* of the cladogram, it has a huge effect on the *timing* of cladogenesis. That is, it requires not only that the divergence of birds from other theropods be in the Triassic but also that much of the divergence of the major clades of theropods be in the Triassic. It must be remembered that there are not many dinosaurs *of any kind* older than *Protoavis* (which is early Norian in age). Thus, much of theropod cladogenesis must be telescoped into a very short period of time in the Triassic, and we should reasonably expect to find Triassic representatives of the ornithomimid, tyrannosaurid, troodontid, and dromaeosaurid clades, among others. The theropod hypothesis for the origin of birds has always had a vexing "time problem" in that the Late Jurassic *Archaeopteryx* predates the Early Cretaceous *Deinonychus* by some 30–40 million years; but this gap is not too egregious and there are reports of even older dromaeosaurid remains that shrink the gap down to 20 million years or less. With *Protoavis* in the picture, however, the time problem balloons to a gap of 95–115 million years!

There are other aspects to the controversy, but these are the major points. Ultimately, each scientist should study the specimens firsthand and come to his or her own conclusions. Few if any of the criticisms noted above have been adequately explored in the scientific literature, although there has been considerable sniping to the popular press. Moreover, such esteemed paleornithologists as Evgeny Kurochkin and D. Stephan Peters have regarded *Protoavis* as a bird in their published accounts of early avian evolution. Having studied the fossils in detail on two occasions subsequent to that spring day in 1986, I can confirm that there are indeed some features that are suggestive of avian status, such as the heterocoelous vertebra, the braincase pneumaticity, and the quadrate (if that identification is correct), yet others are problematic, such as the hand and ankle. Better fossils will certainly clarify the situation, perhaps affirming the avian status of *Protoavis* or perhaps showing that it is some kind of nonavian theropod. In either case, *Protoavis* is both interesting and important.

Although the furor around *Protoavis* is sure to dominate the attention given to this book, *Protoavis* is not the major focus of *The Rise of Birds.* Sankar Chatterjee has spent more than a decade studying the early evolution of birds, and he provides concise accounts of all the important taxa. Especially useful are the discussions of the evolution of the various anatomical systems (e.g., the temporal region of the skull, the hand, the ankle), many of which make little or no reference to *Protoavis*, and a thoughtful analysis of the importance of heterochrony in avian evolution. Also, Chatterjee provides an excellent and detailed analysis of the end Cretaceous extinctions, presenting the data assembled by him and his colleagues for the Shiva Impact Crater in India, which, along with the more famous Chicxulub Crater in Mexico, may hold important clues for the extinction of the conventional dinosaurs as well as many groups of Cretaceous birds. Yet Chatterjee does not end his story with the terminal Cretaceous extinctions; he also examines the "rise of birds" in the Cenozoic world and provides perhaps the most up-to-date cladistic classification of birds. He concludes with an engaging discussion of the continued interactions of humans and birds and of birds' destiny in our hands.

Perhaps the most exciting aspect of this book is the discussion of the origin of avian flight. Traditionally, the hypothesis that birds derive from maniraptoran theropods has been tightly linked to the cursorial origin of avian flight, that is, flight evolving from the ground up. In fact, some workers have viewed a theropod origin of birds and a "trees down" origin of flight as totally incompatible. Chatterjee rejects this notion and instead joins the minority (of which I count myself) who argue that flight in birds arose from small, arboreal, dromaeosaurid-like theropods. But Chatterjee has produced the most sophisticated and detailed model to date, with fairly convincing mechanical analyses and the first real account of the climbing adaptations of dromaeosaurids. This discussion may well wind up being the book's greatest contribution in that it informs us about not only birds but also the functional anatomy of nonavian theropods. Moreover, it is the first formulation of the arboreal theory that draws on the actual morphology of known animals (rather than a vague hypothetical "proavis"). Thus, many of the hypotheses in Chatterjee's model (e.g., that the stiffened tails of small dromaeosaurids functioned as a woodpecker-like prop) make specific anatomical predictions that can be tested through mechanical modeling and comparative analysis.

Avian ancestry and the origin of flight probably will remain as contentious as they always have been. The discovery of new fossils has been a key element of the debate, with *Archaeopteryx*, *Euparkeria*, *Sphenosuchus*, and *Deinonychus* each spawning new notions and moving the debate to a different place. Interestingly, *Protoavis* cannot make this claim. In some ways, this is as it should be if, as Chatterjee argues, *Protoavis* is a more derived bird than *Archaeopteryx* and hence too far removed to be as relevant. *Protoavis* has remained, however, in a rather strange limbo where many scientists have simply ignored it. There has never been a serious, published analysis of the fossils by anyone other than Chatterjee. Perhaps one of the most positive outcomes stemming from the publication of this book will be broader interest in and closer scrutiny of these important fossils. In any case, *The Rise of Birds* surely will be viewed by history as a critical document in the debate on early avian evolution.

LAWRENCE M. WITMER

Preface

In the Hindu epic *Ramayana,* we find a poignant story of resurrection. The fiery sage Gautama curses his beautiful wife, Aholya, and turns her to stone with these words: "You shall remain thus for thousands of years. Only when Rama, the son of Dasharatha, passes this way will you be restored to life." Years pass. Gautama's hermitage has long been abandoned and decays with age, choked by creepers and swallowed by rain forest. Aholya waits alone in the ruins, a stone among crumbling stones. One day, after many millennia, a very young prince Rama comes across the old ashram and sees the petrified image of Aholya, waiting. In reverence, he touches her feet. Immediately, he sees tears where there once were no eyes, flesh where there had been rock. Aholya is alive once more, free from her terrible curse.

Paleontologists have long pursued petrified bones, much like the epic hero Rama. To them fossils are million-year-old Aholyas waiting to be discovered and resurrected. They put muscle and skin on these bones, color them with human passions, and give them life. They recreate the denizens of a lost world in proper chronological and evolutionary context. This book unfolds the story of *Protoavis texensis*—"first bird, from

Texas." *Protoavis* predates *Archaeopteryx* by 75 million years, pushing the emergence of birds back to the Triassic period. Being the world's oldest bird, *Protoavis* provides a new view of avian origins. When I picked up two partial skeletons of *Protoavis* out of the Triassic redbeds of Texas in the summer of 1983, I did not realize that they would generate so much controversy and excitement among scientists. Soon after, avian paleontology went through its own renaissance, as more Mesozoic birds were found during a span of fifteen years than had been discovered during the entire preceding century.

Dinosaurs are so popular that we often neglect their even more fascinating relatives that are still among us. Birds, the true living dinosaurs, deserve considerable respect as successful vertebrates that have evolved, adapted, and survived over a period of 225 million years. The rise of birds in the Pangean world, their flight refinement and global diversification during the continental breakup, their decline at the end of the Cretaceous period, and their explosive radiation during the Cenozoic era are some of the greatest events in the history of vertebrates. Yet their early history is obscure and just beginning to unfold. During the seventy years since the appearance of Gerhard Heilmann's monumental work *The Origin of Birds*, substantive progress has been made in the study of bird fossils and the understanding of the theropod-bird transition. No current book deals exclusively with Mesozoic birds, which compose the first two-thirds of avian history. *The Rise of Birds* is an effort to fill this void. In this book I have attempted a readable and comprehensive account of the early history of the radiation of birds in the shadow of dinosaurs; of their anatomy, function, and evolutionary trends; of the origin of flight; of fossil eggs and embryos, feathers and footprints; of the distribution of birds in the Mesozoic period; of classification and phylogenetic relationships; and of the evolutionary crisis at the end of the Cretaceous. I have also included the explosive evolution of birds during the Tertiary period and their association with humans during the Quaternary to complete my picture of the long odyssey of birds. This book is aimed at an audience beyond the confines of the paleontological community. I give detailed accounts of specific research topics that are currently highly contested or thought provoking.

I am indebted to many colleagues for help, advice, stimulating discussions, and useful insights: Walter J. Bock, Larry D. Martin, Lawrence M. Witmer, Nicholas Hotton, Storrs L. Olson, Alan Feduccia, Erich Weber, J. M. Starck, Gerhard Mickoleit, Wolfgang Maier, Paul Bühler, D. Stephan Peters, Alick D. Walker, E. N. Kurochkin, Luis M. Chiappe, P. L. Zusi, and M. Kent Rylander. The manuscript was reviewed at various stages by Lawrence M. Witmer, James E. Barrick, and Soumya Chatterjee. I gratefully acknowledge their valuable suggestions and editorial skill. The illustrations were rendered by Michael W. Nickell, who gave precious pictorial life to my text. The majority of illustrations were drawn from the actual specimens or by redrawing and sometimes combining drawings found in the original papers describing the material. For permission to use previously published material, I am grateful to the authors and publishers concerned. I acknowledge all of the writers whom I have quoted for their wisdom and inspiration. I thank many of my graduate students for assistance in the field, especially Bryan J. Small and J. Bruce Moring, who serendipitously exposed two delicate *Protoavis* skeletons with a jackhammer. I thank R. C. Miller and Jack Kirkpatrick for allowing access to their property. I am especially grateful to John T. Montford, Donald R. Haragan, Gary Edson, Jane Winer, and Richard E. Peterson of Texas Tech University for their continued support and encouragement of my research. Over the years the research was supported by grants from the National Geographic Society, Smithsonian Institution, Texas Tech University, and University of Tübingen.

Working with the professionals of the Johns Hopkins University Press has been a thoroughly enjoyable and rewarding experience. Richard T. O'Grady initiated the project; David B. Weishampel of the Johns Hopkins University provided the moral and intellectual support; Robert Harington, Douglas Armato, and Ginger Berman led this book to completion. I am most grateful to Linda Forlifer for improving the style and clarity of my writing and to Julie McCarthy for careful editorial assistance. My sincere thanks to all for their patience and guidance.

From *Protoavis* to Pigeon

CHAPTER 1

Surely there is nothing very wild or illegitimate in the hypothesis that the phylum of the Class of Aves has its foot in the Dinosaurian Reptiles—that these, passing through a series of such modifications as are exhibited in one of their phases by Compsognathus, have given rise to [birds].

—Thomas H. Huxley, "On the Animals Which Are Most Nearly Intermediate between the Birds and Reptiles," 1868

Birds in flight symbolize spirits released from the bondage of gravity. From the day that humans first looked up at the skies, birds have summoned a sense of wonder and mystery, enchanting our earth-bound ancestors with their freedom and song. They fly where they please and when they please. The power of flight has opened up to birds a multilayered network of aerial highways and byways, enabling them to reach any place on our planet. Birds are the most successful terrestrial vertebrate, abundant in both numbers of species and populations. Today they live on every continent and occupy virtually all available ecological niches. About 300 billion birds, over 9,000 species, now inhabit the earth, as compared to 3,000 species of amphibians, 6,000 species of reptiles, and 4,100 species of mammals (Gill 1990).

The Genealogy of Birds

Although birds are one of the best-known groups of living vertebrates, their origin, evolution, and early adaptive radiation are poorly documented in the fossil record. The rarity of bird fossils is generally attributed to the extreme fragility, lightness, pneumaticity, and general smallness of their bones. Moreover, the living habits of birds are not conducive to preservation of specimens, as most species prefer arboreal habitat. Except for a few solitary fossils, we have gained essentially no new knowledge of Mesozoic birds during the past century. *Archaeopteryx lithographica* from the Upper Jurassic Solnhofen Limestone of Germany held center stage in avian evolution, being regarded by most researchers as the oldest and most primitive of known birds. Seventy million years after *Archaeopteryx* appeared, birds such as *Hesperornis* and *Ichthyornis,* from the Late Cretaceous of Kansas, had evolved into essentially modern forms, leaving no clues as to their reptilian heritage. It is no surprise that the Mesozoic has been referred to as the Dark Ages of avian history.

During the last fifteen years, the situation has changed dramatically as more and more fossils of Mesozoic birds have been unearthed from different sites around the globe. These Mesozoic birds vary widely in size and possess a wide range of body shapes and ecological adaptations. The new discoveries and the application of cladistics have inspired novel ideas about the antiquity of birds, their evolution, and their phylogenetic relationships. These data have triggered a renaissance in avian paleontology.

Birds, which took to the air during the age of the dinosaurs, are generally classified as Aves, a class separate from the rest of the vertebrates, such as fish, amphibians, reptiles, and mammals. Beauty, grace, feathers, and aerial prowess conceal the true identity and heritage of birds. What is the phylogenetic position of birds among vertebrates? Who were their immediate evolutionary ancestors? These questions have perplexed evolutionary biologists since the time of Darwin. Darwin's friend and champion, Thomas Huxley (1867, 1868a, 1868b, 1870), presented a radical proposal that birds are merely glorified reptiles. He classified reptiles and birds in the same group, Sauropsida. He reasoned that, if birds had not been so outstandingly successful in their aerial adaptation and speciation and if they had remained a relatively small group like pterosaurs, they would now be regarded as an order of reptiles, not a separate class of vertebrates. Huxley was impressed with the stunning similarities in bipedal posture, erect gait, and mesotarsal ankle joint between *Archaeopteryx* and *Compsognathus,* a contemporary theropod dinosaur from the Solnhofen Limestone. For several decades, the famed Eichstätt specimen of *Archaeopteryx* was mistakenly identified as a juvenile individual of *Compsognathus,* indicating how alike these two genera are (Wellnhofer 1974). The fact that theropods and birds always walked bipedally is intriguing and may indicate a common evolutionary history, Huxley argued. Bipedalism is a rare evolutionary event in the history of vertebrates and requires a great deal of balancing and coordination. In birds we see the culmination of this coordination and proprioception. Marsh (1877, 1880) embraced Huxley's proposal of a theropod-bird link when he described Cretaceous toothed birds such as *Hesperornis* and *Ichthyornis.*

In 1926, however, Gerhard Heilmann (1926) swept away the hypothesis of the theropod ancestry of birds in his influential book *The Origin of Birds.* He argued that such small theropods as coelurosaurs, however birdlike in skeletal morphology, could not be considered the ancestors of birds because they lacked clavicles, from which birds derive their furcula. Heilmann was guided by Dollo's law of irreversibility: evolution does not backtrack to recover characteristics. For birds to evolve from theropods, the lost furcula would have to reappear. Faced with such a paradox, Heilmann made a very reasonable suggestion. He argued that both birds and dinosaurs had evolved from a common ancestor. He sought the common ancestor among bipedal "pseudosuchian thecodonts," such as *Euparkeria* and *Ornithosuchus,* which retained the clavicle. From this ancestral stock, two great evolutionary lineages diverged, one leading to dinosaurs and the other to birds. For the next fifty years, Heilmann's pseudosuchian theory enjoyed wide acceptance among evolutionary biologists, with some new variations, such as the crocodilian connection of birds (Walker 1972; Martin 1985).

After almost a century, John Ostrom (1969, 1973, 1976a, 1985a, 1991) revived Huxley's theropod origin of birds and suggested that small theropods, such as dromaeosaurs, come very close indeed to the ancestry of birds. Huxley emphasized mesotarsal ankle structure as the key to the theropod-bird relationship, whereas Ostrom pointed out unique wrist morphology—the semilunate carpal—as the common link between dromaeosaurs and birds. Dromaeosaur fossils are known from the Cretaceous of North America and Asia. These little theropods had slashing toe-claws and were agile and formidable predators, which probably hunted in packs. One of the famous dromaeosaurs from Mongolia, *Velociraptor,* became the undisputed star in Steven Spielberg's movie *Jurassic Park.* In spite of their mean appearances, dromaeosaurs offer important insights into the origin of birds. Ostrom presented an impressive array of skeletal similarities between *Archaeopteryx* and dromaeosaurs, such as a long coracoid, long arms equipped with three fingers, a swivel wrist joint, long slender hind legs, a three-toed foot, and a stiff tail. Except for feathers, *Archaeopteryx* would look like a small dromaeosaur, Ostrom argued. The striking resemblance between *Archaeopteryx* and dromaeosaurs must reflect a common descent, not evolutionary convergence. He emphasized the recent discovery of clavicles in some coelurosaurs, which countered Heilmann's principal objection to a theropod ancestry of birds.

Although Ostrom suggested that birds were the direct descendants of dromaeosaurs, he used traditional classification to separate these two groups; he placed the birds in their own class, Aves, and dinosaurs in Class Reptilia, contrary to Huxley's grouping. However, his former student, Robert T. Bakker (1975), concluded that birds are actually theropod dinosaurs, not just descended from them. Like Huxley, Bakker argued that birds should not have their own separate class because they fly. Bats fly, too, but they are still considered mammals. Birds are as much dinosaurs as bats are mammals.

To understand the bird's place in the dinosaur family tree, we must know the interrelationships of dinosaurs. Dinosaurs began their evolutionary history as small carnivores during the Late Triassic. They became diversified within a relatively short period and then ruled the earth for approximately 160 million years. During that time they adapted to a wide range of conditions and environments and became very successful. All known dinosaurs are divided into two major groups on the basis of pelvic structure: Saurischia and Ornithischia. The Saurischia contains two subgroups: the plant-eating, mostly quadrupedal sauropodomorphs and the carnivorous, bipedal theropods. The Ornithischia includes several subgroups of herbivorous dinosaurs—armored thyreophorans (such as stegosaurs and ankylosaurs), and horned plus duck-billed cerapodans (such as ceratopsians and ornithopods) (fig. 1.1A). The theropods are the most spectacular of all dinosaurs and are linked to the ancestry of birds.

The theropod-bird relationship is strengthened with the application of cladistic analysis. Cladistics differs from older methods of biological classification by using the distribution of evolutionary novelties, called *shared derived characters* or *synapomorphies.* A clade is a group of animals that share uniquely evolved features and therefore a common ancestry. Recognition of a clade or monophyletic group by synapomorphies is the most important step in cladistic hypothesis. Willie Hennig, the German entomologist, first formalized the cladistic method in 1966. During the past two decades, cladistics has been used extensively to infer the phylogenetic relationships among different groups of plants and animals. Using this technique, Jacques Gauthier (1986) provided the first detailed hypothesis of theropod relationships on the basis of skeletal morphology; he recognized several monophyletic groups, or clades, in a hierarchi-

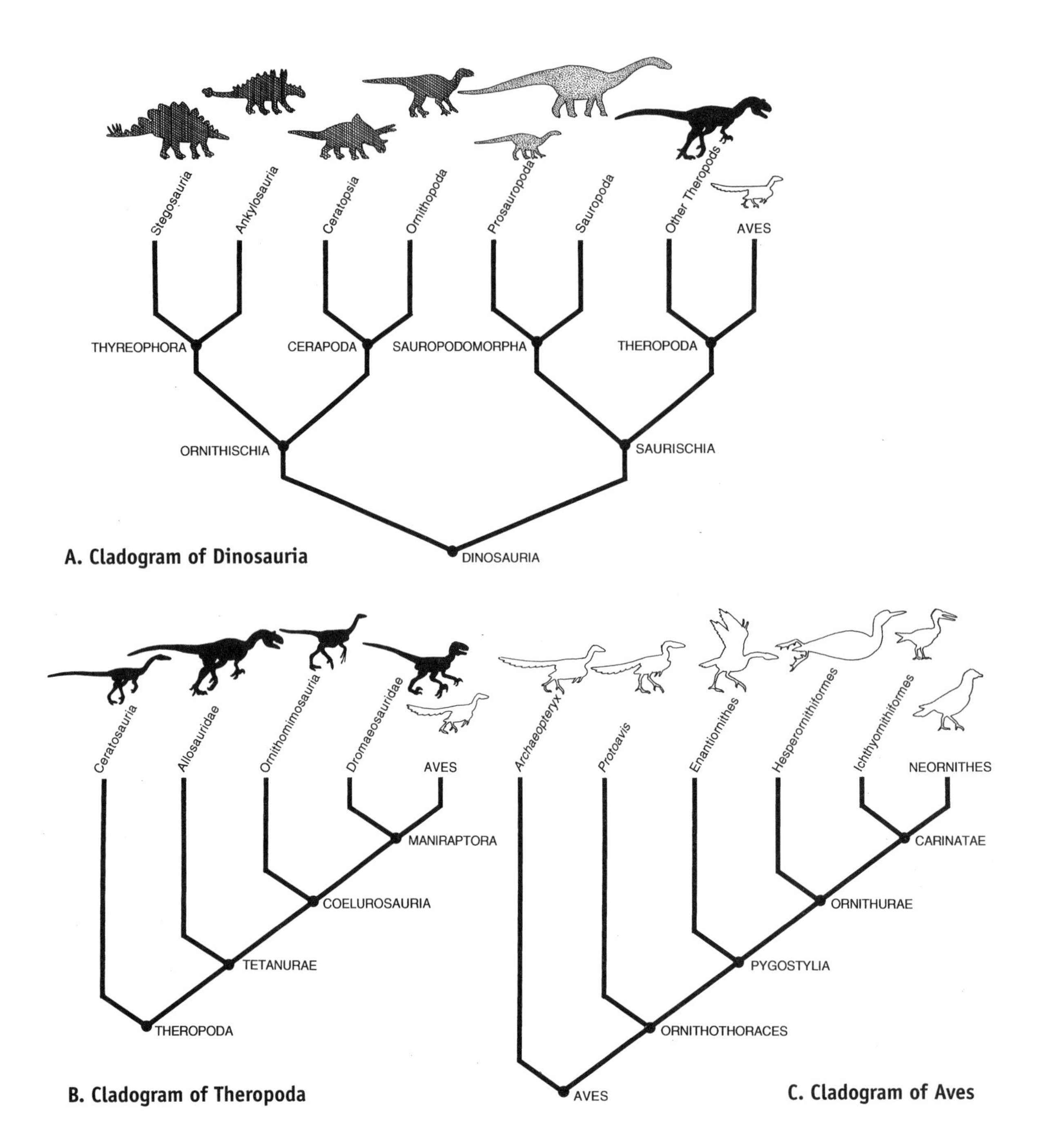

FIGURE 1.1

Cladograms of dinosaurs, theropods, and birds showing their hierarchical relationships in a nested pattern. *A,* cladogram showing the relationships among the key group of dinosaurs. On the *far right* is the clade Theropoda, which includes birds. *B,* cladogram showing the interrelationships of theropods. On the *far right* is the systematic position of birds within the major groups of theropods. *C,* cladogram showing the major groups of Mesozoic birds. *Archaeopteryx* is generally considered to be the most primitive basal taxon of birds, whereas *Protoavis* is included in Ornithothoraces. On the *far right* is the clade Neornithes, or modern birds.

cal pattern. These successive clades are Theropoda, Tetanurae, Coelurosauria, Maniraptora, and Aves (fig. 1.1B). The new phylogenetic relationship suggests that birds are a member of theropods. In fact, birds are now considered not only glorified theropods but also the sole surviving lineage of dinosaurs. The flying-dinosaur image of birds is appealing to the public and also is gaining currency among paleontologists (Weishampel, Dodson, and Osmólska 1990; Sereno and Rao 1992; Chiappe 1995a).

If we look at Gauthier's cladograms (fig. 1.1), it becomes clear that birds should possess all of the characteristics of dinosaurs, saurischians, theropods, tetanurans, coelurosaurs, and maniraptorans in a nested pattern (although some of them may have been lost or modified), but that they are also characterized by a suite of characters uniquely their own. In the phylogenetic scheme, birds are the most derived group of theropods, which acquired many evolutionary novelties in the context of their flight adaptation. The definition and interrelationships of major clades of birds, discussed in a later section, are shown in figure 1.1C.

The Ancestry of Birds

Dromaeosaurs were the closest relatives of birds and shared the most recent common ancestry. The early fossil record of dromaeosaurs is obscure. Thus far, known dromaeosaurs appeared fairly late during the Cretaceous, when birds were already well established. Dromaeosaurs did not continue to become more birdlike. Instead, in their evolutionary course they specialized in killing mechanisms and became considerably larger than birds. The common ancestor of birds and dromaeosaurs has yet to be found in the fossil record. However, we can speculate from cladistic analysis that this hypothetical bird ancestor would be very similar to dromaeosaurs in general morphology. For simplicity, we can refer to this ancestral form as *protodromaeosaurs.* Heilmann (1926) coined the neutral term *proavian* for this hypothetical ancestor. How big was the proavian? We can guess its size from the evolutionary trend in vertebrates, defined in

Cope's law. This law states that, in the course of time, all animals tend to evolve larger body sizes. If Cope's law holds true, the ancestral proavian would be considerably smaller than later dromaeosaur descendants. The proavian may match the size of *Archaeopteryx* and would have given rise to two lineages, birds and dromaeosaurs. Dromaeosaurs are currently the best approximation of the hypothetical bird ancestor and serve as a model when tracing avian ancestry (fig. 1.2).

The Antiquity of Birds

For years, *Archaeopteryx* was considered to be the oldest bird known, but its position has recently been usurped by *Protoavis texensis* from the Late Triassic Dockum Group of Texas, predating *Archaeopteryx* by 75 million years (Chatterjee 1987a, 1991, 1994, 1995, in press; Kurochkin 1995; Peters 1994). Identification of *Archaeopteryx* as a bird is a simple task because *Archaeopteryx* possesses feathers. The recognition of *Protoavis* as a primitive bird requires a thorough knowledge of comparative anatomy of the skeleton because feather impressions were not found with the specimens. Resembling a small nonavian theropod in the rear, *Protoavis* reveals its avian identity in the front portions of the skeleton. It is an excellent example of mosaic evolution, in which some conservative ancestral characters of contemporary nonavian theropods occur with the advanced characters typical of later birds. This mingling of primitive and advanced characteristics seems to have been a common evolutionary pattern in the origination of higher groups of vertebrates.

The primitive characters of *Protoavis* include four metacarpals in the hand, a short ascending process on the astragalus, and a long bony tail. On the other hand, *Protoavis* is fully avian in some significant ways. Its temporal configuration is modified in avian fashion, with the development of streptostylic quadrate and upper jaw mobility. Its braincase is highly inflated, and the orbits are frontally placed. It has heterocoelous (saddle-shaped) centra in the neck, as do modern birds. *Protoavis* has a much more efficient and advanced wing structure than does *Archaeopteryx*. It has a birdlike coracoid and furcula and a keeled sternum for flapping flight. The hand bones show quill nodes for the attachment of primary feathers. The pelvis shows fusion of the ilium and ischium for strength and rigidity, whereas the hindlimbs are reoriented to shift the functional joint from hip to knee.

Detailed cladistic analysis indicates that *Archaeopteryx* is a basal taxon of Aves, whereas *Protoavis* had achieved a structural organization well beyond that of *Archaeopteryx* and is a member of the Ornithothoraces (fig. 1.1C). *Archaeopteryx* thus seems to be a late example of the ancestral type, a "living fossil" in the Jurassic world. The recognition of *Protoavis* as the first bird marks a critical departure from earlier thinking on the origin of birds and the evolution of flight. It provides a more complex picture of morphological diversity early in bird evolution, showing a bushlike adaptive radiation.

However, the discovery of a Triassic bird is not totally surprising. More than a century ago, Yale paleontologist Othniel Charles Marsh (1880) cogently argued that three Mesozoic taxa, *Archaeopteryx, Hesperornis,* and *Ichthyornis,* differ so widely

FIGURE 1.2

Skeletal reconstruction of a dromaeosaur, such as *Velociraptor* from the Upper Cretaceous of Mongolia. Dromaeosaurs were small, agile carnivores and probably behaved like pack-hunting dogs. Dromaeosaurs are currently considered to be the closest relatives of birds. Birds probably evolved from a small, unknown dromaeosaur—the protodromaeosaur; this hypothetical ancestor of birds is called *proavian.*

from one another that the evolution of birds must have taken place at a much earlier time, perhaps at the end of the Triassic. He predicted that Triassic birds with a freely movable quadrate bone would be found to fill the major morphological and evolutionary gaps in avian history. *Protoavis* approaches the predicted structure and size of the ancestral bird envisioned by Marsh. It pushes the avian origin back to the Late Triassic, to the very dawn of the age of the dinosaurs.

Thus, the new avian odyssey begins some 225 million years ago, when *Protoavis* took to the air over tropical Texas forests. This is the beginning of the age of birds. Throughout the Jurassic and Cretaceous, birds diversified, perfected their flight maneuvers, and adapted to various niches during the continental fragmentation. The road from *Protoavis* to pigeon requires a long evolutionary march, with frequent roundabouts and blind alleys. It is paved with the temporary dominance of several different extinct lineages until the Late Cretaceous, when Neornithes (modern birds) emerged. Most Cretaceous birds, such as enantiornithes, hesperornithiforms, *Patagopteryx*, and other less well-known groups, disappeared about 65 million years ago, along with nonavian dinosaurs. Rising Phoenix-like from the ashes of this catastrophe, the neornithine lineage underwent an explosive adaptive radiation of modern forms during the Tertiary.

The Design of an Airframe

With his powerful wings he brought to man the oblation loved by the gods.

—Rig-Veda, ca. 1200 B.C.

CHAPTER 2

All birds have an internal bony skeleton that provides support and keeps the body from collapsing. The bones interconnect at joints; most joints are mobile, but some are fixed. Between mobile joints, the bones are rigid and serve as levers across which the muscles can act. Bones support the muscles used in locomotion, other body movements, and posture; they protect such internal organs as the brain, heart, and lungs; and they house bone marrow, which produces blood. Bone is composed of hydroxyapatite, a hard and durable material that changes little with fossilization and thus provides important clues to vanished soft parts. Tuberosities, depressions, and scars on bones reveal the positions, sizes, and attachments of muscles. The endocranial cavity and cranial nerve foramina reveal the size of the brain and the distribution of the neural network, and the nasal cavity, orbits, and otic capsules provide information about the size and orientation of the sense organs. Teeth, claws, limbs, and girdles are all coordinated with feeding habits and methods of attack and defense, including posture and locomotory patterns. Bones let us see the time dimension in the evolution of vertebrates and, in particular, the rise of birds.

Comparative Anatomy

The primary adaptation that established the birds as a clade was flight. Virtually every distinctive characteristic of bird anatomy has evolved as an adaptation for flying. The skeleton of the dromaeosaur is designed for running and climbing. Dromaeosaurs share several features with birds, including large braincase, large orbits, elongated forelimb, swivel wrist joint, tridactyl manus, large pelvis, reverted pubis, obligatory bipedal posture, and scansorial locomotion. The hand and wrist of dromaeosaurs are so similar and avian in their shape that the arms were normally

FIGURE 2.1

Lateral view of the skeleton of a pigeon (*Columba livia*) showing the light and rigid airframe for flight. Many avian features in the skeleton are visible, such as the kinetic skull; fused bones in the skull, thorax, sacrum, pelvis, carpometacarpus, tibiotarsus, and tarsometatarsus; a flexible neck; a short tail with a pygostyle; a large, keeled sternum; and wings.

held in a folded position, like the wings of a bird. A detailed comparison of the skeletons of dromaeosaurs and birds shows the closeness of these two groups (fig. 2.2). From the dromaeosaur body plan, the avian skeleton has undergone drastic evolutionary modifications related to the flight adaptation. Comparative anatomy provides important clues to the evolution of the airframe.

Like any flying machine, birds must possess two basic attributes to defy gravity: a light, strong structure and efficient power. To minimize weight and maximize power, bones and muscles have been modified. Most bird bones are hollow, light, air filled, and often stiffened internally by struts. Air spaces in the bones link up with the lungs through a network of air sacs. Some bones and teeth, and the long cumbersome tail found in dromaeosaurs, were eliminated to reduce weight; other bones normally sutured were fused for strength and rigidity. Forelimbs and the shoulder girdle were modified into a flight apparatus, and feathers provided the lift and propulsion (fig. 2.1).

Flight is a demanding method of locomotion that requires a great deal of power. This power is provided by two sets of flight muscles, which contract alternately to flap the wings. The collapsible wing supports flight feathers, and the keel of the sternum anchors the powerful flight muscles. A bird must be as light as possible, and flight is more manageable if the remaining weight is concentrated near the center of the body. This is achieved by rearrangement of the internal organs and by posture.

The skeletal features of the pigeon (*Columba livia*) and other recent birds are compared with those of dromaeosaurs to highlight these evolutionary modifications into an airframe (fig. 2.2). Once we understand these skeletal changes between the beginning and end points of avian evolution, the sequence of acquisition of avian features in Mesozoic birds can be traced.

The Skull

The avian skull is modified greatly from the dromaeosaurid design in response to three functional requirements: (1) flight,

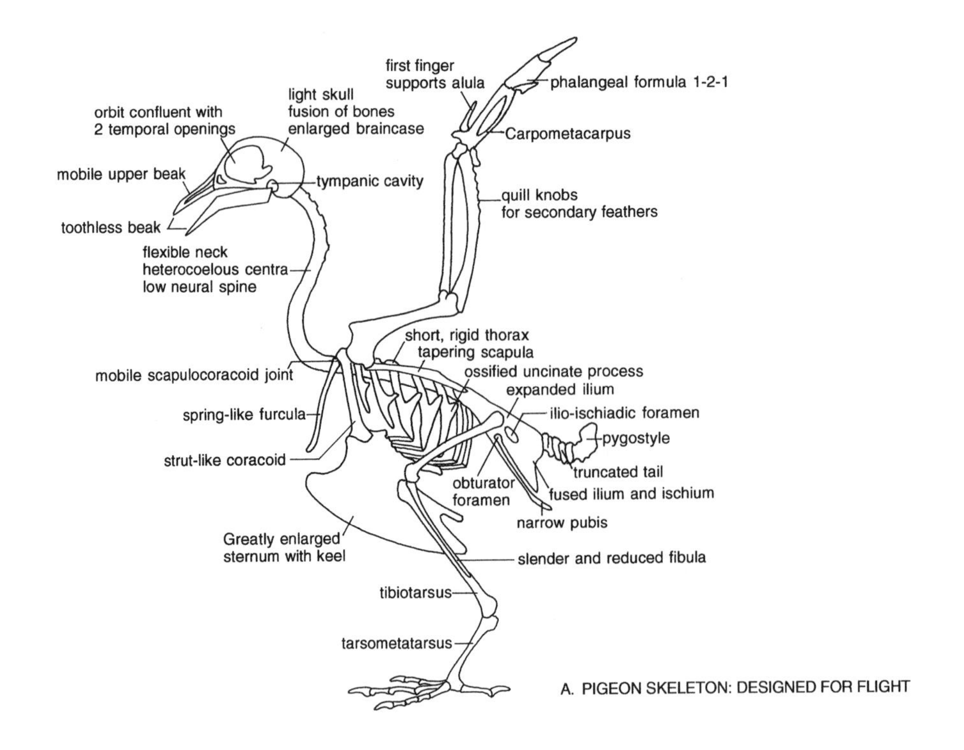

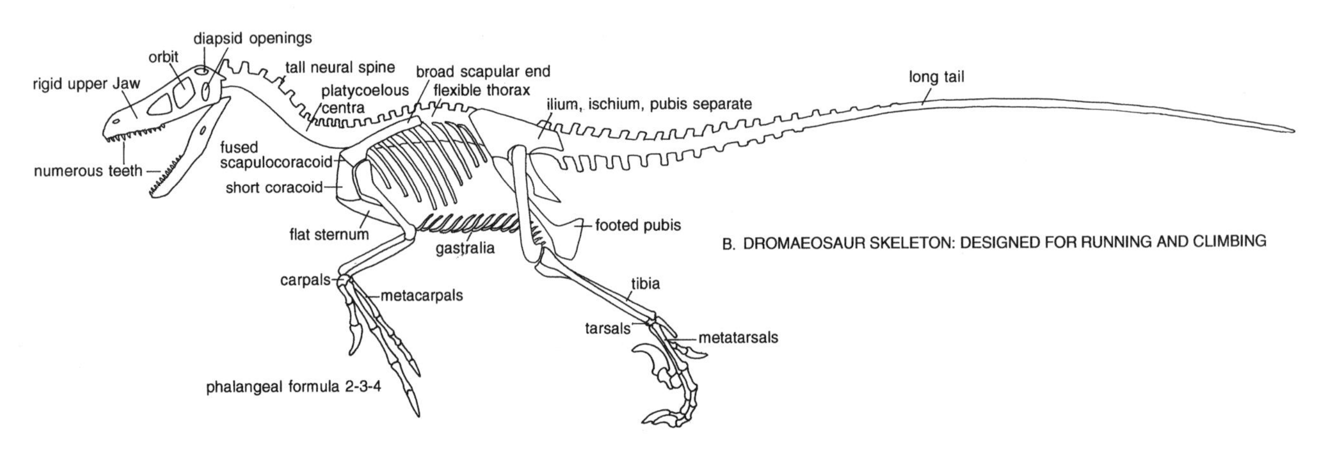

FIGURE 2.2

Comparison of the skeletons of pigeon and dromaeosaur to show the similarities and differences. Many avian features evolved from a dromaeosaur-like body plan in response to flight adaptation in birds. *A,* the skeleton of a pigeon in the left lateral view; *B,* the skeleton of a dromaeosaur in the left lateral view; both *A* and *B* are semidiagrammatic.

(2) enlargement of the brain, and (3) the ability to move the upper jaw, which is linked with altered feeding habits. The skull is lightly built and aerodynamically designed. The pointed beak and rounded braincase act as an airfoil. The weight of the skull has been reduced by thinning of the compact layers of the dermal bones, but the skull remains strong because sutures are fused in adults. Some skull bones have been reduced or lost. The spherical braincase is the optimum shape for holding a highly enlarged brain. The orbits, large depressions on the spherical braincase, accommodate enormous eyes. Birds differ from dromaeosaurs by being able to move the upper jaw relative to the braincase. The triangular, pointed beak is mobile and is an efficient food-gathering device.

The skull of a young chicken (*Gallus gallus*), where the sutures are still open, can be compared with that of *Dromaeosaurus* to trace the architectural changes. The most distinctive departure from the dromaeosaurid skull pattern is the large orbit and the inflated braincase in birds. These modifications become clear when we compare the side view of the skull, especially in the cheek region (fig. 2.3A–B). The postorbital bone and the ascending process of the jugal behind the orbit and between the two temporal openings have disappeared in birds so that the orbit becomes confluent with the two temporal openings. The merger of these three openings gives a very large space for the enormous eyeball, which provides bigger and sharper images. Visual information is extremely important to birds. Another diapsid arch, the squamosal/quadratojugal bar in front of the quadrate, is also eliminated. The freed quadrate becomes mobile, or streptostylic. The teeth are lost so that the jaws become lightweight struts to support the horny beaks. The gizzard in the stomach takes the role of teeth for processing food.

With the loss of the postorbital bone, two important landmarks are visible on the sidewall of the skull behind the orbit: (1) the postorbital process formed by the frontal and the laterosphenoid and (2) the zygomatic process of the squamosal overhanging the quadrate (fig. 2.3B). The quadrate is greatly

skull of dromaeosaur (*Dromaeosaurus*)

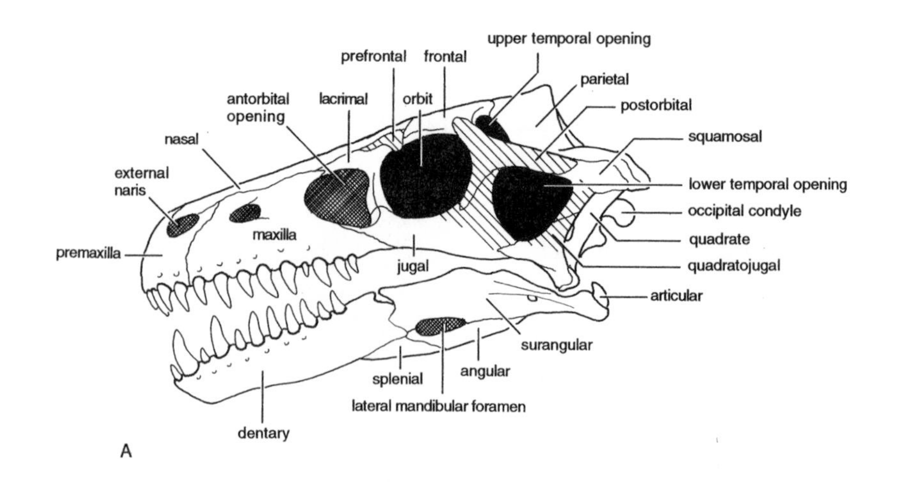

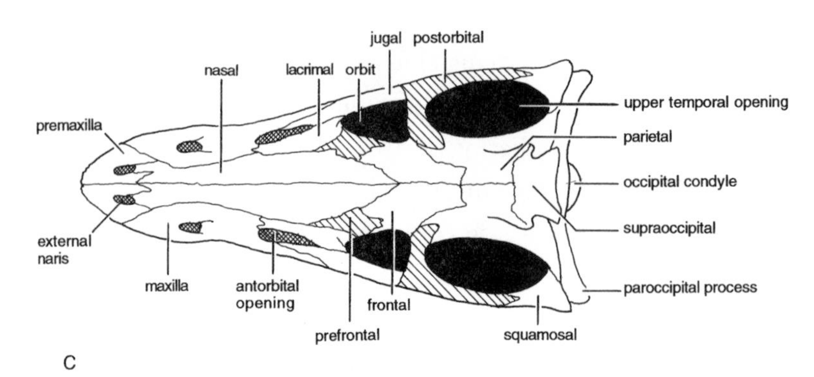

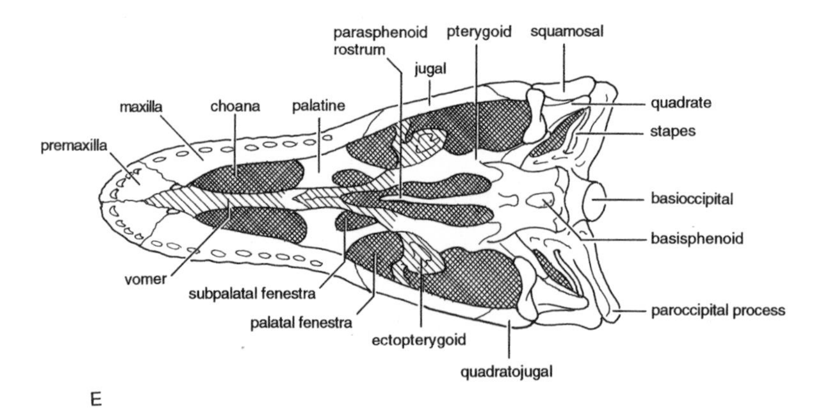

FIGURE 2.3

Comparison of the skull of dromaeosaur and bird (chicken). The most important anatomical innovation in birds is the modification of the temporal region of the skull, where the orbit becomes confluent with the upper and lower temporal openings. This modification allows birds to raise the upper jaw. *A–B,* lateral view; *C–D,* dorsal view; *E–F,* ventral view. Bony regions lost from dromaeosaur to bird skull are represented by diagonally shaded areas (*A,C,E*); new avian features, such as the postorbital process, zygomatic process, and orbital process, are represented by horizontally shaded areas (*B,D*).

skull of chicken (*Gallus*)

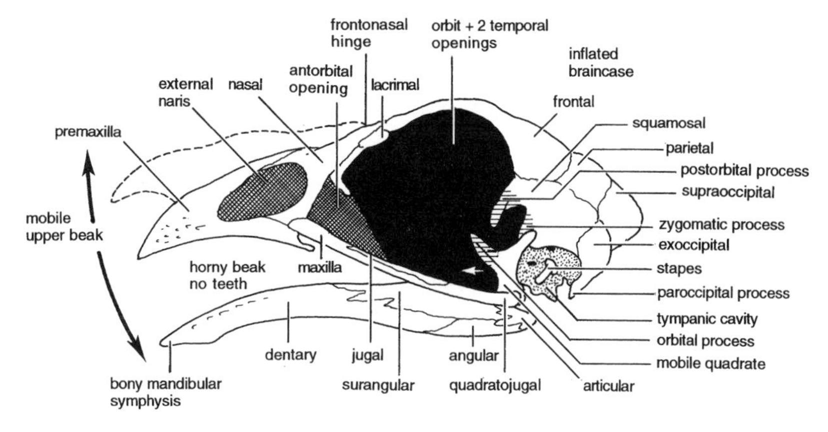

B

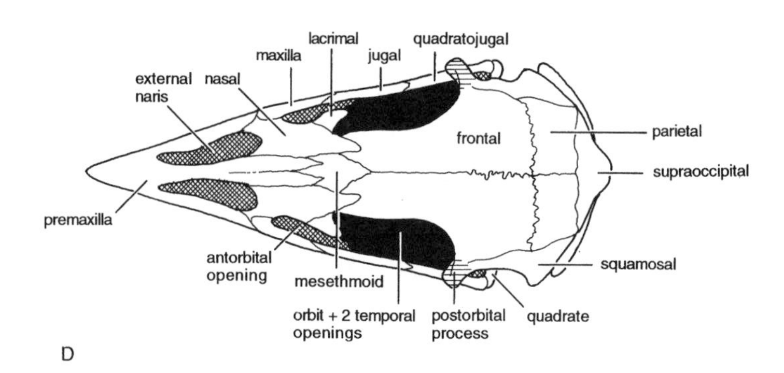

D

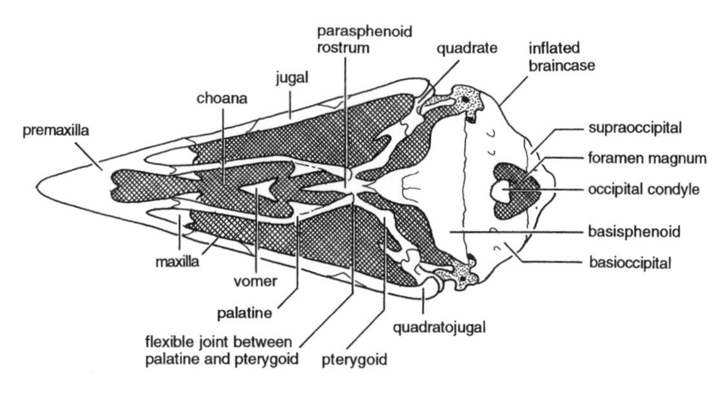

F

modified in birds. Its dorsal head is expanded medially to make an additional contact with the braincase. Medially, near its ventral margin, it receives the pterygoid in a ball-and-socket joint. As a result, it develops a distinct orbital process that acts as a muscle lever. Another bone, the prefrontal, is lost in birds, while the descending process of the lacrimal is reduced to breach the contact with the jugal. With the breakdown of various vertical struts, the conjoined maxilla, jugal, and quadratojugal become a thin horizontal rod, the jugal bar.

In dorsal aspect, noticeable change is seen around the frontonasal hinge. The premaxilla and the frontal become large enough to approach each other along the midline so that the nasal bones are displaced laterally. The nasal bones become very thin and pliable where they contact the frontal bones. Here a new bone, the mesethmoid, is ossified to form a pivot at the frontonasal hinge (fig. 2.3C–D).

In palatal aspect, the ectopterygoid is lost to free the pterygoid from the jugal. The choana is displaced caudally to merge with the palatal fenestra and forms an extensive cleft down the midline. The vomer is atrophied to a sliver of bone, and the palatine is elongated at the expense of the pterygoid. An additional movable joint is established between the pterygoid and the palatine in neognathous birds (fig. 2.3E–F). With the loss and reduction of some bones and the development of a median cleft, the avian palate becomes a delicate, springy framework.

The bird brain has enlarged backward and downward so that the foramen magnum is shifted from the occiput to the ventral side of the skull. Viewed from the ventral aspect, the occipital condyle is visible as a small spherical knob. Anterior to the condyle lies the highly expanded basisphenoid. The enlargement of the brain gives a dome-shaped appearance to the braincase. Some of the dorsal roofing bones, such as the frontal, parietal, and supraoccipital, become inflated to accommodate the large brain. Because of the backward protrusion of the braincase, a large tympanic cavity is created behind the quadrate for the external ear. From the lower and rostral part of the tympanic cavity, a funnel-shaped depression leads to the bony eustachian tube. The inner ear has an elongated cochlear process and enlarged canalicular system for refined hearing and balance.

All six bones in the dromaeosaurid mandible—the dentary, splenial, angular, surangular, prearticular, and articular—are recognized in juvenile birds, but their identity is lost in adult birds. With the loss of teeth, the lower jaw becomes a shallow, delicate structure. The notable change is the development of an ossified symphysis at the dentary to give strength and rigidity to the lower beak.

The Vertebral Column

The vertebral column of a bird has been modified considerably from the dromaeosaurid design. The number of neck vertebrae increased at the expense of the dorsal vertebrae, the thorax became a small rigid box, the synsacrum (a solidly fused series of vertebrae in the pelvic region) was strengthened by the incorporation of more vertebrae, and the long, bony tail was reduced to a few caudals and a pygostyle (a bony plate at the end of the spine). In dromaeosaurs, the cervical centra are platycoelous (slightly hollowed), with well-developed neural spines. In birds, the neck shows extreme mobility in all directions, with the development of heterocoelous (saddle-shaped) centra (fig. 2.4A–F). This long and flexible neck supports an elevated head, so that the beak can be used as a "universal tool." The neural spines are weakly developed in the cervical vertebrae. Behind the axis, the cervical ribs are short and fused with their vertebrae to enclose the canal for the vertebral artery. Ventrally, each centrum bears a pair of projections, or hypapophyses, which may also be fused into a single median ridge.

Fusion of two segments of the vertebral column behind the neck provides stable platforms for the shoulder and pelvic girdles, respectively. Because the thorax forms the fulcrum on which the wings move up and down, it must be short and rigid. The thoracic vertebrae bear complete ribs with uncinate processes, and each process overlaps the next posterior rib to strengthen the

dromaeosaur
(*Deinonychus*)

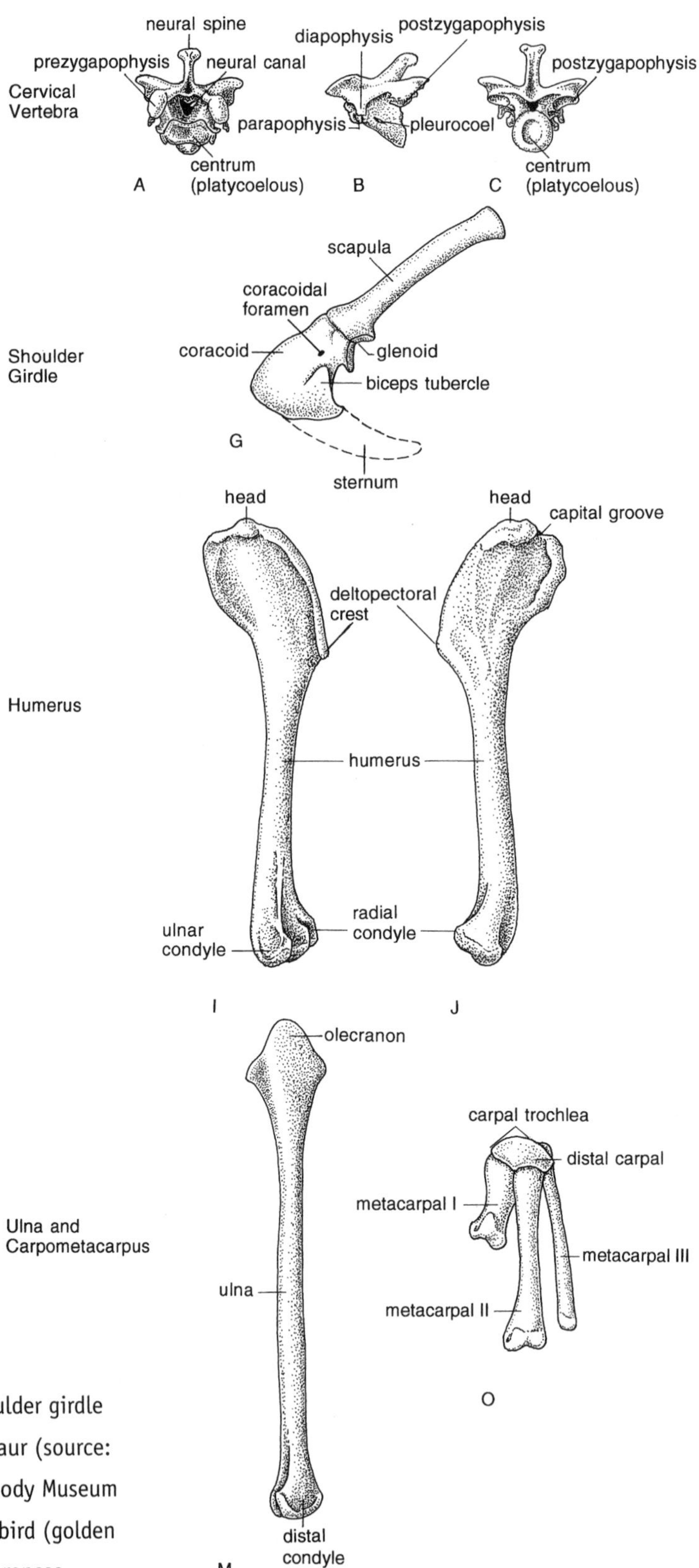

FIGURE 2.4

Comparison of the vertebrae (A–F), shoulder girdle (G–H), and forelimbs (I–P) of dromaeosaur (source: after Ostrom 1969; courtesy of the Peabody Museum of Natural History, Yale University) and bird (golden eagle) to show the similarities and differences.

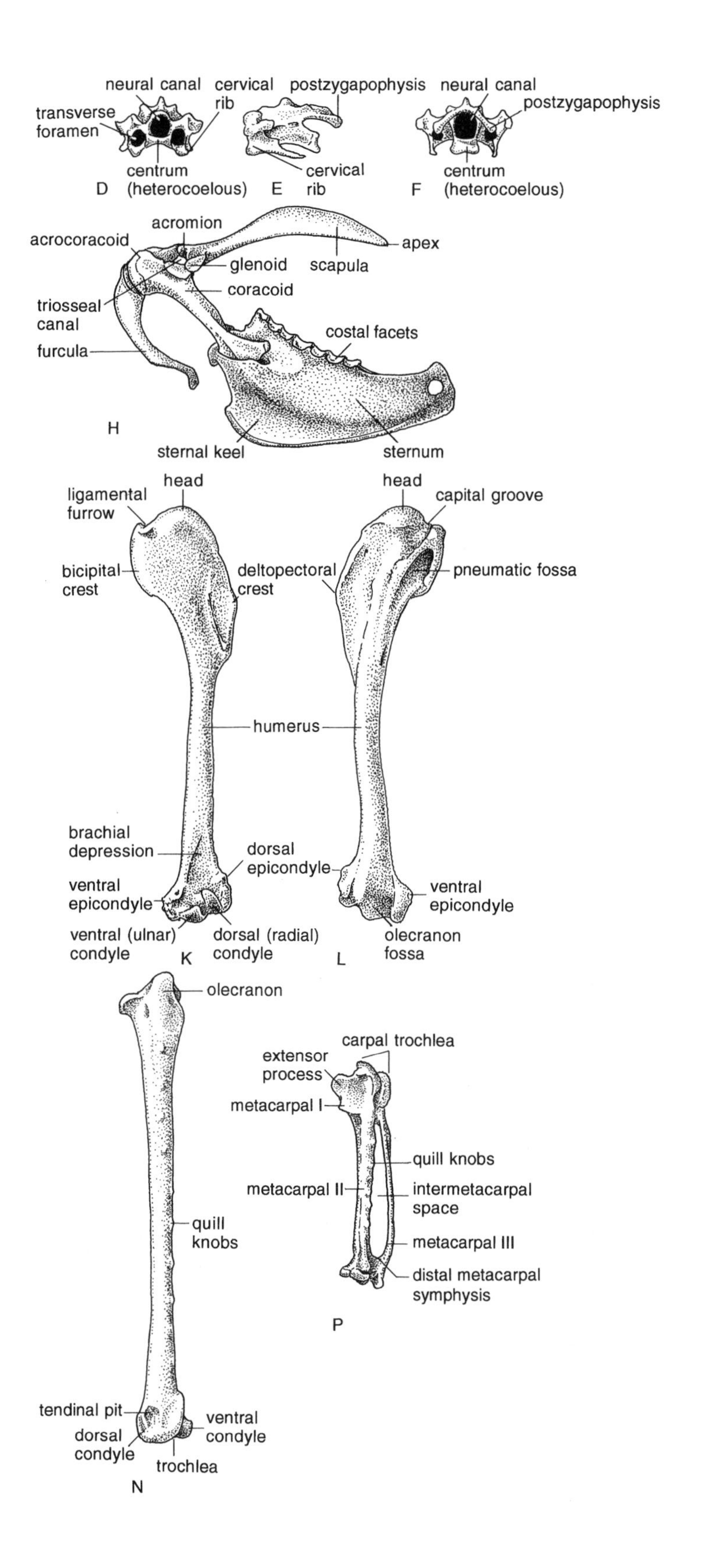

golden eagle
(*Aquila*)

rib cage (fig. 2.1). The gastralia are absent in birds. The neural spines of the thoracic vertebrae are long and are often fused together, except between the hindmost, where there is some flexibility. Bipedal locomotion and the action of the hindlimbs as the landing gear require a strong synsacrum. The synsacrum in birds is extensively developed from the dromaeosaurid condition by the incorporation of additional segments of posterior dorsals and anterior caudals into this structure. Moreover, both the transverse processes and the neural spines also brace the iliac blade. Following the synsacrum are six or seven movable caudal vertebrae that steer the tail for flight and maneuver. The reduction of the long caudal series and the development of the pygostyle are new features in birds. The pygostyle supports the lightweight tail feathers.

The Shoulder Girdle

The shoulder girdle is perhaps the most distinctive feature in the avian skeleton and is built strongly to execute the flight strokes. In dromaeosaurs, the coracoid is relatively small and fused with the scapula to form the glenoid, which faces caudally (fig. 2.4G). The coracoid has developed the biceps tubercle, which may be the precursor to the avian acrocoracoid process (Ostrom 1976b). In birds, the coracoid becomes massive and strutlike to brace the sternum so that it can support the compressive force of the downstroke during flight (fig. 2.4H). The coracoid articulates with the scapula at an acute angle in a flexible joint. The glenoid fossa has been shifted outward and upward to permit the dorsoventral excursion of the humerus. Dorsally, the coracoid develops the acrocoracoid process to complete the opening for the triosseal canal in conjunction with the scapula and the furcula. This canal acts as a pulley, allowing the supracoracoideus muscle to raise the wing. This coracoideus pulley, a major innovation in birds, is used to execute the upstroke. The scapula, a narrow, tapering blade lying parallel to the thorax, reaches almost to the pelvis. The sternum is greatly enlarged and has a prominent keel ventrally for the attachment of large flight muscles. The V-shaped furcula is long, thin boned, and united ventrally at an acute angle. It acts as a flexible spring between the two shoulder joints during flapping flight (Jenkins, Dial, and Goslow 1988). The furcula was recently discovered in an articulated skeleton of a dromaeosaur from Montana allied to *Velociraptor* (David Burnham, personal communication). The sternum of dromaeosaurs is a flat bone like that of flightless birds, without any ventral keel.

The Forelimb

Major modifications of the forelimbs of birds include restriction of the movement of the elbow and wrist joints to one plane, reduction of the number of digits, loss of functional claws, and fusion of the carpometacarpus. In dromaeosaurs, the humerus is longer than the ulna. Moreover, the proximal and distal expansions of the humerus are twisted at an angle (fig. 2.4I–J). In birds, the humerus is shorter than the ulna and the terminal expansions lie in the same plane. Another departure in birds is the development of the bicipital crest and the pneumatic fossa at the proximal end. Unlike dromaeosaurs, birds have distal condyles that are well defined and a radial condyle with a circular profile (fig. 2.4K–L). The radius and ulna are joined with the humerus and the manus in such a way that the wrist and the elbow joints are automatically coupled, allowing folding and unfolding of the wings. The ulna is also distinctive in many birds; a series of quill knobs lie on the outer surface for the attachment of the secondary feathers (fig. 2.4N). Moreover, the distal end has a trochlear surface for articulation with the ulnare. The proximal carpal bones—the radiale and ulnare—are well developed and complex in birds to play important roles in supporting and controlling wing movements (Vasquez 1992). In dromaeosaurs, the two distal carpal bones, distal carpal 1 and 2, often fused to a compound bone, the semilunate carpal. It shows a pulley-like surface, the carpal trochlea, on the proximal end, allowing restricted movement of the wrist joint at the plane of the forearm (fig. 2.4O). This swivel joint at the wrist helps to fold and unfold the hand at the side of the body when not in use. This joint is more refined in modern birds.

However, the carpal and metacarpal bones of dromaeosaurs remain free. In birds, distal carpal 1 is lost. Distal carpal 2 and three metacarpals (I–III) are fused into a single bone, the carpometacarpus (fig. 2.4P). Like dromaeosaurs, birds have three digits, but the number of phalanges has been considerably re-

dromaeosaur
(*Deinonychus*)

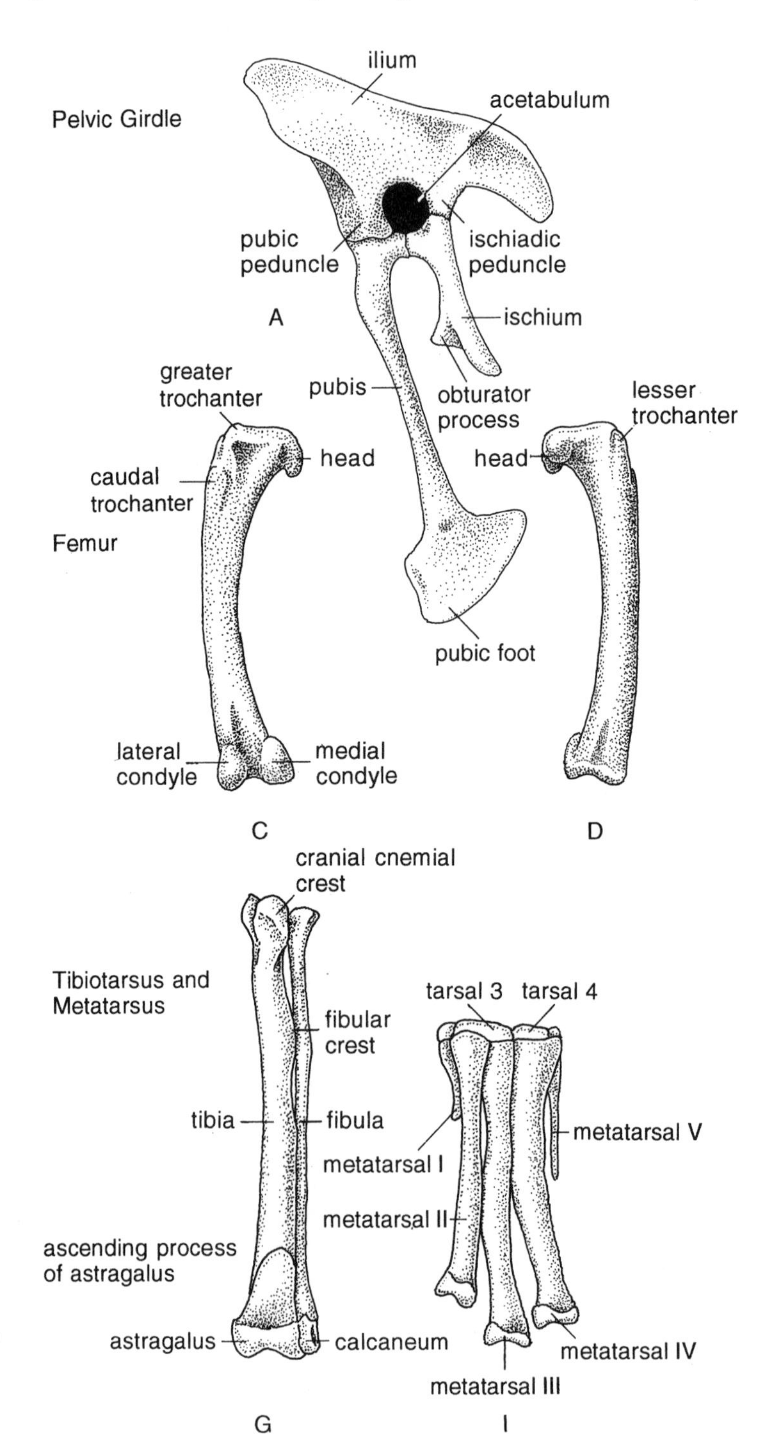

FIGURE 2.5

Comparison of the pelvis (A–B) and hindlimbs (C–J) of dromaeosaur (source: after Ostrom 1969; 1976c; courtesy of the Museum of Comparative Zoology, Harvard University) and bird (golden eagle) to show the similarities and differences.

duced from this primitive pattern of 2-3-4-x-x. The phalangeal formula code is after Padian (1992); O indicates metapodials supporting no phalanges, and x indicates digits that are completely lost.

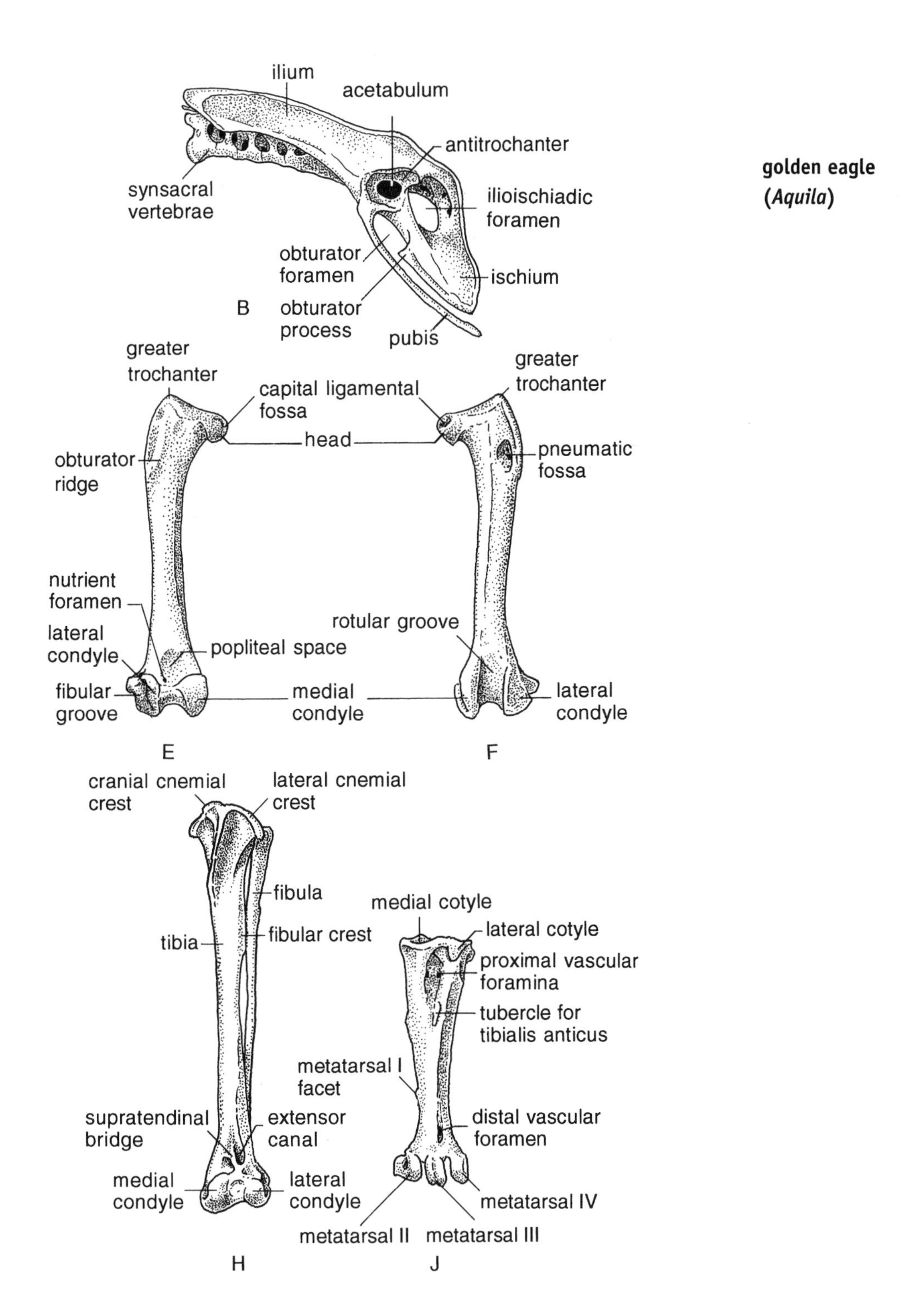

golden eagle (*Aquila*)

The Pelvic Girdle

In dromaeosaurs, three pelvic bones—the ilium, ischium, and pubis—remain free. The pubis is highly elongated and somewhat rotated backward, and it develops a large foot at the distal end (fig. 2.5A). In birds, the pelvic girdle is formed by the fusion of these three bones with the synsacrum and shows extensive remodeling of the dromaeosaurid design (fig. 2.5B). The avian pelvic girdle is greatly expanded to provide a large area for the attachment of hindlimb and abdominal muscles. The ilium is elongated craniocaudally and is strongly fused with the synsacrum to form a rigid structure. Cranially, the two ilia approach each other to form a broad pelvic shield. The ischium, in turn, is fused with the ilium but is interrupted by the ilioischiadic foramen. The pubis is reduced to a thin and narrow strip of bone, rotated backward to run along the ventral edge of the ischium. Cranially, it is fused with the ischium, but is separated by the obturator foramen more distally. A secondary articulation between the femoral head and the acetabulum, the antitrochanter, prevents the abduction of the femur. It also permits the legs to be tucked against the body for flight. A large renal fossa on the medial surface of the ilium and ischium houses the kidney. In the bird, unlike the dromaeosaur, the two halves of the pelvic girdle remain open, without any symphysis.

The Hindlimb

Because the hindlimb is used for bipedal walking and leaping in both birds and dromaeosaurs, the structural differences between these two groups are not pronounced. There are, however, subtle changes in limb position and movement in birds. In dromaeosaurs, the femur moves in a nearly vertical plane at the hip joint during locomotion. Various subsidiary trochanters, such as the fifth and lesser trochanters, are eliminated in birds (fig. 2.5C–D). In birds, the femur is kept subhorizontally and is tucked away under the feathers; the main functional movement has been shifted from the hip to the knee joint. The femur shows some novelties, such as the elevated crest of the greater trochanter and the capital ligamental fossa in the proximal end, whereas distally it develops deep rotular and fibular grooves (fig. 2.5E–F). In dromaeosaurs, the tibia is locked with the astragalus and calcaneum without any fusion (fig. 2.5G). In birds, the tibia is fused with the astragalus and calcaneum to form the distal condyles of the tibiotarsus (fig. 2.5H). Here, it develops a supratendinal bridge over the tendinal groove. Proximally, it shows both cranial and lateral cnemial crests, and the former is an avian innovation. The fibula is greatly atrophied and consists of a slender spicule of bone that fails to reach the ankle. In dromaeosaurs, the five metatarsals are present and remain separate from the distal tarsals. In birds, the distal tarsals are fused with metatarsals II, III, and IV to form a long, rigid, single bone—the tarsometatarsus (fig. 2.5J). Metatarsal I is reduced and articulates with metatarsal II distally; metatarsal V is lost. Proximally, the tarsometatarsus has one or more protrusions, the hypotarsus. The pes becomes anisodactyl where the first digit points backward to form a grasping organ for perching.

The Triassic Treasures of Texas

CHAPTER 3

Tell me, Oh Swan, your ancient story
From which country do you come?
—Kabir, *Songs of Kabir,* a fifteenth-century Indian mystic

The Triassic period of earth's history, from 245 to 208 million years ago, was important in the adaptive radiation and diversification of the world's terrestrial fauna. The Triassic period began after the most catastrophic mass extinction event in geological history. Perhaps as much as 75 to 90 percent of the species in the marine realm became extinct at the end of the Permian period, just before the Triassic began. Among land animals, therapsids (mammal-like reptiles) were the main victims. Plate movements, mountain building, violent volcanism, and regression of the seas produced major changes in the global distribution of continents and oceans as the Permian drew to a close. All of these crises contributed heavily to the breakdown of stable ecological communities and severely disrupted the biosphere. In the aftermath of the Permian extinction, groups of animals that formerly had played minor roles in the ecosystems assumed prominence, and new groups appeared.

By the Late Triassic, archosaurs had replaced therapsids as the dominant vertebrates. During this crucial time in vertebrate evolution, mammals originated from small therapsids, dinosaurs began to diversify, and pterosaurs, kuehneosaurs, and birds took to the air for the first time. Lissamphibians, turtles, and crocodylomorphs also emerged during the Triassic. This faunal turnover in the terrestrial realm revealed the beginnings of many major groups of tetrapods that still persist today.

The Birth and Breakup of Pangea

When the Triassic period began, all of the major continental plates were joined into a single colossal supercontinent called Pangea, which was slowly drifting northward (fig. 3.1). Land animals were able to migrate easily back and forth across the

FIGURE 3.1

During the Triassic period, all continents formed one supercontinent, known as Pangea. In this Pangean world, the first bird, *Protoavis*, lived in the tropical forest of Texas.

land surface of the earth. Pangea comprised two landmasses: Laurasia in the north and Gondwana in the south. A narrow bay of the Tethys Sea, comparable to the present Mediterranean, lay between Indo-Africa and Eurasia. Today, the former position of the Tethys is marked roughly by the Alpine-Himalayan mountain belt. The topography and climate became more uniform during the Triassic because Pangea was disposed evenly above and below the equator. The paleoclimate was subtropical, hot and arid and dominated by monsoonal circulation. Dense forests and swamps full of conifers, cycads, and ginkgoes covered the land from the equator to the high latitudes. No ice sheets covered polar regions. One of the most remarkable features of the Triassic was the widespread emergence of continents and the regression of the seas. Nonmarine sediments composed largely of redbeds were beginning to accumulate in large basins in the interior of the continents. These redbeds were deposited in a complex mosaic of river-deltaic-lake systems in many parts of the world. Today, they are known in India, China, Argentina, Brazil, South Africa, East Africa, Germany, Great Britain, Canada, and the United States. These redbeds have yielded a rich record of Triassic vertebrate fauna. Triassic vertebrates have also been found in Greenland, Australia, and Antarctica.

The unity of Pangea was short-lived. Near the end of the Triassic, Pangea began to break apart, and the continents began to shift toward their present positions on the globe. During this initial fragmentation, a rift opened in the southwest Indian Ocean, moving South America and Africa away from Antarctica-Australia. At the same time, the Atlantic Ocean was beginning to open. When the Atlantic opened, huge fractures in the crust of Nova Scotia, New Jersey, and Connecticut served as the conduits for great outpourings of lava. The ancestral Gulf of Mexico began to form farther south. During the initial stage of rifting, a large meteorite collided with Earth in central Quebec to produce the giant Manicougan Crater, about 100 km across (half the size of Connecticut). This was a time of major environmental disruption and biotic crisis. Another mass extinction struck both marine and terrestrial ecosystems at the end of the Triassic. It eliminated a wide range of land vertebrates. Surprisingly, the dinosaurs (including birds) were unscathed, and they proliferated after this crisis to dominate terrestrial habitats (Chatterjee 1992a).

The Triassic Dockum Group of Texas

Late Triassic history is represented in West Texas by a richly fossiliferous sequence of continental Dockum sediments. Named by geologist W. F. Cummins in 1890 after a small town in Dickens County, Texas, these redbeds crop out around the Southern High Plains and through the Canadian River Valley in the Texas Panhandle (fig. 3.2A). These strata can be physically traced and correlated with the Triassic section exposed northwest in east-central New Mexico, which probably indicates a contiguous depositional basin. Outcrops of the Dockum Group are best exposed at the proximity of the resistant "Caprock" caliche that marks the top of the overlying Tertiary Ogallala Group.

My colleague Thomas Lehman has mapped the Dockum rocks in detail and reconstructed their depositional history. Currently, Dockum Group strata in Texas are assigned to four formations, in ascending order: Santa Rosa, Tecovas, Trujillo, and Cooper Canyon (Chatterjee 1986; Lehman 1994; Lehman, Chatterjee, and Schnable 1992). The Tecovas and Cooper Canyon formations consist primarily of red mudstones intercalated with discontinuous, lenticular bodies of sandstones, whereas the Santa Rosa and the Trujillo are dominantly sandstones and conglomerates. The Santa Rosa-Tecovas sequence is separated from the overlying Trujillo-Cooper Canyon sequence by an unconformity and by mineralogical change; each sequence generally fines upward and indicates different provenance. The thickness of the Dockum ranges from less than 70 meters to as much as 600 meters; the sedimentation began in the Carnian and continued up to the early part of the Norian. The Dockum Basin,

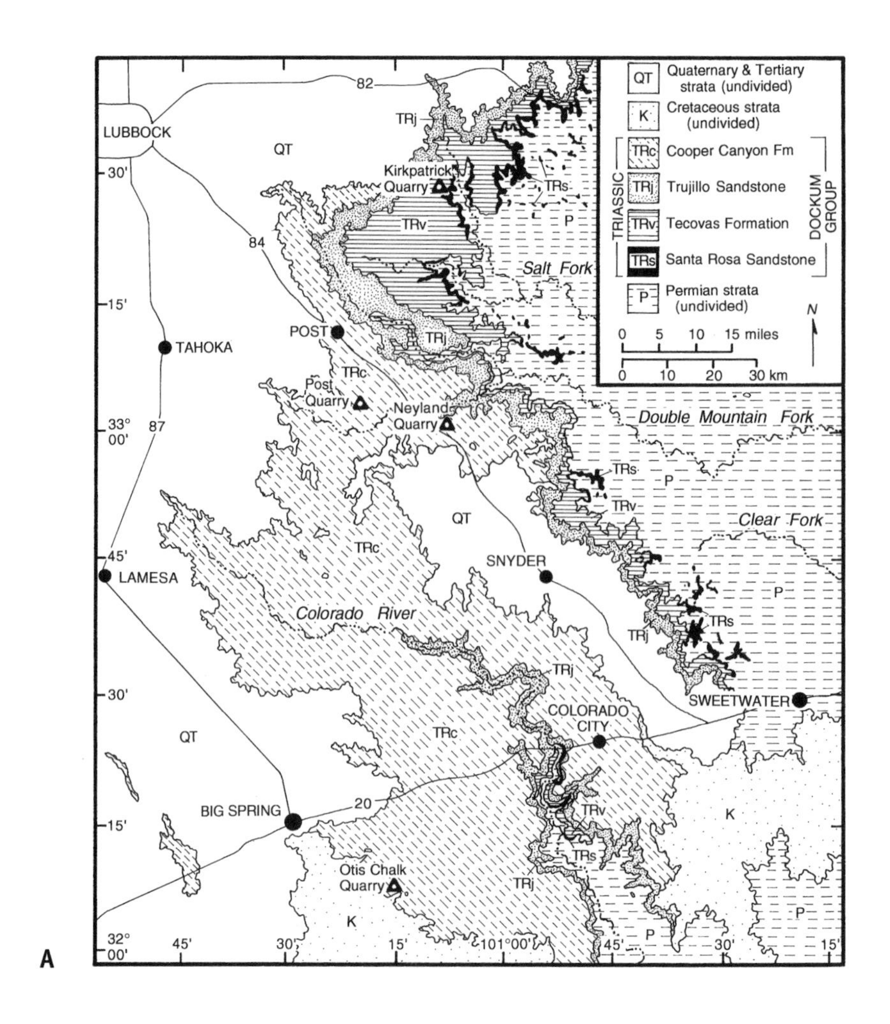

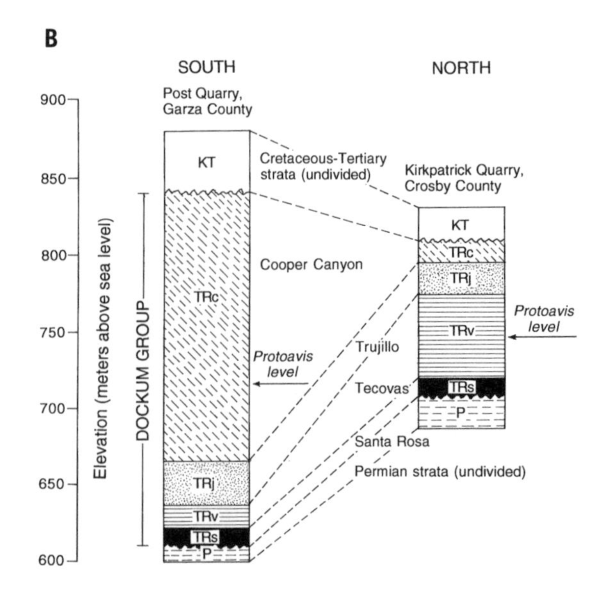

FIGURE 3.2

Stratigraphic sections and exposures of the Upper Triassic Dockum Group of sediments in West Texas. *A,* generalized geological map showing exposures of the Dockum Group in the upper Brazos and Colorado River valleys, West Texas. The locations of several vertebrate fossil sites—Kirkpatrick quarry, Post quarry, Neyland quarry, and Otis Chalk quarry—are shown from north to south in this map. Thus far, *Protoavis* bones have been recovered from Kirkpatrick quarry and Post quarry. *B,* cross section showing correlation of Dockum sediments between the Post quarry and the Kirkpatrick quarry; the stratigraphic level of *Protoavis* bones in two quarries is shown by the *arrows* (source: Lehman 1994).

about 400 km wide and 800 km long, was a major depocenter of Triassic sedimentation when the Gulf of Mexico was opening farther south.

The Dockum Group has been an important source of Triassic vertebrate fossils since 1893, when Edward Drinker Cope first recognized the remains of amphibians and reptiles in these redbeds. Exposures of the sediments are generally good, and there are numerous outcrops in the tributary valleys of the Brazos, Colorado, and Canadian rivers. Three bursts of active collecting and research of the Dockum vertebrates occurred during this century. The first systematic collection was made by E. C. Case of the University of Michigan, Ann Arbor, during the 1920s and 1930s. The second phase of vertebrate collection was undertaken from 1939 to 1941 by the crews of the Work Projects Administration (WPA) at the University of Texas, Austin. One of the areas extensively quarried by the WPA crews was the Otis Chalk area in Howard County, under the direction of Grayson Meade. From the early 1980s until the present, I have systematically explored the area and collected the Dockum vertebrates under the auspices of the National Geographic Society. My students and I have opened three quarries and conducted extensive excavations: the Post and Neyland quarries in Garza County and the Kirkpatrick quarry in Crosby County (fig. 3.2A). The collection now housed at the Museum of Texas Tech University has come into prominence as one of the richest known Triassic vertebrate assemblages in the world.

The Discovery of *Protoavis*

At the beginning of the summer of 1983, I was planning a large-scale excavation at the Post quarry. During the previous three summers we had collected superb skeletons of amphibians and reptiles from a 30-cm-thick, fine-grained mudstone unit of the Cooper Canyon Formation of the Dockum Group. The quarry has produced a prolific array of tetrapods in a single dense layer, including the metoposaur *Apachesaurus,* a new brachyopid, a sphenodontian, a new basal archosaur, a new rhynchosaur, the dicynodont *Placerias,* the tritheledontid *Pachygenelus,* the poposaurid *Postosuchus,* the phytosaurid *Nicrosaurus,* the aetosaurs *Typothorax, Paratypothorax,* and *Desmatosuchus,* an unnamed ceratosaur, the ornithischian *Technosaurus,* and the ostrich dinosaur *Shuvosaurus* (Chatterjee 1983, 1984, 1985, 1986, 1993). Fragmentary remains of such early mammals as *Adelobasileus* were also found here. The Post quarry represents the concentration of carcasses by a mass mortality event. The quarry preserves part of a flood plain adjacent to a meandering channel that was periodically inundated with flooding. The fauna preserved in this quarry were primarily terrestrial animals that probably perished in a flash flood.

This multispecies bone bed occurs near the middle of a steep slope of a 600-meter-high cliff of a Dockum hill (fig. 3.3A). The more we dug, the more fossils we found along the bone bed. Surprisingly, fossils are totally absent from the rest of the section. We had exposed the bone bed as a narrow platform along the strike of the hill for almost 10 meters. We had used picks, shovels, and other hand tools, prepared a detailed quarry map, and collected systematically the delicate fossil remains of the Triassic treasures. We then needed to extend our quarry back into the hill to trace the continuation of the bone bed. This would be a far more difficult task, as we would have to remove quite a large amount of overburden. Ideally, pick and shovel would be the best tools with which to expose the bone bed toward the hill, but this would be a slow process. We had only the summer for field work. Riley Miller, the local rancher who had kindly allowed us to work on his property, offered his bulldozer for this excavation. But the bones were too delicate and the mudstone was too soft for the heavy machinery. Finally, we decided on another plan. We would remove the overburden with a heavy jackhammer to make a wide platform along the slope of the cliff. As soon as we approached the bone bed, we would begin to use pick and shovel, and then trowels, awls, and pins, with whisk brooms and soft brushes to remove dirt from the bones.

We rented a 90-pound Bosch jackhammer commonly used in highway construction and hooked it up to a gasoline genera-

A

B

FIGURE 3.3

A, the Post quarry in Garza County, Texas, where *Protoavis* skeletons were first found. *B,* jackhammer operation during the discovery of *Protoavis* in the summer of 1983. Two delicate skeletons of *Protoavis* were exposed accidentally near the blade of the jackhammer while removing the overburden; *left,* Bryan J. Small; *right,* J. Bruce Moring.

tor. My two students, Bryan Small and Bruce Moring, both superb athletes, became the official jackhammer operators. Mike Nickell and I shoveled away the dirt down the slope to the dry streambed, while Joanne Burley took the field photographs.

For nine days, our work progressed at a rapid pace. On the tenth day, I was watching how easily Bryan and Bruce were removing the large blocks of mudstone (fig. 3.3B). They had already learned how to operate a jackhammer efficiently. We were still in a safe zone, working about a meter above the bone bed. Suddenly, a small block of mudstone, freshly dug out, caught my eyes. Bryan and Bruce stopped the jackhammer. I picked up the mudstone block, cleanly sheared on one side by the thick blade of the jackhammer, and saw a series of small, delicate bones, jumbled up together. In my field notebook of 17 June 1983, I recorded this discovery:

> We found a juvenile skeleton of *Coelophysis* in a mudstone block that fits into my palm. The block broke into 2 pieces exposing more bones. The vertebrae are exceedingly small, less than a centimeter in length, yet the neural arches are fused to the centra. The astragalus-calcaneum look like that of a theropod, but are fused to each other. The fusion of bones in this baby individual is unusual. No skull bones are visible. The specimen needs careful preparation.

We did not know at that time that these bones would be the type specimens of *Protoavis.* What a way to find the world's oldest bird! This was a serendipitous discovery that would eventually generate controversy and excitement among paleontologists and public. Because most of the skeletons were concealed in the matrix, we did not appreciate the importance of this discovery at that time. Moreover, scores of beautiful skeletons of *Coelophysis* were known from the Ghost Ranch of New Mexico. We had all visited Ghost Ranch not long before, when David Berman and his crews from the Carnegie Museum were removing large blocks of *Coelophysis* skeletons. Finding another skeleton of *Coelophysis* did not excite our imagination. We wanted to find something new to science. We put the two mudstone blocks in a cardboard box with a field number of M83/1, brought it back to the lab, and nearly forgot about them. We exposed the bone bed extensively and collected several skeletons of interesting tetrapods from that site in the summer of 1983.

In 1985, while I was studying a little theropod, *Alwalkeria*, from the Triassic of India (Chatterjee 1987b), I retrieved the Dockum specimens for comparison. The Dockum theropods were tiny animals compared to *Alwalkeria;* the bones were delicate, were jumbled together, and required careful preparation under a binocular microscope. It took almost six months to remove the matrix and prepare the material. Soon I began to realize that the Texas specimens represented two individuals of an unknown species never found before. They were not *Coelophysis,* nor did they resemble any known theropods. The most striking feature in the skull is the birdlike quadrate with an orbital process, where the quadrate head fits nicely into a socket of the squamosal. The squamosal is also reduced in avian fashion, with the development of a zygomatic process. It lacks the descending process in front of the quadrate, indicating streptostyly. The jugal and quadratojugal form a horizontal bar. None of these features is known in any nonavian theropod. The temporal region is modified in avian fashion. The braincase is highly enlarged, with numerous pneumatic sinuses. Several postcranial characters, such as the strutlike coracoid, the furcula with a large hypocleidium, the heterocoelous cervicals, and the fused ilioischiadic plate, are present only in birds, not in any nonavian theropods.

I was excited. Was it a bird? For a long time I debated and hesitated to commit myself. Anatomical details of this intriguing animal convinced me that it is more closely related to living birds than is *Archaeopteryx.* Moreover, the animal could fly, which is revealed by its flight apparatus. Many experts on fossil birds came to our museum and confirmed my belief that the newly found material exhibited a suite of distinct avian traits. It became clear to me that, if *Archaeopteryx* is a bird, we must incorporate the Texas specimens as a more derived member of the avian clade. After many years of research and reflection, I finally named and described these animals as *Protoavis texensis* in 1991.

Protoavis was extremely small relative to nonavian dinosaurs. The large individual of *Protoavis* (holotype, TTU P 9200) is a pheasant-sized bird with a long bony tail. Its total length is about 60 cm, corresponding in size to the Solnhofen specimen of *Archaeopteryx.* The smaller individual (paratype, TTU P 9201) probably represents a juvenile individual. It is about half the size of the holotype and matches the size of the Eichstätt specimen of *Archaeopteryx.*

Later, I collected a total of thirty-one isolated postcranial elements of *Protoavis* from the Kirkpatrick quarry, along with other microvertebrates. Here the fossils are concentrated in a thin layer of carbonate granule conglomerate intercalated with red mudstone of the Tecovas Formation. The presence of carbonate granular conglomerate, organic matter, coprolites, and abundant fish fauna (*Ceratodus, Semionotus, Xenacanthus*) in this quarry indicates a quiet pond in a marshlike depositional setting. The preservation of bone is excellent in this quarry. All bird bones collected from the Dockum Group represent only one species, *Protoavis texensis.* They are now housed at the Museum of Texas Tech University. Nearly every skeletal element is represented among these specimens (Chatterjee 1995).

The Paleoenvironments of *Protoavis*

The Dockum sediments and fossils give some idea of the environments in which *Protoavis* might have lived. In Triassic time, Texas was situated close to the equator, with a climate that was warm, probably mostly wet, and dominated by monsoonal circulation (fig. 3.1). The distribution of detrital zircon suggests that a large Mississippi-scale paleoriver—the Chinle-Dockum river—was flowing primarily northwest from western Texas to eastern Nevada, supplied by water from the Amarillo-Wichita Uplift (Riggs et al. 1996). The river fluctuated in size over time, its environs ranged from forest to woodland, and the region was dotted with lakes and ponds.

Plant fossils are excellent indicators of past climate, but the flora from the Dockum sediments is poorly known. Sidney Ash (1972), a paleobotanist, who has identified many plant fossils from the contemporary Petrified Forest of Arizona, also looked at the Dockum flora. He was able to recognize several common plant fossils, including the remains of three ferns (*Cynepteris, Phlebopteris,* and *Clathropteris*), five gymnosperms (*Pelourdea, Araucarioxylon, Woodworthia, Otozamites, Dinophyton*), and one possible cycad (*Sanmiguelia*). The logs of conifer trees such as *Araucarioxylon* are common in the Dockum. Some of these trees were 40 to 50 meters tall. The rarity of plant fossils in the Dockum redbeds is due to the oxidizing environment that destroyed most of their remains. In rare instances, I have encountered beautiful leaf impressions, petrified wood, and lignite layers, as well as conchostracans (small aquatic crustaceans) in green mudstone facies that may represent reducing pond deposits.

From the remains of the Dockum flora, Ash concluded that the climate was moist and warm. He argued that the living relatives of the Dockum ferns and gymnosperms now mostly inhabit humid tropical areas and that the large specimens of *Araucarioxylon* suggest an abundant water supply. The pollen assemblages demonstrate that a moderately diverse flora was present in the Dockum environments and that the climate was moist. There were three distinct plant communities in the Dockum: (1) a flood plain swamp community of ferns and cycads, (2) a lowland closed canopy of forest of *Araucarioxylon*, and (3) an upland gymnosperm community. These plants supplied the food and camouflage for the Dockum herbivores, including forms such as aetosaurs and trilophosaurs (fig. 3.4).

The Dockum fauna gives us further clues about the Late Triassic environment. The presence of large bodies of water is supported by the locally abundant aquatic animals, such as bivalves, conchostracans, fish, metoposaurs, and phytosaurs. Within the Dockum ecosystems, three principal habitat subzones for the vertebrates have been identified: (1) aquatic—rivers, lakes, and ponds inhabited by fish, metoposaurs, and phytosaurs; (2) lowland—margins of rivers, lakes, and ponds occupied by the brachyopids, rhynchosaurs, protorosaurs, trilophosaurs, and squamates; and (3) upland—divides between two streams, interfluves, populated by aetosaurs, poposaurs, dinosaurs, and therapsids (fig. 3.4).

In the Dockum terrestrial communities, two major groups of archosaurs thrived: the pseudosuchians and the dinosaurs (including birds). When the dinosaurs first appeared on the Triassic scene, they were relatively small, overshadowed by the pseudosuchians. Some of the Triassic pseudosuchians show stunning convergences with later dinosaurs. For example, aetosaurs show ankylosaur-like body armor and limb structures, whereas *Postosuchus* exhibits tyrannosaur-like posture and carnivorous adaptations. These superficial similarities indicate that the Triassic pseudosuchians had nearly identical functional requirements and, thus, evolved structural adaptations similar to those of later dinosaurs.

Early dinosaurs, such as ceratosaurs, *Technosaurus*, and *Shuvosaurus*, have been recovered from the Post quarry (Chatterjee 1984, 1993). These animals lived in a dense riverine forest of conifer trees with an understory of ferns that formed the base of the terrestrial ecosystems (fig. 3.5). The Dockum food chain started with the plant-eating animals and progressed through a series of larger carnivorous animals. Surprisingly, at the top of the food chain was the poposaurid *Postosuchus,* which was not a theropod, but a pseudosuchian (fig. 3.4). Evolution and resource availability played major roles in the formation of the Dockum community. A diverse variety of food was abundant in the Dockum environment, allowing an array of different types of animals to coexist with little interference (Chatterjee 1985, 1992a).

In the continuing process of natural selection in the Dockum environment, *Protoavis* emerged as a small feathered theropod that leaped into the air to exploit a new frontier of ecospace. This was a giant leap that heralded the age of birds and their conquest of the skies. Although both birds and nonavian dinosaurs appeared simultaneously during the latter part of the Triassic and coexisted throughout the Mesozoic, they adapted different strategies and explored different niches for obtaining food and shelter. Nonavian dinosaurs tended toward large to gigantic body size in terrestrial habitat and became the dominant land animals in the Jurassic and Cretaceous periods. Unlike birds, all nonavian dinosaurs disappeared at the end of the Cretaceous calamity. Birds maintained a small body size because of their flight constraints and exploited a wide range of ecological

FIGURE 3.4

The food chain of the terrestrial components of the Dockum paleocommunities. *Postosuchus* was at the top of the food chain.

niches beyond the reach of nonavian dinosaurs. They survived two major mass extinctions in the Mesozoic and enjoyed bursts of adaptive radiation during the Tertiary. In living birds you can hear the distant murmurs of a pioneer Texan dinosaur who ventured into a new frontier to conquer the sky about 225 million years ago.

FIGURE 3.5

Artist's representation of a Dockum landscape, showing some of the Late Triassic archosaurs in the Dockum quarry: the pseudosuchian *Postosuchus* at the *center*, the early ornithischian dinosaur *Technosaurus* in the *left corner*, the early ornithomimosaurid dinosaur *Shuvosaurus* at the *right*, and the earliest known bird *Protoavis* in the sky.

A Portrait of *Protoavis*

By George, this must be the trail of the father of all birds!

—Arthur Conan Doyle, *The Lost World,* 1912

CHAPTER 4

Before we start to look at different kinds of Mesozoic birds and to appreciate the origin and diversity of this group, we need to look more closely at the anatomy of *Protoavis.* Being the earliest bird, its skeleton provides the general body plan that has been modified to meet the rigorous demands of flight. The skeleton of *Protoavis* has a basically avian structure but retains several primitive characteristics of nonavian theropods, an example of mosaic evolution. The systematic hierarchy of *Protoavis* is as follows:

Aves Linnaes, 1758
Ornithothoraces Chiappe and Calvo, 1994
Protoavidae Chatterjee, 1991
Protoavis Chatterjee, 1991
Protoavis texensis Chatterjee, 1991

The Anatomy of *Protoavis*

THE SKULL

As in most Mesozoic birds, the cranial bones are not fused in *Protoavis.* They are thin and delicate, and they tend to become somewhat disarticulated during fossilization. The skull is relatively small: 74 mm long, 38 mm wide, and 32 mm deep as restored in the holotype (fig. 4.1). The short snout, enormous orbit, and inflated braincase have crowded the cheek region. Many of the bars behind the orbit and between the two temporal openings have disappeared. The skull shows several avian hallmarks in lateral aspect, including a modified diapsid condition, the postorbital process, the zygomatic process, and the orbital process that are lacking in *Archaeopteryx.* The teeth are retained at the tip of the jaws, but the posterior teeth are lost. The edentulous region was evidently encased by horny sheath, as in modern birds. The three premaxillary teeth oppose the two dentary teeth; each tooth is small, compressed sidewise, and lacks serration. The naris is bounded by the premaxilla and the nasal. The maxilla is edentulous and has an ascending process in front of the antorbital fenestra that underlies the nasal. The prefrontal is absent, as in other birds. The frontal is an elegant, inflated bone that surrounds the orbit. The orbits are very large, about one-third of the skull length, and are forwardly placed, suggesting a large component of stereoscopic vision. The postorbital bone is lost. In its place, the frontal and laterosphenoid form a postorbital process for the attachment of the postorbital ligament. Rostrally, the frontal becomes narrow and flat to form a flexible hinge with the nasal. Behind the frontal, the parietal is transversely arched and contacts the squamosal laterally. The lacrimal forms a vertical bar between the orbit and the antorbital opening and has a sliding joint with the jugal. The jugal and quadratojugal form a slender jugal bar without any vertical struts. The quadratojugal, in turn, develops a flexible pin joint with the quadrate. Because of the elimination of the diapsid arches and the squamosal/quadratojugal bar, the quadrate becomes streptostylic. The squamosal is considerably reduced; it shows a distinctive zygomatic process and the lack of a descending process in front of the quadrate. It has developed a separate ventral cotyle for the quadrate head to form the ball-and-socket joint.

In the palate, the choana has been shifted considerably backward so that the palatal components of the maxillae and the intervening vomers contribute to the secondary palate, as in primitive birds (fig. 4.1C). The vomers are long and fused rostrally to lie between the premaxillae. Caudally, they are separated and meet the pterygoids. The palatine is a long, slender bone that forms the lateral margin of the choana. The pterygoid is reduced considerably rostrally. It runs backward and outward to the ventral condyle of the quadrate, to which it is loosely attached. The ectopterygoid is lost, as in birds, to make the palate more mobile. The quadrate has a single, spherical head but shows several avian attributes: a free orbital process, a ventral condylar articulation with the pterygoid, a lateral cotylus for the quadratojugal, a tricondylar articulation with the mandible, and the acquisition of streptostyly.

The braincase (fig. 4.2A) is inflated and shows neurosensory specialization possibly associated with balance, coordination, muscular control, and flight. The occipital condyle is fairly small relative to the foramen magnum. A distinctive epiotic bone is present on the posterior aspect of the occiput and is separated from the supraoccipital by a sinus canal. Medially, a large and deep floccular recess occurs between the epiotic and prootic; this recess is surrounded by a bony tube for the rostral vertical semicircular canal (fig. 4.2B). The otic capsule is built in avian fashion and shows three foramina clustered together in the tympanic recess: the fenestra ovalis in the front, the fenestra pseudorotunda behind, and the caudal tympanic recess at the caudodorsal corner. Behind the fenestra pseudorotunda, a large metotic strut is added to the exoccipital so that the vagus foramen (X) has been diverted backward and emerges through a single foramen in the occiput. Lateral to it lies the exit for the hypoglossal (XII) nerve. A parabasal notch lateral to the vagus foramen gives passage for the internal carotid artery. This notch

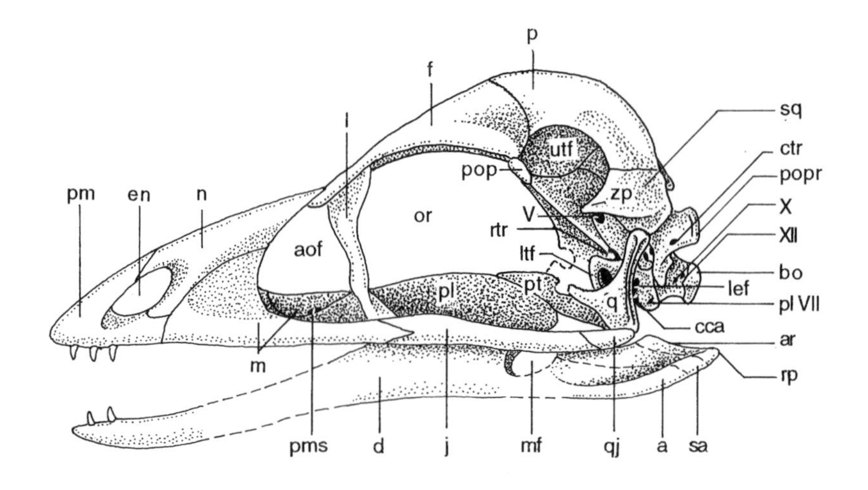

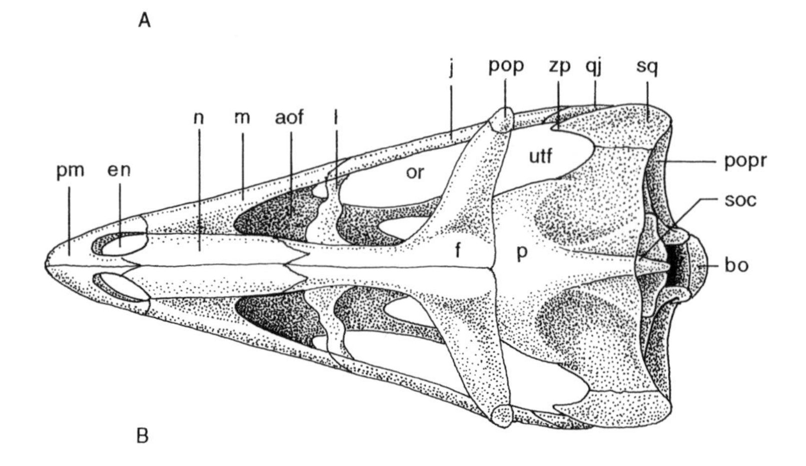

FIGURE 4.1

Protoavis texensis; composite restoration of the skull representing the size of the holotype, TTU P 9200. *A,* left lateral view; *B,* dorsal view; *C,* ventral view; *D,* occipital view (source: Chatterjee 1991, 1995, n.d.). Abbreviations: *a,* angular; *ams,* anterior maxillary sinus; *aof,* antorbital fenestra; *ar,* articular; *bo,* basioccipital; *bpt,* basipterygoid process; *bs,* basisphenoid; *cca,* foramen for cerebral carotid artery; *ch,* choana; *cp,* cultriform process; *ctr,* caudal tympanic recess; *d,* dentary; *dtr,* dorsal tympanic recess; *en,* external naris; *eo,* exoccipital; *eov,* foramen for external occipital vein; *ep,* epiotic; *f,* frontal; *fm,* foramen magnum; *fo,* fenestra ovalis; *fpr,* fenestra pseudorotunda; *j,* jugal; *l,* lacrimal; *lef,* lateral eustachian foramen; *ltf,* lower temporal fenestra; *m,* maxilla; *met,* metotic strut; *mf,* mandibular foramen; *n,* nasal; *oa,* occipital artery foramen; *opr,* orbital process; *or,* orbit; *p,* parietal; *pbn,* parabasal notch; *pl,* palatine; *pm,* premaxilla; *pms,* posterior maxillary sinus; *pop,* postorbital process; *popr,* paroccipital process; *pr,* prootic; *ps,* parasphenoid; *pt,* pterygoid; *ptf,* posttemporal fenestra; *q,* quadrate; *qj,* quadratojugal; *rp,* retroarticular process; *rtr,* rostral tympanic recess; *sa,* surangular; *sc,* sinus canal; *sma,* foramen for sphenomaxillary artery; *soc,* supraoccipital; *sq,* squamosal; *utf,* upper temporal fenestra; *v,* vomer; *vef,* ventral eustachian foramen; *zp,* zygomatic process; Roman numerals indicate the foramina for cranial nerves.

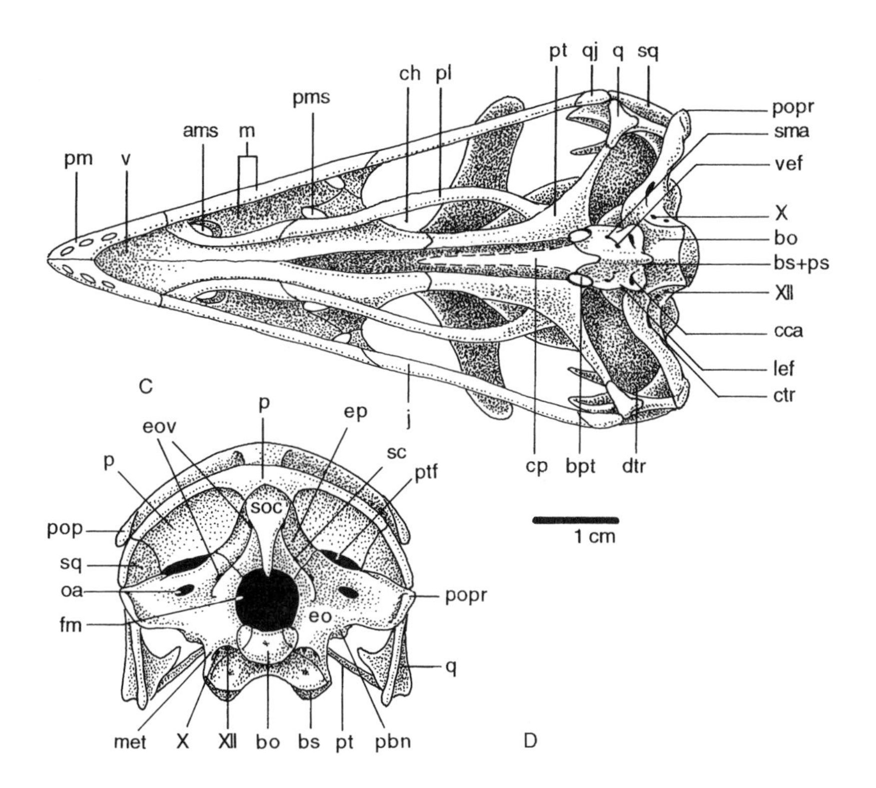

continues forward as a bony parabasal canal. The middle ear cavity region is highly pneumatized, and all of the air sinuses of birds—rostral, caudal, dorsal, quadrate, and articular tympanic recesses—are present. The functions of these tympanic recesses are not well understood but may be linked to better detection of low-frequency sound (Witmer 1990). As in living birds, the rostral tympanic recess in *Protoavis* shows contralateral communication that might provide critical directional information in three-dimensional space (Rosowki and Saunders 1980). The basioccipital-basisphenoid complex is extensively permeated by pneumatic cavities. The bony eustachian tube is enclosed within the parasphenoid; its lateral opening is highly enlarged, whereas its ventral opening lies medial to the basipterygoid process.

The lower jaw (fig. 4.1A) is slender, tapers slightly forward, and is very shallow for most of its length. The mandibular symphysis is ossified, and the postdentary bones are fused. The external mandibular foramen is small and elongated and is bor-

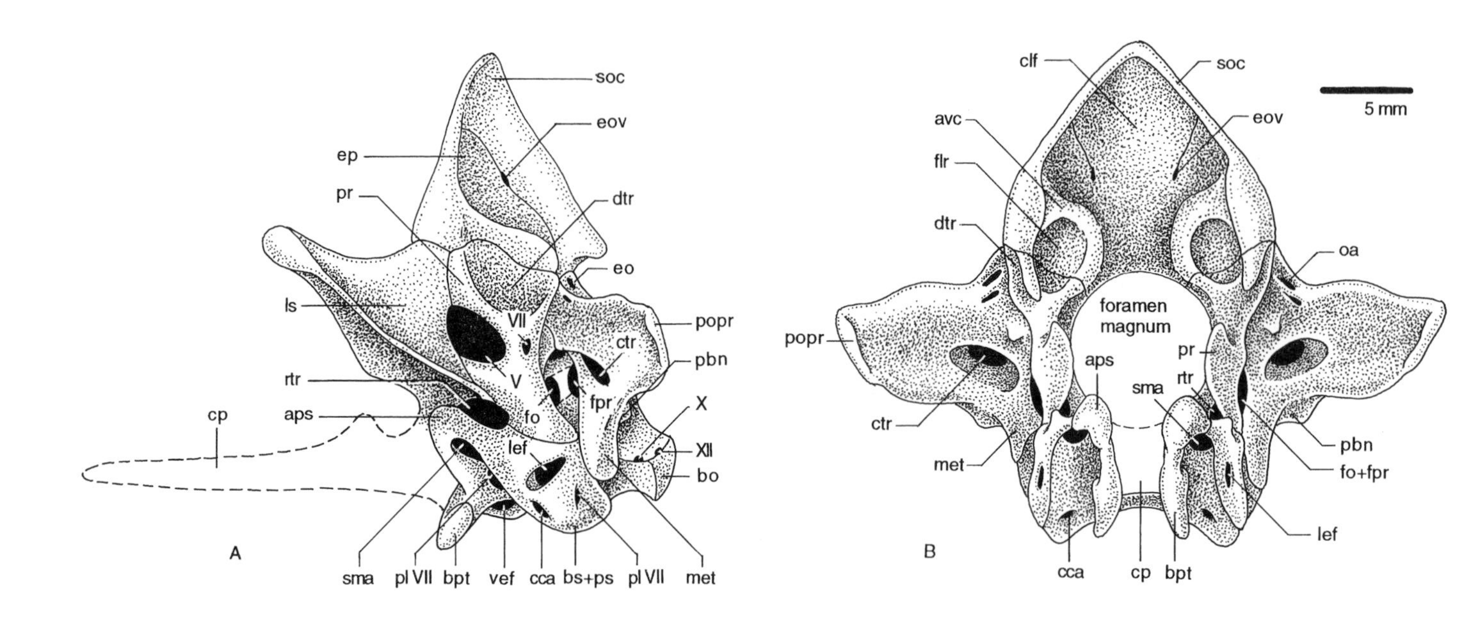

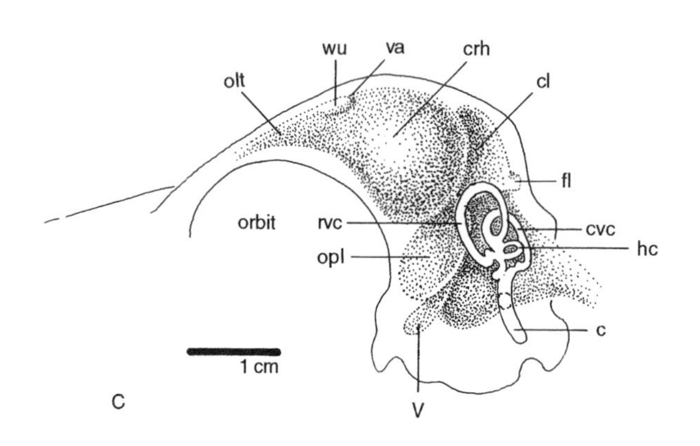

FIGURE 4.2

Protoavis texensis; composite restoration of the braincase and endocast, representing the size of the holotype. *A*, left lateral view; *B*, cranial view; *C*, restoration of brain from endocast, lateral view (source: Chatterjee 1991, 1995, n.d.). Abbreviations: *aps*, alaparasphenoid; *bo*, basioccipital; *bpt*, basipterygoid process; *c*, cochlear duct; *cca*, foramen for cerebral carotid artery; *cl*, cerebellum; *clf*, cerebellar fossa; *cp*, cultriform process; *crh*, cerebral hemisphere; *ctr*, caudal tympanic recess; *cvc*, caudal vertical canal; *dtr*, dorsal tympanic recess; *eoa*, foramen for external occipital vein; *ep*, epiotic; *euf*, foramen for eustachian canal; *fl*, floccular lobe; *flr*, floccular recess; *fm*, foramen magnum; *fpr*, fenestra pseudorotunda; *hc*, horizontal canal; *ic*, foramen for internal carotid artery; *lef*, lateral eustachian canal; *ls*, laterosphenoid; *met*, metotic strut; *oa*, foramen for occipital artery; *olt*, olfactory tract; *opl*, optic lobe; *pbn*, parabasal notch; *popr*, paroccipital process; *pr*, prootic; *rtr*, rostral tympanic recess; *rvc*, rostral vertical canal; *sma*, foramen for sphenomaxillary artery; *soc*, supraoccipital; *va*, vallecula; *vef*, ventral eustachian foramen; *wu*, Wulst; Roman numerals indicate the foramina for cranial nerves.

dered by a rostral fossa. The surangular shows a lateral process for the attachment of the postorbital ligament. The retroarticular process, formed mostly by a posterior extension of the articular bone, is well developed.

THE VERTEBRAL COLUMN

It is estimated that there are twelve cervical, nine dorsal, six sacral, and twenty caudal vertebrae in *Protoavis.* The heterocoelous centra and hypapophyses in the anterior cervical region of *Protoavis* (fig. 4.3A) are highly derived features among Mesozoic birds. These features are restricted to few Cretaceous taxa, such as Enantiornithes, Hesperornithiformes, *Patagopteryx,* and the Antarctic loon *Polarornis.* Each prezygapophysis in this region faces forward and forms a convex rolling surface in lateral aspect. The curvature of the prezygapophyseal facet approximates a circular arc, permitting a continuous range of movement of the neck without obvious osteological stops. The long and flexible neck with the development of heterocoelous vertebrae permitted *Protoavis* a wide range of movement of the head, which functioned as a universal tool. The neural spines are atrophied. In the following presacral vertebrae, the centrum is amphicoelous (with both the anterior and the posterior surfaces being concave) to platycoelous (flat or concave ventrally and convex dorsally); it has a large hollow core in cross section with a paper-thin bony shell around it in camellate fashion. The absence of pleurocoel on the centra may be attributed to the juvenile nature of the specimen (9200), but pneumatic fossae on the centra and pneumatic chambers in the neural arch are present throughout the column. In several isolated specimens, pleurocoels are preserved. The neural canal is highly enlarged relative to the centrum, presumably indicating a higher demand for nerve signal traffic. The trunk appears to be short, horizontal, and rigid without any sign of fusion. The sacrum is poorly preserved; the first sacral shows the characteristic expanded ribs. *Protoavis* had a long bony tail similar to that of *Archaeopteryx,* without any development of pygostyle at the end. The caudal

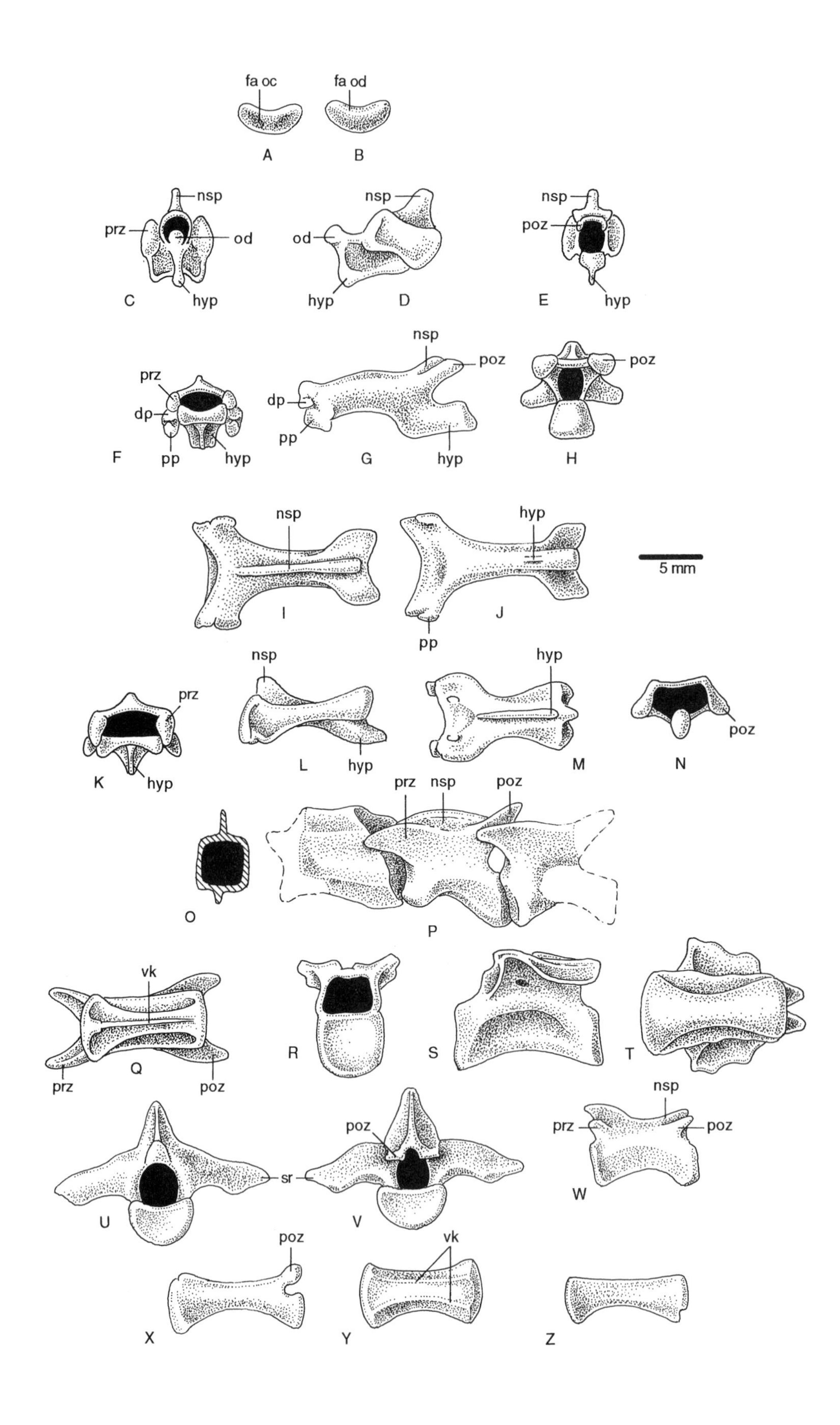
fa oc
fa od
A
B
nsp
prz
od
hyp
C
D
E
poz
dp
pp
F
G
H
I
J
5 mm
K
L
M
N
O
P
vk
Q
R
S
T
sr
U
V
W
X
Y
Z

centra are weakly amphicoelous and maintain a uniform length throughout the column but decrease in diameter posteriorly. The terminal tail vertebrae become highly slender and delicate rods. They lack both zygapophyses and intersegmental mobility; they probably functioned together as a stiff unit to control the pitch during flight.

THE SHOULDER GIRDLE

The shoulder girdle of *Protoavis* was designed on the same mechanical principle as that of modern birds, with the development of a strutlike coracoid, a triosseal canal for the supracoracoideus pulley, a springlike furcula bearing hypocleidium, and a keeled sternum (fig. 4.4A–B). The scapula is a long and tapering bone that meets the coracoid in a flexible manner. A pneumatopore is present near the acromion, where the scapula joins the furcula. The glenoid cavity faces outward and upward, which permits dorsoventral movement of the humerus. The coracoid is long and stout and bears both procoracoid and acrocoracoid processes. The sternum has a ventral keel for the attachment of two sets of flight muscles, namely the pectoralis and supracoracoideus (fig. 4.4M). Dorsally, the sternal basin is

FIGURE 4.3

Protoavis texensis, vertebral column, representing the size of the paratype TTU P 9201. *A* and *B,* cranial and caudal views of the atlas intercentrum. *C–E,* cranial, lateral, and caudal views of the axis. *F–J,* cranial, lateral, caudal, dorsal, and ventral views of the third cervical; note the heterocoelous centrum and hypapophysis. *K–N,* cranial, lateral, caudal, and ventral views of the fourth cervical. *O,* cross section of the ninth cervical showing the hollow core. *P,* lateral view of the ninth, tenth, and eleventh cervicals. *Q,* ventral view of the tenth cervical. *R–T,* rostral, lateral, and ventral views of the first dorsal. *U* and *V,* rostral and caudal views of the first sacral. *W,* lateral view of the fourth caudal. *X,* lateral view of the thirteenth caudal. *Y,* ventral view of the sixth caudal. *Z,* lateral view of the twentieth caudal. (Source: Chatterjee 1995.) Abbreviations: *dp,* diapophysis; *fa ac,* facet for occipital condyle; *fa od,* facet for odontoid; *hyp,* hypapophysis; *nsp,* neural spine; *od,* odontoid; *pp,* parapophysis; *poz,* postzygapophysis; *prz,* prezygapophysis; *sr,* sacral rib; *vk,* ventral keel.

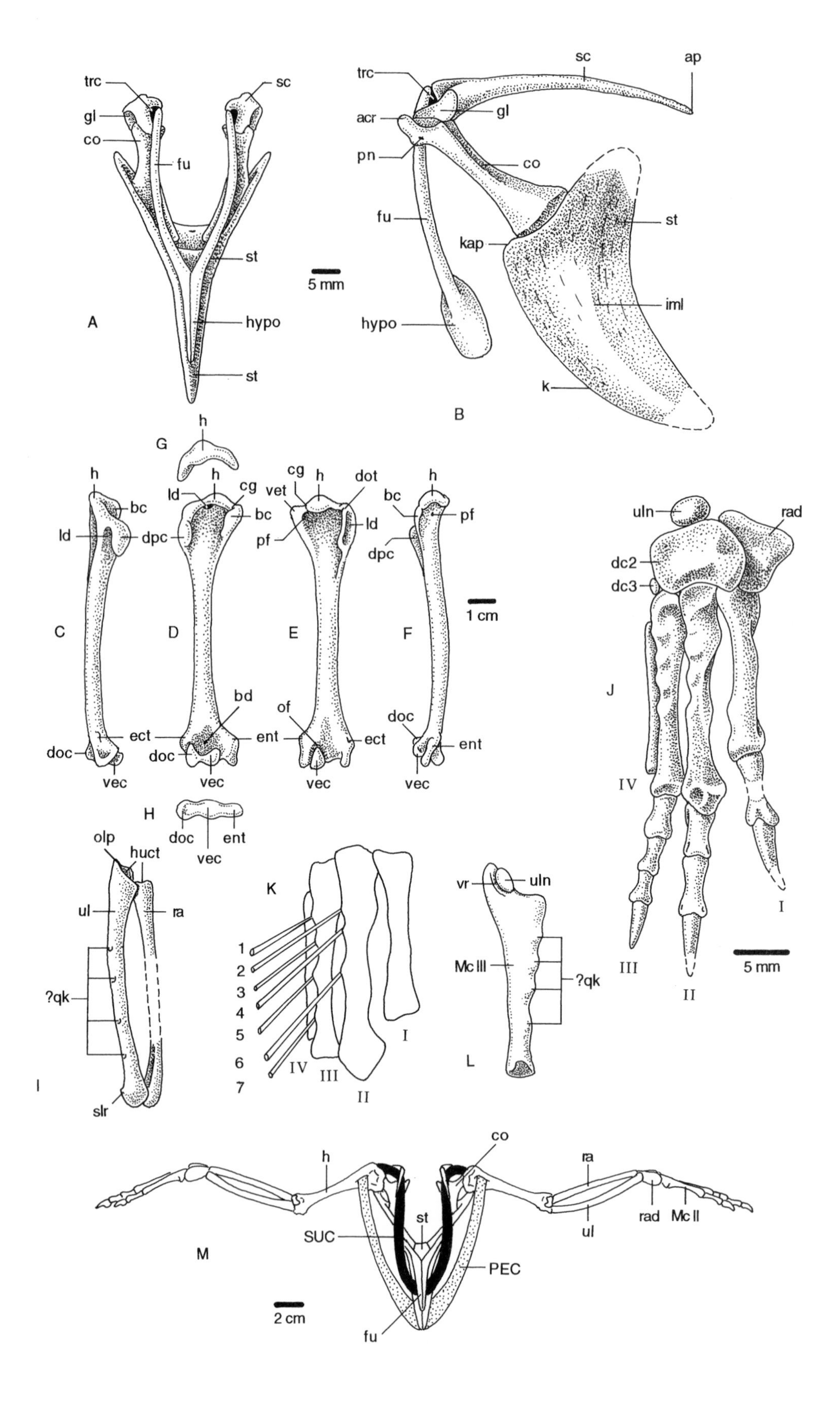

trc
sc
gl
co
fu
st
hypo
st
A
5 mm
trc
sc
ap
gl
acr
pn
co
fu
st
kap
iml
hypo
k
B
G
h
h
ld
cg
bc
h
bc
ld
dpc
cg
h
dot
vet
pf
ld
h
bc
pf
dpc
1 cm
C
D
E
F
bd
of
ect
ent
ect
doc
ent
doc
doc
vec
vec
vec
vec
H
doc
vec
ent
uln
rad
dc2
dc3
J
IV
III
II
I
5 mm
olp
huct
ul
ra
?qk
I
slr
K
1
2
3
4
5
6
7
IV
III
II
I
vr
uln
Mc III
?qk
L
co
h
ra
st
SUC
PEC
rad
Mc II
ul
M
2 cm
fu

hollow and is pierced by several pneumatic foramina. The furcula is very similar to that of a chicken (*Gallus*) in morphology, with a large hypocleidium. It would act as a spring between two shoulder girdles, facilitating inflation and deflation of the clavicular sac during flight (Jenkins, Dial, and Goslow 1988).

THE WING

Like living birds, *Protoavis* acquired two independent and specialized methods of locomotion, flying with the forelimbs and walking with the hindlimbs. *Protoavis* was an obligatory biped, as indicated by the development of the swivel wrist joint, which would restrict the hand movement for terrestrial locomotion. The forelimb developed a linkage system at the elbow and wrist joints and was modified into a collapsible wing. The humerus (fig. 4.4C–H) of *Protoavis* is remarkably avian. It is a strong, tubular bone where the two expanded ends lie in the same plane. Proximally, there is a pronounced head that fits into the

FIGURE 4.4

Protoavis texensis, composite restoration of shoulder girdle and forelimb, representing the size of the holotype. *A* and *B,* cranial and lateral views of the shoulder girdle. *C–H,* dorsal, cranial, caudal, ventral, proximal, and distal views of the right humerus. *I,* dorsal view of the right radius and ulna. *J,* dorsal view of the right manus. *K,* outlines of metacarpals showing the quill knobs for attachment of primaries. *L,* caudal view of metacarpal III showing its interlocking articulation with ulnare by ventral ridge. *M,* shoulder girdle and extended wings in cranial view to show the action of the flight muscles. (Source: Chatterjee 1995.) Abbreviations: *acr,* acrocoracoid; *ap,* apex; *bc,* bicipital crest; *bd,* brachial depression; *bt,* biceps tubercle; *cg,* capital groove; *co,* coracoid; *dc,* distal carpal; *doc,* dorsal condyle; *dot,* dorsal tuberosity; *dpc,* deltopectoral crest; *ect,* ectepicondyle; *ent,* entepicondyle; *fu,* furcula; *gl,* glenoid fossa; *h,* head; *hu,* humerus; *huct,* humeral cotyle; *hypo,* hypocleidium; *iml,* intermuscular lamina; *k,* keel; *kap,* keel apex; *ld,* ligamental depression; *mc,* metacarpal; *of,* olecranon fossa; *olp,* olecranon process; *PEC,* pectoralis muscle; *pf,* pneumatic fossa; *pn,* pneumatic foramen; *qk,* quill knobs; *ra,* radius; *rad,* radiale; *sc,* scapula; *str,* semilunate ridge; *st,* sternum; *stb,* sternal basin; *SUC,* supracoracoideus muscle; *trc,* triosseal canal; *ul,* ulna; *uln,* ulnare; *vec,* central condyle; *vr,* ventral ridge; Roman numerals indicate the digits in manus.

glenoid. In cranial aspect, a ligamental depression occurs just below the head for the attachment of the supracoracoideus muscle. Farther dorsally the deltopectoral crest is expanded and has a ridgelike projection for the insertion of the pectoralis muscle. At the ventral side, the bicipital crest, an unusual advanced feature in a Triassic bird, is weakly developed. Both dorsal and ventral tuberosities are encountered near the proximal end, but they are relatively subdued in comparison to those in modern birds. A shallow pneumatic fossa occurs below the ventral tuberosity but is not pneumatized by the foramen. Here the head is separated from the ventral tuberosity by a capital groove. The shaft is long, hollow, and flattened cranially. The distal expansions show two well-defined, asymmetrical condyles for the radius and the ulna, respectively. The dorsal or radial condyle is larger and more pronounced, elongated parallel to the axis of the bone. The ventral or ulnar condyle is almost spherical and continues onto the posterior surface, where it is bordered by a deep olecranon fossa. Both ectepi- and entepicondyles are well developed on the dorsal and ventral borders of the distal end.

The radius is a straight cylindrical bone that is slimmer than the ulna. Proximally, it shows a concavity for the reception of the humerus; distally, it is expanded to receive the radiale. The ulna is a curved bone with an olecranon process at the proximal end. The shaft is damaged and bears a series of faint bumps similar to quill knobs (fig. 4.4I), but their identity is uncertain. The distal end is slightly expanded into a trochlear surface and shows a weak semilunate ridge on the dorsal edge for articulation with the ulnare.

The manus is beautifully preserved in *Protoavis.* It shows four separate carpal and four metacarpal bones without any sign of fusion (fig. 4.4J). Four metacarpals are also present in embryonic birds, but in adults the first three metacarpals are fused with the distal two carpal bones to form the carpometacarpus, whereas the fourth metacarpal disappears.

The wrist is intact in *Protoavis.* The radiale is ovoid in outline and considerably larger than the ulnare. Proximally, it has a distinct facet for the radius; distally, it has a rolling surface for the distal carpal 2. The ulnare is a small disc, lying between the ulna and the distal carpal 2. The smallest carpal bone, proximate to the metacarpal III, is identified as the distal carpal 3. The distal carpal 2 is the largest carpal element and has a well-designed joint surface. Proximally, it has a pulley-like groove, the carpal trochlea, which extends along the whole width to form the swivel joint. This allows the hand to move along the plane of the wing surface. Distally, the semilunate bone has a single socket for reception of the metacarpal II. The hand has three functional digits, with a vestige of metacarpal IV. Metacarpal I shows a subdued extensor process. Metacarpal II is the most robust and longest bone in the series. Metacarpal III has a ventral ridge at the proximal end to interlock with the ulnare. Both metacarpals II and III appear to fuse at the ends to form a large intermetacarpal space. They bear a series of quill knobs for the attachment of primary feathers. There are about seven quill knobs, arranged alternately on the dorsal surfaces of metacarpals II and III. Unlike modern birds, *Protoavis* shows that both metacarpals might have supported primary feathers in alternate fashion, while metacarpal IV provided lateral support for the shaft of the feathers. Metacarpals I–III retain phalanges and terminal claws, with the formula being 2-3-4-0-x.

THE PELVIC GIRDLE

The pelvis of *Protoavis* shows derived features not reported in many Mesozoic birds: (1) the ischium is rotated parallel to the ilium and fused distally to enclose the ilioischiadic foramen, (2) the ischium and pubis are open ventrally without any symphysis, (3) there is a renal fossa, (4) the ischiadic peduncle of the ilium is short and reduced, and (5) the pubis is relatively short, without any sign of distal expansion or "foot."

The ilium expands into a broad, flaring blade with long preacetabular and postacetabular processes; the former is hooked and bears a prominent ventral projection for the attachment of the femoral protractor muscle (fig. 4.5A–C). The opisthopubic pelvis and the cranial elongation of the preacetabular process may be linked to the subhorizontal attitude of the femur. The acetabulum is very shallow, circular, and completely perforated to receive the inturned head of the femur. The medial surface of the ilium is very complex and shows the renal fossa on the postacetabular area to accommodate the kidney. The fossa is bordered medially by the ilioischiadic pila, as in some modern birds, to reinforce the ilioischiadic plate. The renal fossa is a diagnostic feature of birds.

The ischium is a narrow, bony plate directed backward to fuse with the ilium; the line of fusion is visible on the lateral aspect, but the fusion is obscured medially by the ilioischiadic pila. Proximally, the bone is forked cranially around the acetabulum; the pubic process is a narrow rod, whereas the ischiadic process is broad and stout, bearing a weak antitrochanter process. Immediately below the pubic process is a prominent obturator process on the ischium.

The pubis, approximately equal to the ilium in length, is considerably longer and slimmer than the ischium. It projects downward and backward at a 45° angle to the horizontal and is separated from the ischiadic shaft by a triangular gap. In modern birds, the pubis lies alongside the ischium and contacts the obturator process to enclose the obturator foramen. The proximal end of the pubis is expanded and differentiates into a large iliac and a narrow ischiadic process around the acetabulum. A pectineal process on the pubis near the acetabulum is the origin of the ambiens muscle. The shaft is long and narrow without any distal expansion or symphysis, unlike the condition in *Archaeopteryx.*

THE HINDLIMB

The hindlimbs of *Protoavis,* as in many Mesozoic birds, are intermediate in morphology between those of nonavian theropods and modern birds. The femur is a cylindrical bone with a slight craniocaudal curvature. The head is oval in outline and broadly convex to fit into the acetabulum. The greater trochanter is weakly developed and contacts the antitrochanter on the acetabulum. From the greater trochanter a prominent obturator ridge curves distally as an oblique rugose crest on the caudal aspect of the shaft. The posterior trochanter, a muscle scar recognized in dromaeosaurs and *Archaeopteryx* by Ostrom (1976a), is also present in *Protoavis.* It occurs immediately below the greater trochanter. Unlike modern birds, *Protoavis* shows the primitive presence of the lesser trochanter. It projects as a conical boss below the head on the cranial surface and is set off from the shaft by a prominent shelf. The shaft is hollow, thin-walled, and oval in cross section. Distally, the femur flares into medial and lateral condyles for articulation with the tibia. The lateral condyle is highly pronounced and is separated from the fibular condyle by the fibular groove. The distal end is pierced by two nutrient foramina, one at the popliteal space and the other at the site of flexor attachment. I have observed these foramina in several species of modern birds.

The proximal head of the tibia is characteristically avian, with the development of both lateral and cranial cnemial crests (fig. 4.5J–K). The cranial cnemial crest is an avian feature that is absent in *Archaeopteryx.* A well-defined fibular crest to which the fibula is attached lies behind the lateral cnemial crest. The distal end is slightly inflated and is notched for the ascending process of the astragalus. Unlike modern birds, *Protoavis* has a tibia that is not fused with the proximal tarsal bones to form the tibiotarsus.

The fibula is a narrow, slender rod approximately equal in length to the tibia; it does not show any sign of fusion or reduction. A similar primitive fibula is known in several Mesozoic birds (e.g., *Archaeopteryx* and *Sinornis*). The proximal end is spatulate and is closely appressed to the fibular crest of the tibia. The shaft has an unusual tibial crest along the medial margin; this crest wraps tightly around the tibia. The distal end is oval and fits nicely into a corresponding socket of the calcaneum.

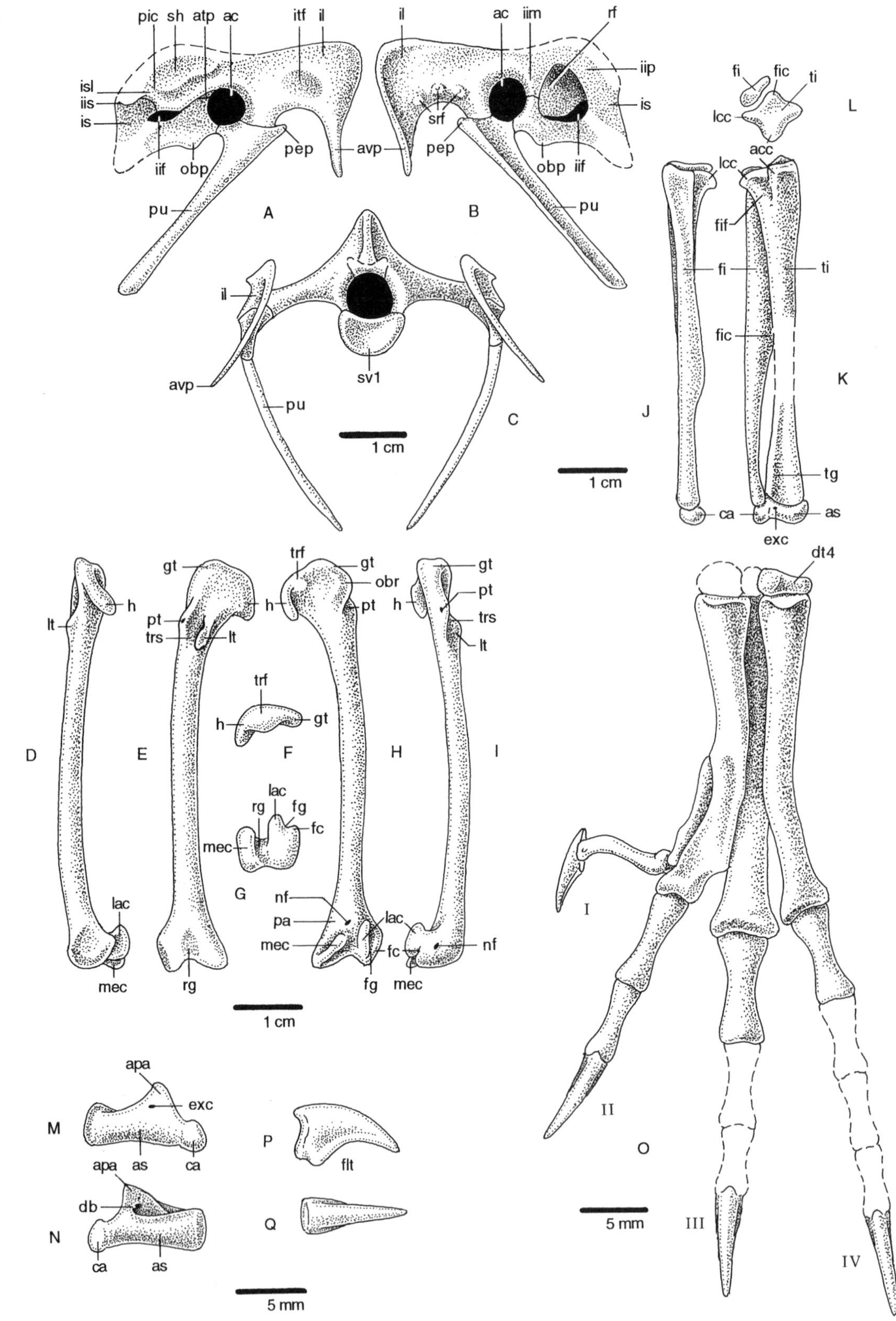
pic
sh
atp
ac
itf
il
isl
iis
is
iif
obp
pep
avp
pu
A
il
ac
iim
rf
iip
is
srf
pep
obp
iif
pu
B
il
avp
pu
sv1
C
1 cm
fi
fic
ti
lcc
acc
L
lcc
fif
fi
ti
fic
K
J
tg
ca
as
exc
1 cm
dt4
gt
trf
gt
obr
gt
h
lt
h
pt
trs
lt
h
pt
h
pt
trs
lt
trf
h
gt
D
E
F
H
I
lac
rg
fg
fc
mec
G
nf
pa
lac
mec
fc
nf
lac
mec
rg
fg
mec
1 cm
I
II
O
III
IV
5 mm
apa
exc
M
as
ca
P
flt
apa
db
N
ca
as
Q
5 mm

The pes is anisodactyl where the hallux is large, opposable, and fully reversed (fig. 4.5O). Metatarsal I is reduced to a splint and articulates with the distal end of metatarsal II. The long opposable hallux situated on the lower side of metatarsal II clearly indicates the development of a grasping foot. The central metatarsals (II–IV) are stout, elongated, and tightly appressed proximally in such a fashion that metatarsal III is proximally overlapped by metatarsal II and metatarsal IV in the cranial aspect. The distal trochlear surface of metatarsal III is symmetrical, while those of metatarsals II and IV are oblique and divergent. Metatarsal V is absent. The first four digits terminate in large, compressed, recurved claws with well-developed flexor tubercles, which are indicative of climbing or perching (fig. 4.5P–Q). *Protoavis* apparently possessed a capacity for tree climbing as well as bipedal walking.

FIGURE 4.5

Protoavis texensis, composite restoration of the pelvic girdle and hindlimb, representing the size of the holotype. *A–C,* lateral, medial, and cranial views of the right pelvic girdle. *D–I,* medial, cranial, proximal, distal, caudal, and lateral views of the right femur. *J–L,* lateral, caudal, and proximal views of the right tibia and fibula. *M–N,* cranial and caudal views of the left astragalocalcaneum. *O,* anterior view of the left pes. *P–Q,* lateral and dorsal views of a claw of the third digit. (Source: Chatterjee 1995.) Abbreviations: *acc,* anterior cnemial crest; *ac,* acetabulum; *apa,* ascending process of astragalus; *avp,* anteroventral process of ilium; *ca,* calcaneum; *db,* dorsal basin; *dt,* distal tarsal; *exc,* extensor canal; *fc,* fibular condyle; *fg,* fibular groove; *fi,* fibula; *fic,* fibular crest; *flt,* flexor tubercle; *gt,* greater trochanter; *h,* head; *iif,* ilioischiadic foramen; *iim,* intermedial iliac crest; *iip,* ilioischiadic pila; *iis,* ilioischiadic suture; *il,* ilium; *ilp,* iliac process; *is,* ischium; *isl,* ischiadic lamina; *isp,* ischiadic process; *itf,* iliotrochanteric fossa; *lac,* lateral condyle; *lcc,* lateral cnemial crest; *lt,* lesser trochanter; *mec,* medial condyle; *mt,* metatarsal; *nf,* nutrient foramen; *obp,* obturator process; *obr,* obturator ridge; *pa,* popliteal area; *pep,* pectineal process; *pic,* posterior iliac crest; *pt,* posterior trochanter; *pu,* pubis; *rf,* renal fossa; *rg,* rotular groove; *srf,* sacral rib facets; *sv1,* first sacral vertebra; *ti,* tibia; *tic,* tibial crest; *tg,* tendinal groove; *tro,* trochlea; *trs,* trochanteric shelf; Roman numerals indicate the digits in pes.

The Brain and Sense Organs

Although soft parts are not preserved in the fossil vertebrates, the braincase contains various internal molds that can reveal a great deal of information about the morphology of the neurosensory organs it housed. The fossil record may provide important clues to the evolutionary pattern of brain, behavior, intelligence, and senses of extinct vertebrates.

The vertebrate brain is an extremely complex organ whose different components have been fashioned by 500 million years of natural selection. The structure of the brain can be seen simply by making an internal cast of the cranial cavity. This internal cast, or endocast, reveals the external morphology and size of the brain, especially for birds and mammals, thus providing crucial information about behavior and intelligence. In rare cases, natural endocasts are preserved in the fossils. Endocasts have been made for some theropods in a phylogenetic context to discover when and why brain enlargement might have taken place. In the theropod lineage, as will be discussed later, the enlargement of the brain may be critically linked to arboreal adaptation that demanded three-dimensional orientation in space. Nervous systems have changed in the course of evolution, in conjunction with adaptive specialization of the rest of the body.

BRAIN MORPHOLOGY

The brain of *Protoavis,* like that of modern birds, seems to have filled the cranial cavity almost completely; thus, the endocast provides a broad picture of the size and shape of the brain. It is differentiated into three segments from front to back in the avian fashion: the cerebrum, optic lobes, and cerebellum. The brain is large and deep, oriented horizontally to form an arc around the back of the orbit (fig. 4.6A). At the front of the cerebrum, the olfactory lobes are short and reduced, reflecting poor development of the sense of smell. The cerebral hemispheres are expanded with the development of the visual Wulst, bordered by a shallow vallecula laterally (fig. 4.6B). The Wulst shows a small bulge on the dorsal surface. Cerebral enlargement indicates integration of the senses and enhanced intelligence in

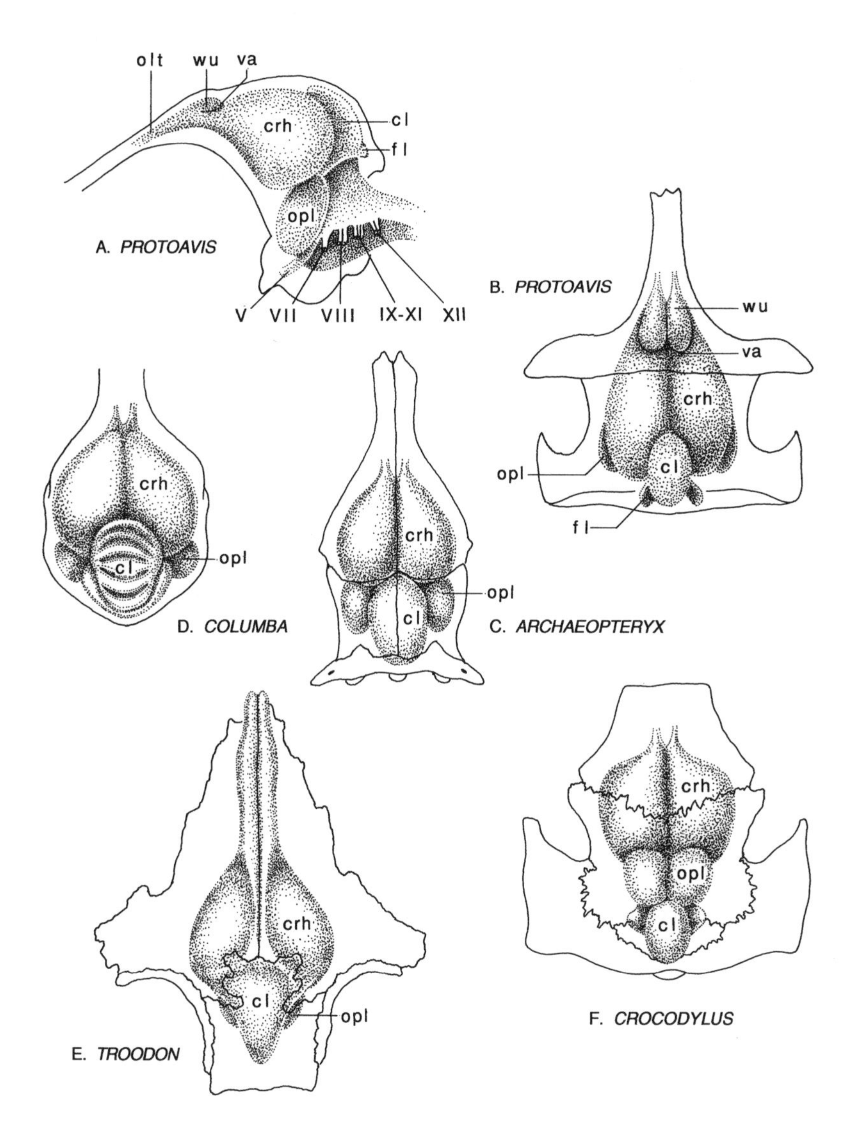

FIGURE 4.6

Comparative brain morphology of crocodile (*F*), non-avian theropod (*E*), and birds (*A–D*); these fossil brains are reconstructed from endocasts. *A–B, Protoavis;* note the development of large cerebral hemispheres, Wulst, and vallecula, as well as displacement of the optic lobes ventrally and laterally, as in birds; the olfactory lobes are reduced considerably. *C, Archaeopteryx*; *D,* pigeon (*Columba*); *E, Troodon*; *F, Crocodylus.* Note the considerable brain enlargement and the rearrangement of the optic lobes in *Troodon* and birds relative to crocodiles. (Source: Bühler 1985; Chatterjee 1991.) Abbreviations: *cl,* cerebellum; *crh,* cerebral hemisphere; *fl,* floccular lobe; *olt,* olfactory tract; *opl,* optic lobe; *va,* vallecula; *wu,* Wulst; Roman numerals indicate the foramina for cranial nerves.

Protoavis. The enlargement of the cerebrum has led to its contact with the cerebellum dorsally, thus displacing the optic lobes ventrally and laterally. The large optic lobes must have played an important role in processing visual information. The cerebellum is largely associated with balance, muscular coordination, and posture, all so important to flying animals. In *Protoavis,* the cerebellum is fairly large and erect, with a dorsal swelling, and is housed between the parietal and the supraoccipital. A large and prominent floccular lobe extends from the lateral side of the cerebellum. This feature may be linked to bipedalism, for which balance is important. The flocculi are generally absent or minute in quadrupedal dinosaurs. The ventral swelling of the medulla into the pons indicates interconnection of the cerebrum with the cerebellum to mediate the complex motor activity of early birds. Below the cerebellum the medulla extends caudally and is the site of origin for the last eight (V–XII) cranial nerves. Only those cranial nerves responsible for smell and vision are located rostral to the medulla.

The gradual modification of brain morphology from primitive archosaurs to modern birds is shown in figure 4.6. Among archosaurs, the crocodiles (fig. 4.6F) and most nonavian dinosaurs show a primitive pattern of brain morphology where the forebrain, midbrain, and hindbrain are narrow, elongate, and serially arranged. Brains of the nonavian coelurosaur and pterosaur approach avian morphology, with cerebral expansion and lateral displacement of the optic lobes (fig. 4.6E). This spatial rearrangement of the brain components may indicate similar neurosensory specialization associated with visual acuity, balance, and coordination in a three-dimensional world.

ENCEPHALIZATION AND INTELLIGENCE

Brain size varies considerably among vertebrates, depending on their level of activity, mode of life, and intelligence. Like other organs, the brain is large or small in different animals according to whether body size is large or small. For example, the elephant brain is almost four times larger than ours. That does not mean that elephants are more intelligent than humans. Absolute brain size is not a reliable correlate of intelligence. What is needed is a method of estimating relative brain size in relation to body size.

For more than a century, it has been known that the relationship between brain weight and body weight is not linear but exponential; brain weight increases more slowly than body weight from a small to large animal within a single group. Edward Drinker Cope, a great nineteenth-century paleontologist, observed that, in an evolutionary lineage, the animals were steadily increasing in size. Most evolutionary brain enlargement is merely the result of the entire animal growing bigger. Encephalization, a measure of relative brain size, refers to the special process wherein the brain expands more rapidly than would be expected from the growth of the rest of the body. Harry Jerison, a pioneer in paleoneurology, used endocasts to study in detail the encephalization process of extinct animals. In 1973, he synthesized his ideas in his seminal book *Evolution of the Brain and Intelligence.* Jerison devised a method that compensates for the effect of body size, allowing independent comparisons of brain size among related groups. He used the ratio of actual brain weight to the expected weight as an estimate of relative brain size, which he termed the encephalization quotient (EQ). For reptiles, especially nonavian archosaurs, Jerison's encephalization quotient was

$$\text{EQ} = \text{brain weight}/[0.005 \times (\text{body weight})^{0.66}]$$

whereas for higher vertebrates (birds and mammals) he estimated the encephalization quotient as

$$\text{EQ} = \text{brain weight}/[0.12 \times (\text{body weight})^{0.66}]$$

To determine the encephalization quotients for the extinct vertebrates, one must estimate both the brain size and the body size. The brain size is determined from the volume of the endocast, whereas the body size is inferred from the estimated model based on skeletal reconstruction. Using Jerison's method, I have estimated the EQs of selected species of pterosaurs, nonavian dinosaurs, and birds. The selected data are plotted in figure 4.7A, superimposed on Jerison's "brain:body maps" for living reptiles and birds. The estimates of brain and body size of *Protoavis* give a relative brain size that falls entirely within the avian polygon and is a clear departure from the reptilian level. By contrast, pterosaurs were clearly reptilian. The EQs of pterosaurs, nonavian theropods, and birds are also plotted graphically in figure 4.7B for comparison. The estimated EQ value of *Protoavis* is 0.41, whereas that of *Archaeopteryx* is 0.34. The EQs of early birds fall within the lower range of living birds but lie well outside the range of nonavian theropods and pterosaurs. Brain size increased greatly in the line of evolution toward birds as the cerebrum and cerebellum enlarged and increased in complexity.

I have calculated the EQs of many living birds. Smaller birds (body weight of <100 g) in general have proportionately larger EQs (approximately 1) than larger birds. In my survey, the crow is the intellectual elite as its EQ approaches 1.6. The ostrich, on the other hand, is not so bright; it has a relatively small brain showing the lowest EQ (0.15). Most nonavian dinosaurs had brains no greater in relative size than crocodiles. Some nonavian coelurosaurs, such as *Troodon*, began to show enlargement of brains (EQ = 0.25). A similar trend of brain expansion is expected in dromaeosaurs, but I do not have any EQ data.

It appears that such early birds as *Protoavis* and *Archaeopteryx* possessed somewhat larger brains than those of pterosaurs and nonavian dinosaurs. What was the selection pressure that led to the astonishing cranial expansion of early birds? Jerison speculated that the evolution of birds from their reptilian contemporaries was the result of their invasion of a new adaptive zone—complex, three-dimensional, arboreal niches. He noticed that, although both birds and pterosaurs developed the ability to fly, which demanded balance and coordination, the pterosaur brain was still reptilian in relative size, whereas birds were more encephalized. From this observation he argued that flight, per se, did not necessarily produce selection pressure toward avian

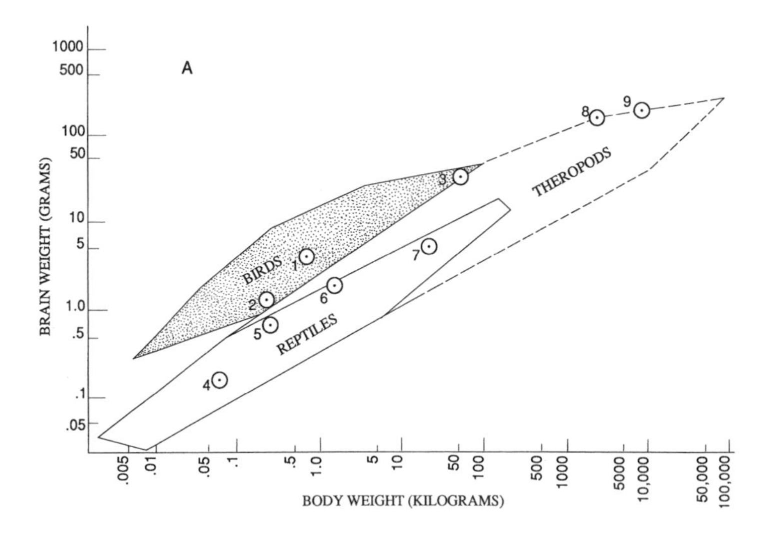

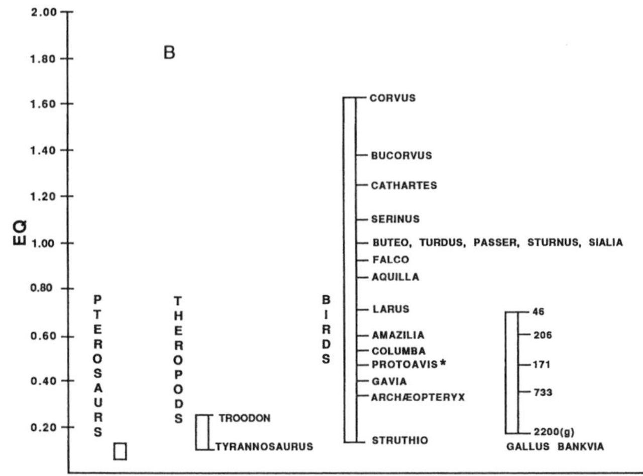

FIGURE 4.7

A, brain:body maps are the minimum complex polygons that can be drawn to enclose a set of points representing brain size plotted against body size. Number points are for early birds (*1, 2*), nonavian theropods (*3, 8, 9*), and pterosaurs (*4, 5, 6, 7*). *1, Protoavis; 2, Archaeopteryx; 3, Troodon; 4, Pterodactylus; 5, Rhamphorhynchus; 6, Scaphognathus; 7, Pteranodon; 8, Allosaurus; 9, Tyrannosaurus.* The ratio of brain size to body size in early birds and *Troodon* falls within the avian polygon, indicating considerable enlargement of brain; the same ratio in pterosaurs and most nonavian dinosaurs is in the reptilian domain. (Source: Jerison 1973; Hopson 1980; Chatterjee 1991.) *B,* encephalization quotients (EQs) for pterosaurs, theropods, and birds. Except for coelurosaurs such as *Troodon,* dinosaur and pterosaur brains are generally small, in the reptilian range. The EQ values of *Archaeopteryx* and *Protoavis* lie in the lower range of living birds, whereas the EQ value of ostriches lies at the bottom. The *right-hand bar* shows that body weight is inversely correlated with EQs during ontogeny; the smaller the animal, the larger the relative brain size. (Source: Chatterjee 1991.)

brain enlargement. The degrees of encephalization of two different aerial vertebrates may be related to their different styles of adaptation.

The pterosaurs habitually flew over shallow regions of the sea and fed on fish and other slippery animals. They lived in colonies on cliffs near sea coasts. Their activity was limited to a two-dimensional world, whether it was a vertical cliff or a horizontal expanse of water. The low EQs of pterosaurs may be correlated with their adaptation to a two-dimensional world, which is neurologically less demanding.

In contrast, the early birds, such as *Protoavis* and *Archaeopteryx*, inhabited woodland niches, most nearly like those of tree-dwelling primates, in a three-dimensional world. The judgment of distance is far more complex in this environment. The confusingly mottled background of leaves, branches, and other foliage at different levels of trees provided a strong selection pressure for enlargement of the brain to process complex three-dimensional audiovisual information. These early birds were depending more and more upon sight and sound and less and less upon smell. Life in the trees would have promoted the development of stereoscopic vision and hearing acuity, as well as the ability to orient in three-dimensional space. Jerison thus reasoned that the evolution of enhanced vision and hearing in woodland niches is the key factor for the sudden enlargement of the brain in early birds.

But why did the early birds adapt to the arboreal life in the first place? There may be several reasons. First, arboreal habitat provided safer and more secure niches and conferred protection from the contemporary ground-dwelling reptiles. Second, trees were the launching pad for early flights. Third, large flying insects, an almost untapped source of animal proteins, may have been another stimulus toward the arboreal and aerial habitats of early birds.

Jerison believed that there is a general correspondence between relative brain size and intelligence. Intelligence is a quality of mental acuteness and comprehension unique to a species.

It has a nonphysical as well as a physical reality, so it is hard to study. The brain is simply the organ that processes intelligence. Intelligence probably indicates adaptation to new environments, the ability to learn, and the capacity to create a perceptual world. Intelligence is related to the complexity of interactive behavior that can be studied and measured. Jim Hopson (1980), who has studied dinosaur brains, suggested that relative brain size may be linked to daily activity and thermoregulation. For example, most nonavian dinosaurs had relatively small brains even by reptilian standards and were probably less active than are living endotherms. On the other hand, the very large brain of nonavian coelurosaurs associated with cursorial adaptation (adaptation for running) and agility may indicate an endothermic level of activity. Hopson found a correlation between high EQ and endothermy. *Protoavis*, with an EQ close to 0.4, was probably metabolically as active as some living birds.

THE EYE

The orbits are so large and deep in *Protoavis* that the right and left cavities nearly touch each other. The enormous size of the orbit, about one-third the length of the skull, clearly indicates a large eyeball, about 20 mm in diameter, which would provide larger and sharper images (fig. 4.8A). There is a prominent caudal vertical wall to support the large eye. Like most living birds, *Protoavis* must have been a visually oriented animal. It had not only large eyes but also well-developed optic lobes. Flying animals demand high resolving power for better perception of movements and for avoidance of collisions.

Walls (1963) has discussed the correlation between food habits (predator and prey species) and the position of the eyes in avian skulls. In predatory birds such as hawks, eagles, and owls, the orbits are frontally placed, giving them excellent binocular vision. In prey species, on the other hand, the orbits are laterally placed, allowing the widest field of view to keep watch for impending danger. In *Protoavis*, the orbits are frontally placed with a large overlapping field. The snout is sharply

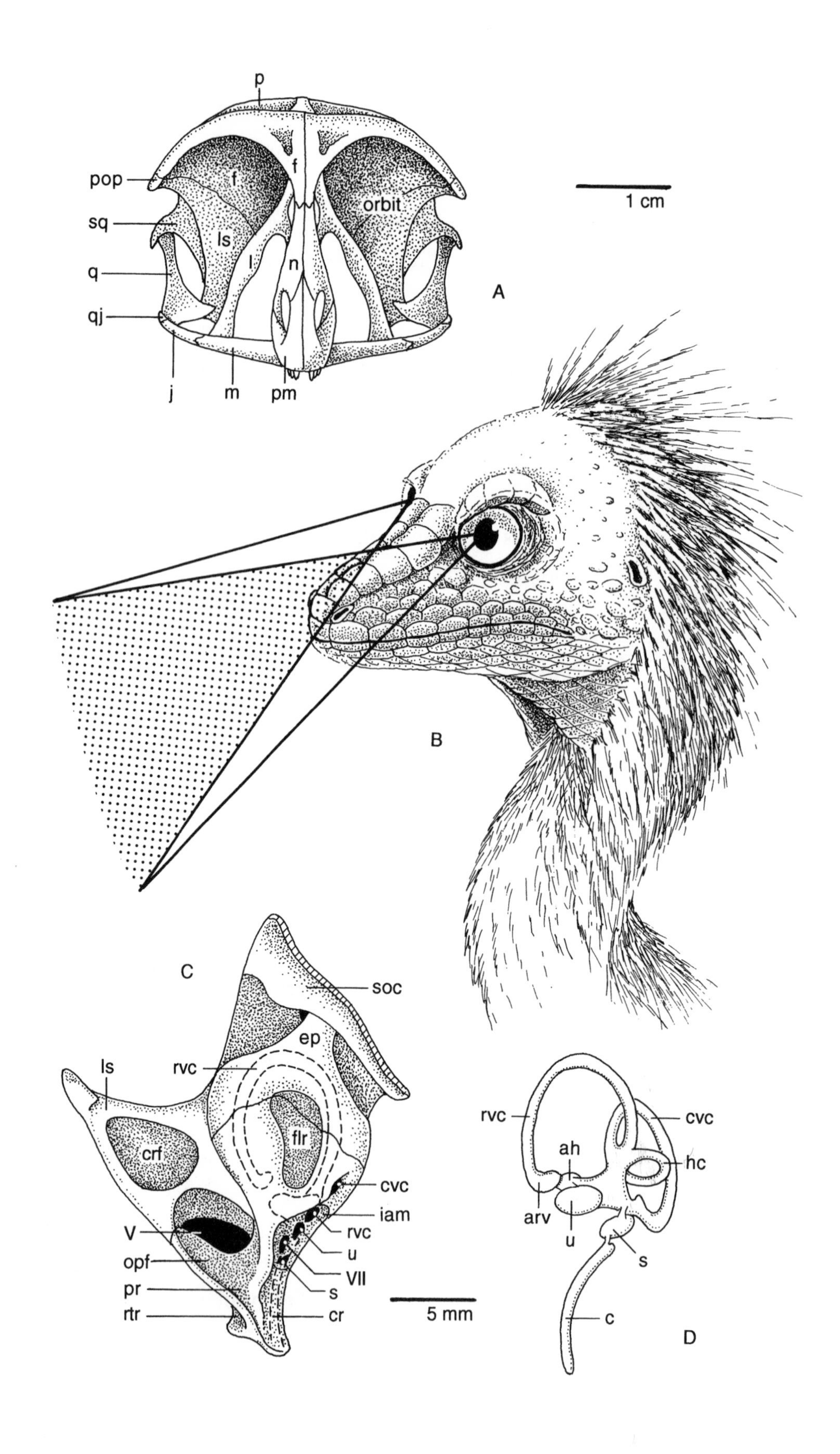
p
f
pop
f
sq
ls
orbit
1 cm
l
n
q
A
qj
j
m
pm
B
C
soc
ep
ls
rvc
flr
crf
cvc
iam
V
rvc
u
opf
VII
pr
s
rtr
cr
5 mm
rvc
cvc
ah
hc
arv
u
s
c
D

tapered to clear its field of vision. John Pettigrew, an Australian neuroscientist, has worked on the position and significance of visual Wulst in the forebrain region in living birds (Pettigrew 1979; Pettigrew and Frost 1985). In owls the Wulst is highly developed, with a trigeminal bulge that may indicate stereopsis—highly precise, binocular depth perception. In *Protoavis,* we see the beginning of the development of a bulge on the Wulst region (fig. 4.6B). The frontal position of the eyes, coupled with the development of bulging in the Wulst, may indicate that *Protoavis* had achieved stereoscopic vision (fig. 4.8B). Stereoscopic vision would offer the possibility of some judgment of distance to help locate food more accurately. Pettigrew also noticed that the position of the vallecular groove relative to the olfactory bulbs is highly variable among recent birds and may be related to differences in feeding style and visual acuity. In *Protoavis,* the Wulst is proximate to the olfactory bulb, as in some recent birds such as plovers (*Vanellus*) and bee-eaters (*Merops*), without any development of trigeminal expansion. This cerebral morphology indicates a visual feeding style rather than probe feeding.

THE EAR

Protoavis had improved not only the visual but also the auditory system to supplement the sense of smell. The otic capsule was large, and spaces within it indicate that an elongate cochlea

FIGURE 4.8

Reconstruction of the sense organs of *Protoavis. A,* restoration of the skull of *Protoavis* in the cranial view to show the frontal position of the orbits and the stereoscopic view. *B,* life restoration of the head of *Protoavis;* with both eyes facing forward, *Protoavis* shows strong adaptations of stereoscopic vision, as in modern nocturnal birds of prey. *C,* internal view of the right side of the braincase showing the floccular recess (*flr*), the C-shaped bony tube for the rostral vertical canal (*rvc*), and the long tubular cochlear recess (*cr*). *D,* restoration of the inner ear of *Protoavis* (source: Chatterjee 1991). Abbreviations: *arv,* ampulla for the rostral vertical canal; *c,* cochlea; *cvc,* caudal vertical canal; *hc,* horizontal canal; *rvc,* rostral vertical canal; *s,* sacculus; *u,* utriculus; Roman numerals indicate the foramina for cranial nerves; for other abbreviations, see figures 4.1 and 4.2.

had evolved. This is the part of the inner ear that gives birds their keen sense of hearing. The middle ear region of *Protoavis*, as discussed earlier, is built in an avian fashion. It shows two foramina, the fenestra ovalis for the reception of stapes in the front and the fenestra pseudorotunda behind (fig. 4.2A). The bony eustachian tube leading from the tympanic cavity to the throat helps to equalize air pressure on both sides of the eardrum. The middle ear region is highly pneumatized where the rostral, dorsal, and caudal tympanic recesses can be seen. The rostral tympanic recess shows contralateral communication. This feature is attributed to mechanisms of sound localization in birds. The tympanic recesses may have increased the sensitivity of the middle ear for detecting low-frequency sound.

The inner ear of *Protoavis* is reconstructed from the bony labyrinth and the cochlear recess (fig. 4.8C). On the medial surface of the braincase, the deep floccular recess is the most important landmark for the orientation of the canalicular system. An inflated bony tube around the floccular recess indicates the size and location of the rostral vertical semicircular canal. This is the longest part of the labyrinth. Two other semicircular canals, the caudal vertical and horizontal canals, are shared between the prootic and the opisthotic (fig. 4.8D).

The reception of auditory sensation is confined in birds to the papilla basilaris of lagena, which is elongated to form a bony tubular cochlear recess. This cochlear recess is well developed in *Protoavis* in the lower part of the prootic and opisthotic (fig. 4.8C). The elongated cochlea suggests enhanced auditory reception of *Protoavis*, resulting in improved discrimination of sound frequency. In modern amphibians (frogs), reptiles (crocodiles, geckos), and birds, refined hearing is linked to vocalization. It is likely that *Protoavis* was vocal and presumably could hear its own voice for communication.

Cranial Kinesis

Major features that differentiate avian skulls from those of dromaeosaurs are modifications in the temporal region concomitant with the development of cranial kinesis. In dromaeosaurs, the diapsid arches are intact (fig. 4.9A). In birds, three diapsid arches surrounding the lower temporal opening (the dorsal, ventral, and caudal bars) are eliminated so that the quadrate becomes streptostylic (fig. 4.9B). Cranial kinesis allows birds to move the upper jaw with respect to the braincase. In modern birds, two main types of kinesis are recognized relative to the position of the dorsal line of flexure and the nature of nasal opening. In prokinesis, bending occurs at a single transverse axis across the frontonasal hinge, so that the entire upper jaw moves as a unit. Prokinesis is associated with holorhinal naris (the posterior outline of the opening is fairly rounded), which makes deformation within the upper jaw impossible (fig. 4.9B). In rhynchokinesis, the dorsal flexion zone of the upper jaw has displaced forward so that its rostral part can be moved. Different forms of rhynchokinesis are characterized by the location, number, and extent of the hinges on the dorsal bar. Rhynchokinesis is generally associated with schizorhinal naris (posterior margin of the opening forms a deep slit) (fig. 4.9C). Prokinesis is a primitive adaptation, whereas rhynchokinesis is a derived feature.

Cranial kinesis is powered by the movement of the quadrate. In birds, the quadrate can move in a variety of directions relative to the braincase. In *streptostyly* the quadrate swings forward and backward. In *opisthostyly* the quadrate moves only caudally from its resting position (fig. 4.10A–B). In *parastyly* the quadrate moves in a transverse direction (Chatterjee 1991).

In *Protoavis,* we see the beginning of the development of prokinesis. Like modern birds, *Protoavis* has lost the rostral, dorsal, and caudal arcades of the lower temporal opening, as well as the ectopterygoid, making the jugal bar, quadrate, and pterygoid mobile (fig. 4.10C–D). The ventral end of the lacrimal has developed a sliding joint with the jugal bar. Both the quadratojugal and the quadrate ramus of the pterygoid have minimized contacts with the quadrate to enhance mobility. These articulations form pin joints restricted to the foot of the quadrate. The orbital process in *Protoavis* serves as an effective

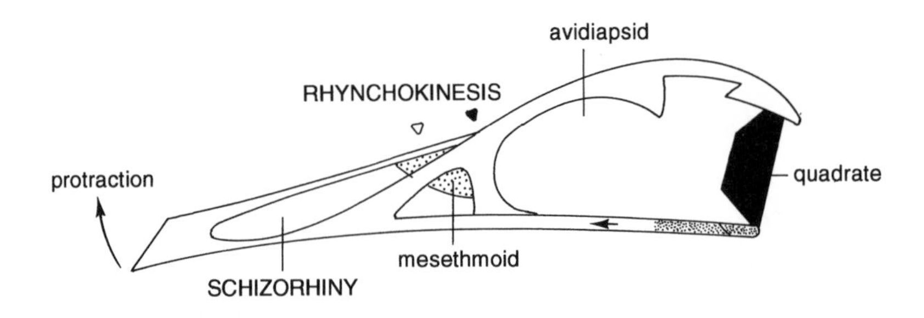

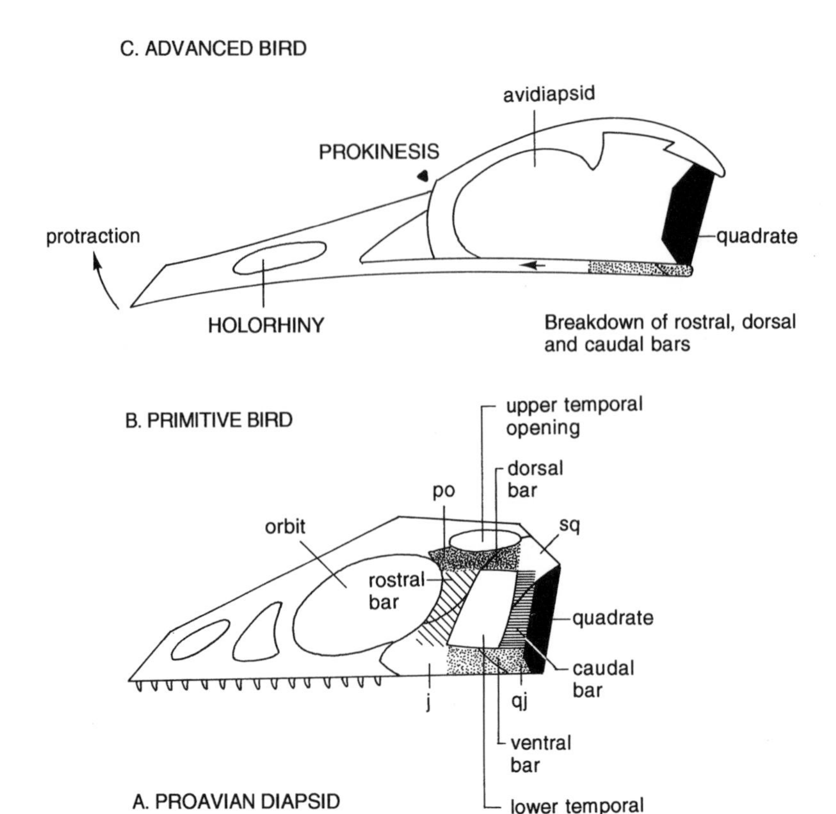

FIGURE 4.9

Evolution of the avian temporal region and skull kinesis in response to streptostyly. *A,* in proavian diapsids such as dromaeosaurs, the lower temporal opening is framed by four bony bars (rostral, dorsal, caudal, and ventral), which form as a blocking device; as a result the quadrate cannot move forward. *B,* prokinetic birds with holorhinal nostrils acquired streptostyly and skull kinesis by loss of the rostral (postorbital/jugal), dorsal (postorbital/squamosal), and caudal (squamosal/quadratojugal) bars; the quadrate becomes mobile and is able to raise the upper jaw as it is pushed forward. *C,* rhynchokinesis evolved after the loss of teeth from the premaxilla/maxilla bar, the development of the schizorhinal nostril, and forward extension of the mesethmoid under the dorsal bar; the *solid pointer* indicates the craniofacial hinge; the *open pointer* indicates the additional bending axis on the dorsal bar in rhynchokinesis.

lever arm for the protractor quadrate muscle. The spherical head of the quadrate fits into a concavity of the squamosal to form a ball-and-socket joint, allowing the quadrate to swing freely in any direction.

Like modern birds, *Protoavis* has four functional units and four principal joints on each side of the skull (fig. 4.10E). The functional units are (1) the upper jaw, (2) the jugal bar + the pterygoid-palatine bar, (3) the quadrate, and (4) the braincase. These four units form a four-bar crank chain. If the braincase is held stationary and the quadrates are swung forward, the upper jaw is raised. The upper jaw is the mobile unit for procuring food, whereas the braincase is the stationary unit for housing neurosensory organs. These two units are joined by various links. The quadrate is the vertical link and the main crank device. Its head forms a fulcrum on the undersurface of the squamosal and moves forward and backward at its foot so that the force can be transmitted to the beak through a pair of horizontal links (fig. 4.10E). The jugal bar forms the lateral link, whereas the pterygoid-palatine bar acts as the median link. The quadrate is flexibly attached by pin joints to these horizontal links on either side of its foot. Because the jugal bar and the pterygoid-palatine bar share a similar mechanical function, as a push rod between the upper jaw and the quadrate, they are considered functionally as a single link to simplify the model. There are three flexible or bending zones that permit movement of the upper jaw. The dorsal bending zone lies between the nasal and frontal bones, the lateral bending zone is between the maxilla and jugal, and the palatal bending zone occurs across the vomer/palatine contact. In modern birds, the bending zones become very thin and pliable with the fusion of skull bones. In *Protoavis,* these zones are connected by thin and flexible joints (Chatterjee 1991).

Two pairs of muscles act together to raise the upper jaw and depress the lower jaw (fig. 4.10C–D). The orbital process of the quadrate acts as a lever arm for the protractor pterygoidei et quadrati muscle. The action of the muscle is to pull the

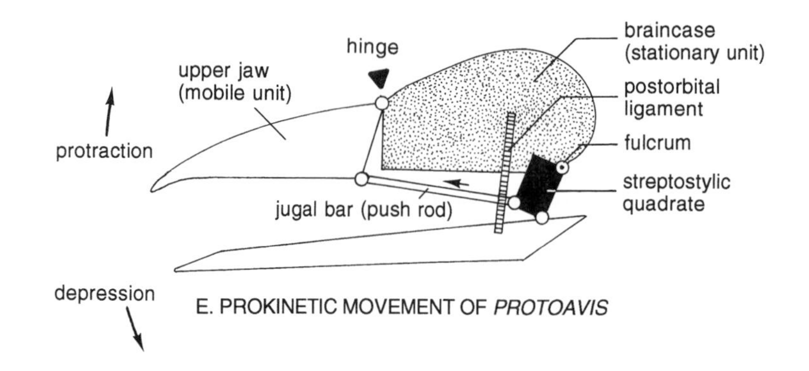
hinge
upper jaw
(mobile unit)
braincase
(stationary unit)
postorbital
ligament
fulcrum
streptostylic
quadrate
protraction
jugal bar (push rod)
depression
E. PROKINETIC MOVEMENT OF PROTOAVIS

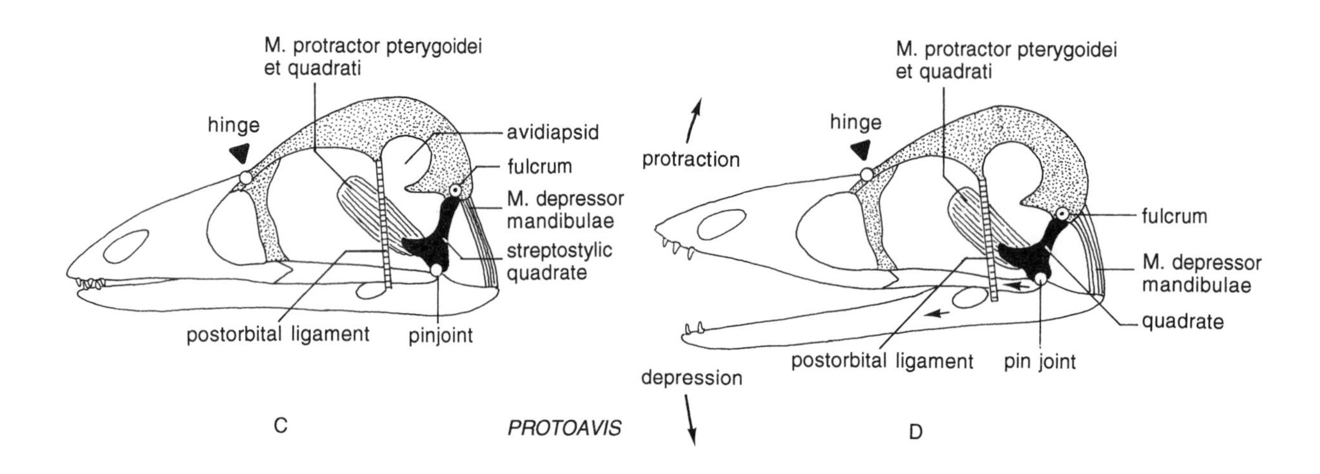
M. protractor pterygoidei
et quadrati
hinge
avidiapsid
fulcrum
M. depressor
mandibulae
streptostylic
quadrate
postorbital ligament
pinjoint
M. protractor pterygoidei
et quadrati
protraction
hinge
fulcrum
M. depressor
mandibulae
quadrate
postorbital ligament
pin joint
depression
C
PROTOAVIS
D

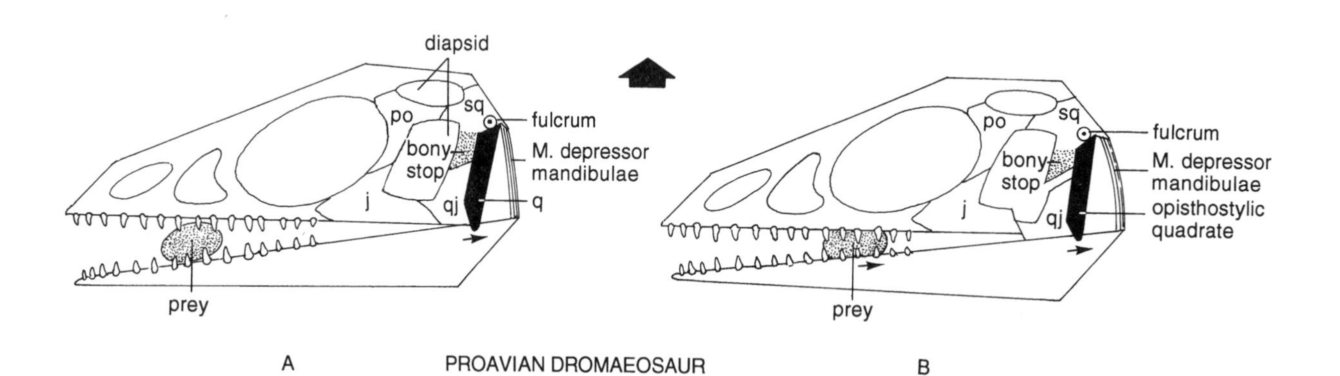
diapsid
po
sq
fulcrum
bony
stop
M. depressor
mandibulae
j
qj
q
prey
po
sq
fulcrum
bony
stop
M. depressor
mandibulae
j
qj
opisthostylic
quadrate
prey
A
PROAVIAN DROMAEOSAUR
B

quadrate forward. As the quadrate foot moves forward, this force is imparted to the jugal bar, which, in turn, pushes the upper jaw forward. Since the upper jaw is flexibly attached at the bending zones, the forward push of the jugal bar rotates the upper jaw dorsally. Why does the foot of the quadrate move forward when the lower jaw is depressed? Walter Bock provided an explanation. He suggested that in most birds there is a mechanical linkage between the elevation of the upper jaw and the depression of the lower jaw (Bock 1964). Typically, such coupling is effected by the postorbital ligament. This ligament is made of collagen and is only slightly extensible. It runs from the postorbital process of the braincase to the external process of the lower jaw. As soon as the lower jaw is opened by the depressor mandibulae muscle, it tightens the postorbital ligament, pushing the foot of the quadrate forward and thus raising the upper jaw (fig. 4.10). Because both postorbital and external processes

FIGURE 4.10

Evolution of cranial kinesis in birds. *A* and *B,* in the proavian dromaeosaur the quadrate head has developed a ball-and-socket joint with the squamosal. The quadrate cannot move forward because of the blocking action of the squamosal/quadratojugal bar in front of the quadrate, which prevents streptostyly. However, the combined quadratojugal-quadrate can move backward (opisthostyly), which, in turn, would permit caudal movement of the lower jaw relative to the skull during the bite; this backward movement of the jaw during prey capture would assist in slicing and shifting the flesh back toward the throat, ready for swallowing. *C,* in *Protoavis,* with the acquisition of an avidiapsid condition (i.e., confluence of the orbit with the two temporal fenestrae), the squamosal/quadratojugal bar in front of the quadrate is eliminated, and a pin joint is developed between the quadratojugal and the quadrate. The quadrate becomes streptostylic and can move forward and backward. *D,* during contraction of the depressor mandibulae muscle, the foot of the quadrate moves forward; this rotates the upper jaw dorsally. The elevation of the upper jaw and the depression of the lower jaw is coupled by the postorbital ligament (Bock 1964). *E,* functional interpretation of prokinesis in the skull of *Protoavis,* based on the four-bar crank mechanism. The skull has four functional units and four principal joints on each side of the skull. The braincase is the stationary unit. As the jaw is depressed, the quadrate is swung forward and the upper jaw is raised. Cranial kinesis allows a larger gape to facilitate high intake of food.

are present in *Protoavis,* it is likely that the postorbital ligament functioned like that in living birds for jaw coupling.

Kinesis in birds increases the gape to allow a high intake of food. It also allows sophisticated manipulation of food in the mouth. Prokinesis is not possible among nonavian dinosaurs because of the presence of the squamosal/quadratojugal bar, which acts as a blocking device for streptostyly (fig. 4.9A). In birds, when the jaws are closed, further opisthostylic quadrate movement creates a wedge that opens backward in the mouth, thus preventing food from escaping. This opisthostylic quadrate movement also allows the upper and lower jaws to close quickly (fig. 4.10D). Opisthostyly is a primitive feature and is developed in small nonavian theropods, such as ceratosaurs and coelurosaurs, to allow more efficient capture of food and to push prey back toward the throat, ready for swallowing (fig. 4.10A–B). In some birds, when the jaw is lowered, the quadrate moves sidewise (parastyly), which, in turn, moves the caudal part of the lower jaw sidewise. This lateral spreading of the lower jaw increases the diameter of the throat, which helps when swallowing large pieces of food. Parastyly is common among large theropods (i.e., allosaurs and tyrannosaurs), where the lateral mandibular spreading occurs along a vertical flexible joint between dentary and postdentary bones.

Mode of Life

The adult *Protoavis,* as restored from available skeletal material, is a small, gracile bird, about the size of a pheasant (*Phasianus*), with a long bony tail (fig. 4.11). The skull is attached to an elongated, S-shaped neck. The trunk is short, the girdles are robust, and the limbs are well built for locomotion. The estimated overall length is 60 cm, which is comparable to the Solnhofen specimen of *Archaeopteryx. Protoavis* is relatively strongly built, like predatory birds. Its estimated weight is about 600 g, heavier than *Archaeopteryx.*

Protoavis lived in the tropics of Texas and mainly in trees. It was a predatory bird, which is indicated by its carnivorous teeth at the tip of the jaws; the posterior parts of the jaws were edentulous and covered by horny sheath. The brain itself became relatively larger than that of the contemporary dinosaurs and more elaborate in design. The outstanding feature of the skull is the relocation of the enormous orbits to the front, as in owls and nocturnal primates. Animals that are active at night must be able to see in dim light. Their eyes are very large to gather in as much light as possible. The orbits of *Protoavis* are so frontally placed that they must have been used for hunting at dusk or in the dark (fig. 4.8B). The eyes were set in a bony cup that held them separate from the brain. The earliest birds probably survived as small, nocturnal animals to avoid direct interaction with their diurnal reptilian contemporaries. To locate food or prey in the hazardous three-dimensional world of the trees, they developed sharp eyes in preference to keen noses.

Unlike modern raptors, *Protoavis* lacked sharp talons, the actual killing instrument, as well as a compact, hooked beak for chopping and ripping flesh into pieces. There are no obvious adaptations in the skeleton of *Protoavis* for dismembering larger prey. It is likely that *Protoavis* took smaller and juvenile tetrapods, invertebrates, and insects available in the Dockum biota. *Protoavis* was a swift, agile, arboreal predator. Predation

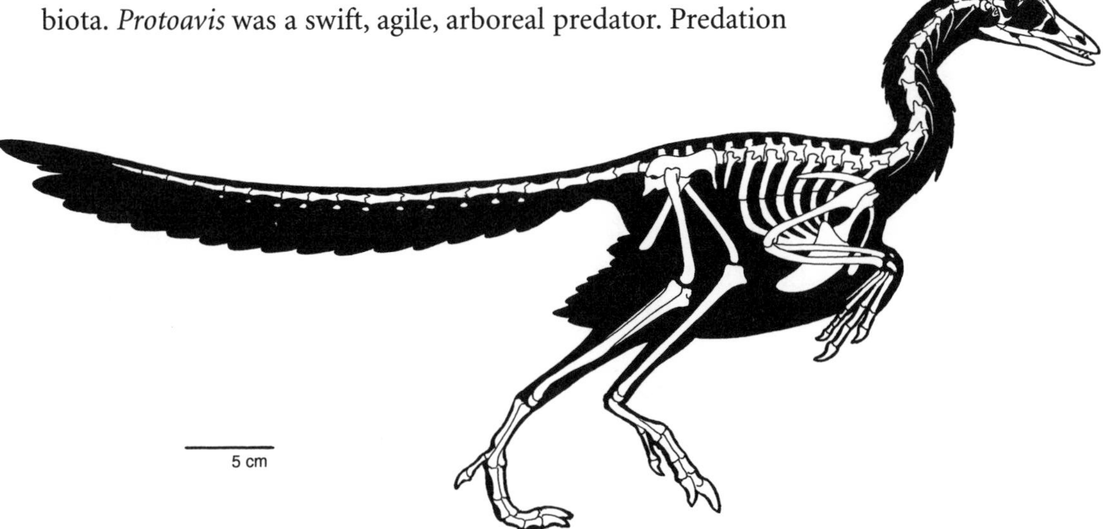

FIGURE 4.11

Skeletal restoration of *Protoavis texensis*. In life, the adult *Protoavis* would be about the size of a pheasant (*Phasianus*), 60 cm long, if its long bony tail is taken into account.

in *Protoavis* was an extension of theropod habit, but the predatory weapon was the beak, as its forelimbs were used for flight. Heterocoelous neck vertebrae permitted a wide range of motion of the head in the performance of these tasks. The jaws acquired kineticism to increase gape and facilitate a high intake of food.

Protoavis acquired hearing acuity that might be associated with vocal sounds for communication. In woodland and forest environments, where it may be difficult to see, sounds may be an effective way of mediating social interactions. Its claw morphology shows that *Protoavis* could climb trees. Trees would have been used in many ways by early birds—as an escape refuge, as a hiding place, as a nesting site, or as a resting site. Like modern birds, *Protoavis* probably lived in a complex habitat, spending some time feeding on the ground but using trees for hunting, sleeping at night, and nesting.

Bipedal Locomotion

Like modern birds, *Protoavis* acquired two independent and specialized methods of locomotion, flying with the forelimbs and walking with the hindlimbs. *Protoavis* was an obligatory biped, and the forelimbs played no role in supporting weight. Maintaining proper balance on two legs for support and locomotion was a major innovation in nonavian theropods. Many of these bipedal hallmarks persisted in early birds. For example, the hip joint was adapted to an erect posture with perforation of the acetabulum and a completely inturned head. The hindlimbs are slender but strongly built and show features of cursorial adaptation. The distal segments of the hindlimb became longer than the proximal to increase stride length and enhance speed. The metatarsals are elongated and closely bound together. The mesotarsal ankle joint suggests digitigrade posture. *Protoavis* walked on its digits, with the heel and ankle bones carried off the ground. The fifth digit was lost. The reversed hallux forms a prop to support the digitigrade pes.

The increase in the area of the pelvic girdle, the backward rotation of the pubis, and the reduction of the tail indicate a subtle change in the posture of early birds from the dromaeosaurid design. The principal effect of this reorganization of pelvis and tail would be to shift the center of gravity forward. In nonavian theropods, the femur rotated back and forth considerably at the hip joint for propulsion. In *Protoavis*, the leg bones were held in a Z configuration, as in modern birds. In this position the femur was kept subhorizontally and its distal end moved up and down slightly. The main functional movement was transferred to the knee joint, where the tibia swung forward and backward. Several osteological features suggest this new posture in *Protoavis*. The preacetabular process of the ilium was elongated cranially and provided a large area for the attachment of the iliotibialis cranialis muscle. This muscle would protract the femur in the subhorizontal position. With the development of the cranial prolongation of the ilium, the pubis was relieved of its role in femoral protraction and could rotate backward. The principal femoral retractor, the caudofemoral longus muscle, was considerably reduced, as is evident from the loss of the fourth trochanter and the truncation of the tail (Gatesy 1990). The tail, an important balancing organ in dromaeosaurs, was considerably reduced in *Protoavis*. To counterbalance the long tail, the pubis was rotated backward so that heavy viscera could be shifted caudally under the pelvis and over the legs. The ischium and pubis became open ventrally to make room for viscera. With the development of the subhorizontal attitude of the femur, the center of gravity was positioned near the distal end of the femur. The tibia now played a major role in propulsion, and the knee joint showed greater motion than the hip joint. The development of both cranial and later cnemial crests in *Protoavis* may reflect their important role in knee movement.

Flight Capability

Protoavis was small, with the adult weighing about 600 g. Small size confers favorable weight/surface relationships for flight, allowing a bird to move through the air with smaller wings and lower air speed. *Protoavis* shows many derived features associated with flapping flight that are not encountered in *Archaeop-*

teryx. Of particular importance are the keeled sternum, hypocleidium-bearing furcula, strutlike coracoid, large flight muscles, and triosseal canal for the supracoracoideus tendon (fig. 4.4). The glenoid cavity faces dorsolaterally, which permits unrestricted movements of the humerus in all directions. The humerus is an exceptionally strong bone and shows all the bumps and ridges for the attachment of the flight muscles. The specialized linkage system of elbow and wrist joints would stiffen the entire extended wing into a plane to resist twisting when exposed to air pressure during flight. The presence of feathers is inferred indirectly from the development of quill knobs in the hand. The furcula and the keeled sternum must have accommodated large flight muscles. The flexible furcula would act as a spring between the shoulder joints during the flight strokes. The presence of a triosseal canal suggests that the supracoracoideus muscle must have functioned effectively as a wing elevator and that the animal could possibly take off from the ground. The enlargement and partial fusion of pelvic bones, a relatively short horizontal femur with antitrochanter articulation, and mesotarsal ankle joint all may indicate that the hindlimb was modified as a landing gear.

The elbow and wrist joints of *Protoavis* were coupled automatically without much muscular effort and were restricted to a single plane, as in modern birds. This complex linkage system between elbow and wrist joint is central to the normal functioning of the avian wing, not only in its folding and unfolding but also in maintaining the proper position of the skeletal elements during flight. *Protoavis,* however, did not achieve the strength, rigidity, and reinforcement that are required to withstand the strain imposed by sustained flight. Rick Vasquez (1992, 1994) discussed in detail the complex wrist movements during the flapping flight of modern birds. The wrist bones of *Protoavis* show some interesting though primitive morphology, which may provide clues to the early evolution of flapping flight. The radiale is very pronounced and would lock the manus in the extended position. This mechanism prevents the manus from supinating during gliding or maneuvering. Vasquez noticed that, in modern birds, the ulnare is a large, complex bone that interlocks with the ulna and the ventral ridge of metacarpal III. This arrangement prevents the manus from hyperpronating during the downstroke. In *Protoavis,* the ulnare is a small, spherical bone that articulates with a ventral ridge of metacarpal III but lacks the complex interlocking device known in modern birds. It is doubtful that such a small ulnare in *Protoavis* could prevent the manus from hyperpronating during slow, maneuvering flight. In all probability, *Protoavis* was designed for powered flight for a short distance at a cruising speed. As in *Archaeopteryx,* the long tail of *Protoavis* provided stability and enabled it to control its motion with respect to pitch and roll. The animal probably kept a straight and level course without much turning, climbing, or diving. Cruising flight is more primitive and less strenuous than slow flight. With further improvement of the wrist, fusion of the skeleton at critical joints, and loss of tail in the Cretaceous birds, more sophisticated flight evolved.

The Jurassic Birds

CHAPTER 5

It took the whole of Creation
To produce my foot, my each feather:
Now I hold Creation in my foot.
—Ted Hughes, *Crow,* 1960

During the Jurassic period (about 208–144 million years ago), rifts in the continental crust produced the Atlantic Ocean. Laurasia began to separate from Gondwana, creating an equatorial seaway between them, leaving great deposits of evaporites along the Gulf of Mexico, and forming much of the coastline bordering the Tethys Sea. Climates became warm worldwide, and moist winds from newly created oceans brought rain to inland deserts, supporting lush vegetation of cycads and conifers. Dinosaurs became giants, spread throughout the Jurassic world, and adapted a wide range of environments from the equator to the polar regions. Above them flew small birds and pterosaurs. The plesiosaurs and ichthyosaurs shared the shallow seas with marine crocodilians.

Bird skeletons from the Jurassic world are relatively rare; the most celebrated is *Archaeopteryx* from Germany. Recently, another contemporary taxon, *Confuciusornis*, has been reported from China. A tantalizing bird skeleton associated with feather impression has been discovered in Korea recently but awaits description and diagnosis (fig. 5.1). Various avian footprints from North America and Africa indicate that birds were more diverse during the Jurassic than is generally believed.

Archaeopteryx of Altmühl

Until recently, our knowledge of the origin of birds and the evolution of flight has been based entirely upon *Archaeopteryx lithographica* from the Upper Jurassic Solnhofen Limestone of Bavaria, Germany. This fine-grained rock, which is extensively quarried for lithographic stone, was evidently deposited in a shallow lagoon of a tropical sea. *Archaeopteryx* has been held as

FIGURE 5.1

Late Jurassic continental configuration showing the locations of three avian taxa: *Archaeopteryx* from Germany, *Confuciusornis* from China, and an unnamed Korean bird.

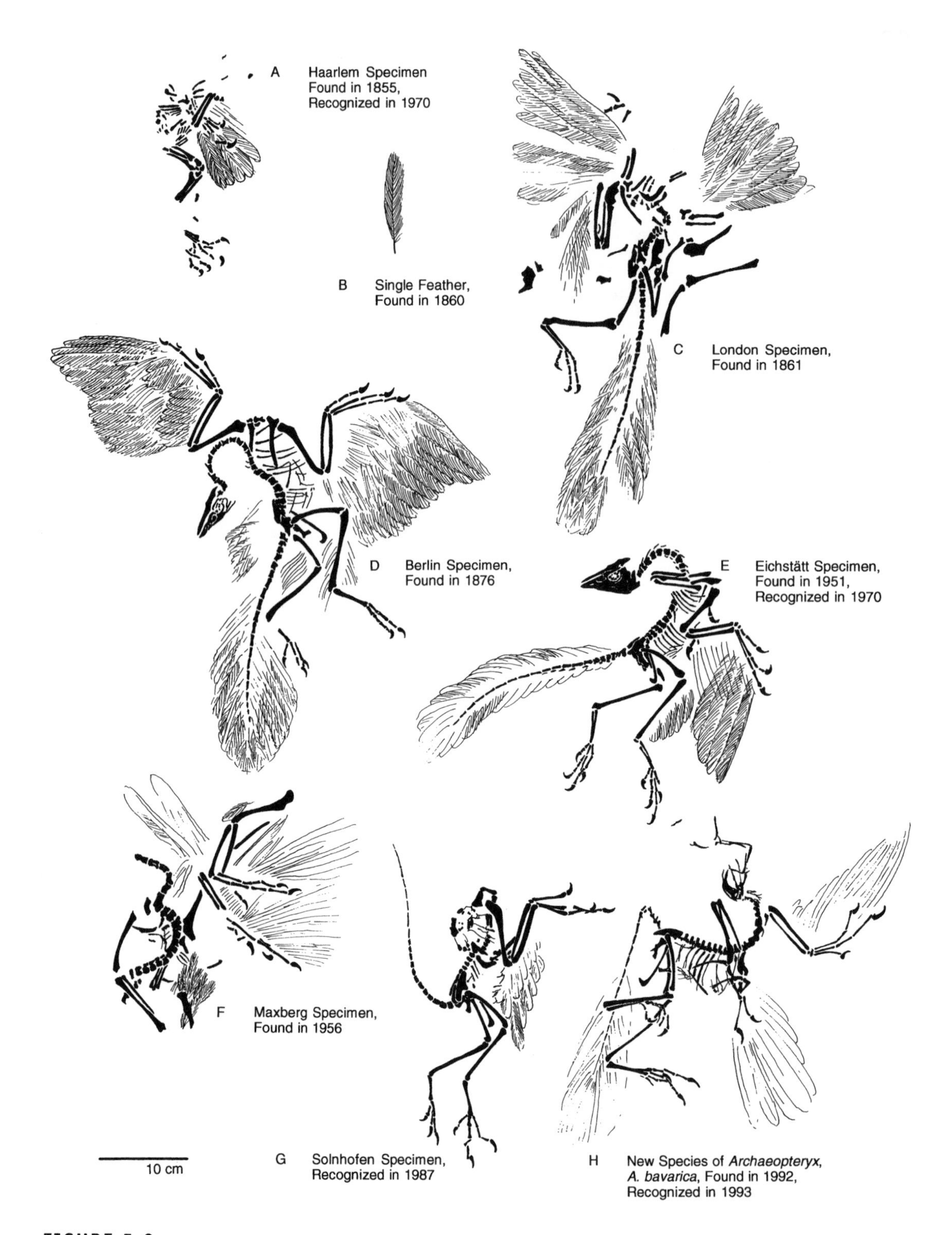

FIGURE 5.2

Seven fossil skeletons and a solitary feather of *Archaeopteryx,* including the popular name and the year of discovery or recognition (source: modified from Wellnhofer 1990, 1993).

the crown jewel in paleontology since the beginning of the Darwinian evolution. It has been studied, analyzed, scrutinized, criticized, and debated for more than a century. Recently, it was ridiculed as a fraud by two eminent astronomers, Hoyle and Wickramasinghe (1986), without any basis. Because of its evolutionary importance, *Archaeopteryx* was the subject of an international symposium at Eichstätt, Germany, in 1984. An attempt was made to resolve some of the controversies, but there was little consensus about its status, ancestry, relationships, mode of life, and flight capabilities.

The story of *Archaeopteryx* begins in 1860 with the discovery of the impression of an isolated feather in the pastoral Altmühl Valley, near the town Solnhofen, not far from Munich (fig. 5.2B). The feather was 60 mm long and 11 mm wide, asymmetrical with two unequal vanes separated by a quill. The specimen was reported in 1861 by the German paleontologist Hermann von Meyer (1861a). One month later, von Meyer (1861b) reported the discovery of a second specimen from the same Solnhofen limestone, an almost complete skeleton associated with distinct feather impressions. He recognized a curious admixture of both reptilian and avian features in the skeleton and coined a neutral name to designate this specimen, *Archaeopteryx lithographica*, meaning ancient wing from the lithographic limestone.

Controversy started immediately as J. Andreas Wagner (1862) of the University of Munich, who had previously described the splendid specimen of *Compsognathus* from the same limestone strata, proclaimed that *Archaeopteryx* was not a bird, but a feathered reptile. He substituted a new name for it, *Griphosaurus problematicus*, with a reptilian emphasis, relating the new specimen to the fabulous griffin lizard. Two years later this specimen was sold to the Museum of Natural History of London for 700 British pounds. The London specimen (fig. 5.2C) includes almost the entire skeleton. The skull bones are scattered but preserve a natural endocast. The specimen was described in detail by Sir Richard Owen (1863), an arch enemy of

Darwin, as an aberrant bird without any evolutionary implication. Thomas Huxley (1868a, 1868b), on the other hand, emphasized the importance of *Archaeopteryx* as a transitional form between reptiles and birds. The reptilian features include teeth, a long bony tail, gastralia, and sharp claws in the wing. The avian characteristics are feathers, the furcula, and a reverted pubis. Surprisingly, Huxley used *Compsognathus*, not *Archaeopteryx*, as a missing link between dinosaurs and birds. He viewed *Archaeopteryx* as a side branch of the main line of avian evolution, having little to do with the origin of modern flying birds. He considered *Archaeopteryx* somewhat irrelevant to the issue of bird origins. Darwin remained almost silent regarding *Archaeopteryx* and never used it in defense of his theory of evolution.

The third specimen of *Archaeopteryx* was reported by E. Häberlein in 1877 and is now housed at Berlin's Humboldt Museum of Natural History. The Berlin specimen (fig. 5.2D) is by far the best; the skeleton is articulated in a natural pose with an intact skull and extended wings. The fossil was exquisitely preserved, including fine structural details of feathers on its wings and tail.

This was the beginning of a series of discoveries of *Archaeopteryx* specimens from the Solnhofen Limestone that would captivate paleontologists and evolutionists for more than a century. Today, the species is represented by one isolated feather and six partial or complete skeletons that vary greatly in size and represent different developmental stages. The specimens are identified on the basis of the institution that either originally or presently houses them. These are known as the London, Berlin, Eichstätt, Solnhofen, Maxberg, and Haarlem specimens, respectively (fig. 5.2). The Solnhofen specimen (fig. 5.2G) is by far the largest, about the size of the holotype of *Protoavis*. Recently, a seventh skeleton of *Archaeopteryx* was described by Peter Wellnhofer (1993). This is the smallest of the specimens, yet it shows two features unknown in other skeletons: an ossified sternum and interdental plates on the inner side of the lower jaw. Wellnhofer concluded that it was an adult specimen representing a new species, *Archaeopteryx bavarica* (fig. 5.2H)

THE ANATOMY OF *ARCHAEOPTERYX*

In spite of the excellent preservation of several skeletons, detailed anatomical information on *Archaeopteryx* is not yet available. The main reason for this deficiency is that none of the specimens has been completely prepared and freed from the limestone matrix (Martin 1995a). Our knowledge of the anatomy of *Archaeopteryx* is thus limited to its two-dimensional profile. The skull of *Archaeopteryx* is difficult to interpret because of extensive crushing. Peter Wellnhofer, a leading authority on *Archaeopteryx,* made the first accurate restoration of the Eichstätt skull in 1974. Alick Walker (1985) provided a detailed interpretation of the braincase of the London specimen. Later, I modified the skull restoration on the basis of additional evidence gleaned from the London and Berlin specimens. The seventh specimen, *A. bavarica,* provides additional cranial information (Elzanowski and Wellnhofer 1996).

The skull of *Archaeopteryx* is dromaeosaur-like in general structure. It retains many primitive features, such as interdental plates, ectopterygoid, epipterygoid, rostroventral wing of the prootic, postorbital and squamosal processes of the squamosal, ascending process of the jugal, and distinct sutures (fig. 5.3A). Numerous carnivorous teeth are present in the rostral half of the jaws, and these teeth continue at the midpoint of the antorbital fenestra. The teeth are conical, unserrated, and widely spaced, with distinct necks and replacement pits; they are bordered by a series of inderdental plates at the lingual margin. The nasal opening is large and elliptical, and the maxilla takes part in the formation of the rim, as in primitive archosaurs. The antorbital opening is triangular and shows two internal foramina in front of it. The nasal bones act as median roofing elements between the premaxillae and the frontal bones. The frontals show inflated topography for housing of the cerebrum. The

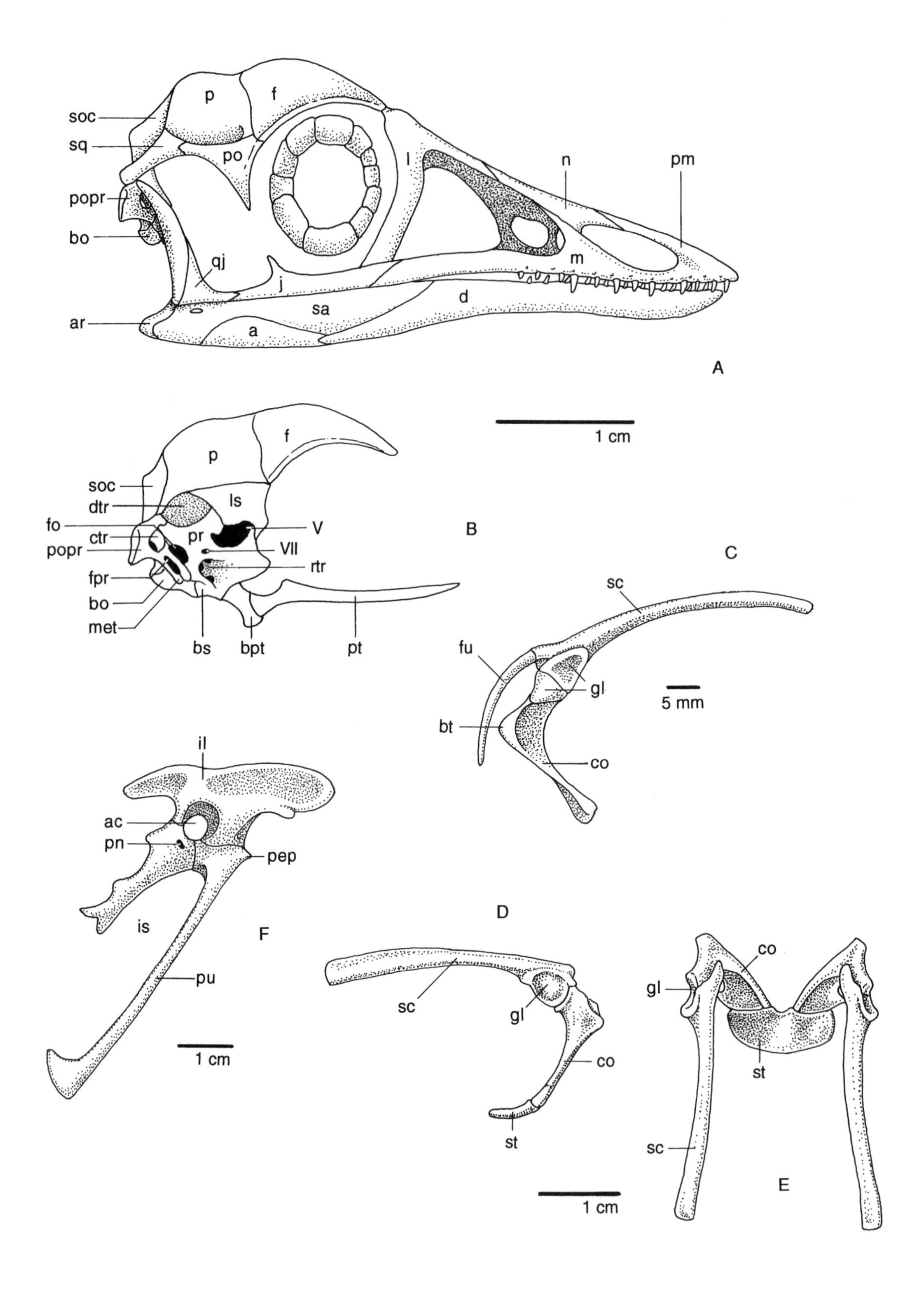

soc
sq
popr
bo
ar
p
f
po
l
n
pm
qj
j
m
sa
d
a
A
1 cm
B
soc
dtr
fo
ctr
popr
fpr
bo
met
p
f
ls
pr
V
VII
rtr
bs
bpt
pt
C
sc
fu
gl
5 mm
bt
co
il
ac
pn
pep
is
F
pu
1 cm
D
sc
gl
co
st
co
gl
st
sc
E
1 cm

parietals have a median crest. The orbit is large and circular, containing a ring of sclerotic plates, and is somewhat laterally placed. There is a partial modification of the temporal region. The postorbital is present, but the ascending process of the jugal seems to be atrophied to breach its contact with the postorbital. Because of this change, the lower temporal opening communicates with the orbit, while the upper temporal arcade remains intact. This is the first step toward achieving an avian temporal configuration (fig. 5.3A).

The squamosal-quadratojugal bar appears to be present in front of the quadrate, but their connection is probably lost, as seen in ceratosaurs. This bar would act as a blocking device to prevent the streptostylic movement of the quadrate. The quadratojugal is L-shaped and makes a long sutural contact with the quadrate. The palate is poorly known, but it retains the ectopterygoid bone. The palatine is birdlike and is differentiated into three flanges: the premaxillary process rostrally, a hook-shaped choanal process, and a long pterygoid wing. The quadrate is very similar to that of a nonavian theropod, with a single head contacting squamosal and paroccipital process. It bears a large pterygoid flange but lacks the avian orbital process. Its ventral surface shows bicondylar articulation with the lower jaw. The lower jaw is slender and lacks coronoid and lateral mandibular fenestra. The inflated frontal and modified temporal regions seem to be two avian attributes in the skull roof of *Archaeopteryx.*

The braincase of *Archaeopteryx* is more derived and exhibits many features of primitive birds. It has developed all three tym-

FIGURE 5.3

Composite restoration of skeletal elements of *Archaeopteryx. A,* reconstruction of *Archaeopteryx* skull (based on the Eichstätt specimen and the new species *A. bavarica*). The temporal region of the skull is poorly understood; however, the orbit may be confluent with the lower temporal fenestra (semi-avidiapsid condition). *B,* braincase; *C,* shoulder girdle; *D* and *E,* lateral and dorsal views of right shoulder girdle of *Archaeopteryx bavarica,* showing a small, ossified sternum (source: after Wellnhofer 1993). *F,* pelvic girdle. For abbreviations, see figures 4.1–4.5.

panic recesses (rostral, dorsal, and caudal) in the same topographic positions as in modern birds (fig 5.3B). The otic capsule shows the fenestra ovalis and fenestra pseudorotunda, separated by a thin crest of opisthotic. The dorsal end of the opisthotic displays a typical concavity leading to the caudal tympanic recess. There is a development of the metotic strut behind the fenestra pseudorotunda so that the vagus canal (cranial nerve X) has diverted back to the occiput. The occiput shows a sinus canal between the epiotic and supraoccipital. The brain is fairly large, and the EQ is 0.34, which is in the range of some living birds.

John Ostrom (1976a) has shown many similarities in the postcranial skeletons of *Archaeopteryx* and dromaeosaurs. The postcranial skeleton of *Archaeopteryx* is described in detail by Wellnhofer (1974, 1993). The neck is long and flexible, and the thorax is comparatively short and compact. The vertebral column is like that of nonavian theropods. There seem to be nine cervical, fourteen dorsal, five sacral, and twenty or twenty-one caudal vertebrae. The centra are amphicoelous throughout the column and lack hypapophyses. Some posterior dorsal vertebrae show small pleurocoels. In the tail region, the zygapophyses are highly elongated, and the chevrons behind the 7th caudal vertebra are prominent. The tail is unique, in that it has a row of rectrices on either side. The thoracic ribs lack uncinate processes, and a series of gastralia behind the coracoids are present in all specimens.

The scapulocoracoid (fig. 5.3C) is reminiscent of the condition in dromaeosaurs, in which the scapula is long, slender, and straplike; the coracoid is subrectangular, with the development of a biceps tubercle below the glenoid. These two bones are joined by a suture in *Archaeopteryx*, unlike dromaeosaurs. Moreover, the glenoid has shifted its position from backward to outward. The biceps tubercle—the precursor to avian acrocoracoid—is more pronounced in *Archaeopteryx* and projects forward. However, the coracoid of *Archaeopteryx* is primitively built. It lacks all of the avian features, such as strutlike configuration, procoracoid, acrocoracoid, and triosseal canal. The development of a strong, robust, U-shaped furcula is the main departure from the dromaeosaurid design. Although the sternum is absent in all specimens of *A. lithographica*, it has been identified as a small, flat, rectangular bony plate in *A. bavarica* (fig. 5.3D–E)

The forelimbs were elongated and obviously functioned as wings. They show long flight feathers extending out from the hand and lower arm bones. The humerus is primitive, with a large deltopectoral crest, but lacks the avian bicipital crest. The two expanded ends lie oblique to each other. The distal condyles and epicondyles are not well differentiated. The radius and ulna are long and slender and display the same proportions as in dromaeosaurs. The tridactyl manus with its hooklike claws is exceptionally large and represents about 40 percent of the length of the forearm. The wrist has the distinctive semilunate distal carpal, as in dromaeosaurs, to receive metacarpals I and II (Ostrom 1976a). This swivel wrist joint allowed folding of the hand in the plane of the forearm when not in use. The manus is three-fingered: metacarpal I is the shortest, II the longest, and III of intermediate size; metacarpals IV and V were lost. The phalangeal formula is 2-3-4-x-x, and the terminal claws are thin and recurved, indicating adaptation for perching and climbing (Feduccia 1993).

The pelvis is more advanced than in dromaeosaurs. The configuration of the pelvis of *Archaeopteryx*, especially the orientation of the pubis, has generated much controversy in recent years. Ostrom has restored the pelvis in a nonavian, theropod fashion, with the pubis extending vertically, but most paleontologists agree that the pubis has been rotated considerably backward, as is seen in the Berlin specimen. The ilium, ischium, and pubis are separate. The ilium has a large, expanded preacetabular process and a stout pubic peduncle (fig. 5.3F). The ischial morphology is unusual, showing a foramen proximally below the acetabulum and a bifurcation at the distal end, which may indicate cartilaginous extension. The pubis is twice the length of the ischium. It has a pectineal process proximally and shows a distinct foot at the distal end, as in nonavian theropods. Medially, the pubes are fused at the symphysis.

The hindlimb appears somewhat more advanced than in dromaeosaurs. The femur has an inturned head, well-defined greater and lesser trochanters, a posterior trochanter, and a distal expanded end without any fibular groove. The femoral shaft is somewhat bowed craniocaudally. The tibia, longer than the femur, has a lateral cnemial crest but no cranial crest. The fibula is a thin, splintlike bone and is as long as the tibia. The proximal tarsal elements are not fused with the tibia to form the tibiotarsus. The astragalus has a long and narrow ascending process, and the calcaneum is reduced. The foot is four-toed and anisodactyl. The central toes (metatarsals II, III, and IV) show a partial fusion at the proximal end in the Solnhofen specimen. The phalangeal formula is 2-3-4-5-x; the terminal claws show digital flexors, which may indicate grasping or perching capability. A skeletal restoration of *Archaeopteryx* is shown in figure 5.4.

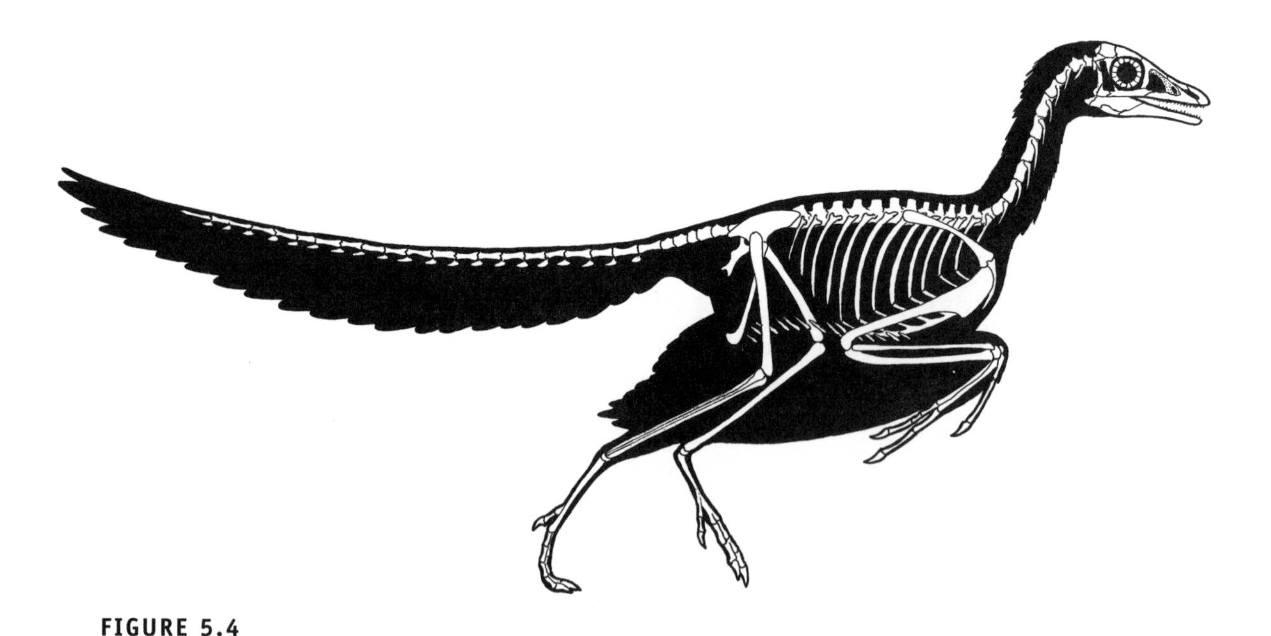

FIGURE 5.4

Skeletal restoration of *Archaeopteryx,* mainly based on the London specimen.

THE PALEOECOLOGY OF *ARCHAEOPTERYX*

What sort of bird was *Archaeopteryx*? What was its habitat? Was it arboreal or cursorial? Could it fly? These are some of the contested issues that have been debated for more than a century. To understand this controversy, we need to know in which paleoenvironment *Archaeopteryx* lived. During the Late Jurassic time, Pangea was breaking up; the Tethyan seaway appeared as the northern and southern continental landmasses separated. Arid conditions were accompanied by deposits of evaporites that formed along much of the coastline bordering the Tethys Sea. Animal communities were becoming separated and endemic. The Solnhofen region during this time, as reconstructed by Günter Viohl (1985), was a quiet, tropical lagoon lying behind coral islands on the northern shores of Tethys (fig. 5.5A). The lagoon was extremely salty and anoxic and did not support life.

Archaeopteryx was obviously an island bird; it evolved in isolation on several Central European Islands, such as London-Brabant Massif, Central German Swell, and Bohemian Massif. Many of the rich biota from these islands, including cycads, conifers, ginkgoes, insects, lizards, pterosaurs, *Compsognathus*, and *Archaeopteryx*, were probably drowned by occasional storms and carried to this lagoon to be preserved in exquisite detail. These animals and plants were preserved in the lagoon, far from their original life habitat. The preservation of articulated skeletons of *Archaeopteryx* with feathers and pterosaurs with skin impressions certainly indicates unusual circumstances of fossilization in the Solnhofen setting, perhaps autochthonous burial. Were the pterosaurs and *Archaeopteryx* downed in flight and blown into the lagoon? Or did they die and become mummified at the beach, then to be transported to deep water, settling eventually into the bottom of the lagoon? These remarkably preserved fossils at Solnhofen evoke many unusual explanations. The water was calm and stagnant; there was no scavenging or bacterial decomposition at the bottom of the lagoon as a consequence of poisonous planktonic bloom. This unusual setting was conducive to the preservation of such delicate, soft structures as feathers, insect wings, wing membranes of pterosaurs, hairlike setae of crustaceans, and medusae of jellyfish. *Archaeopteryx*, a crow-sized bird, lived in the forests of coral islands in Central Europe and was frozen in time into this lithographic limestone by accident.

Although traditionally *Archaeopteryx* has been envisioned as an arboreal bird, Ostrom (1973, 1976a, 1985a) suggested that it was a ground dweller similar to quail and roadrunners. Ostrom was so influenced by the anatomical similarity between *Archaeopteryx* and dromaeosaurs that he interpreted *Archaeopteryx* as a terrestrial cursor without any flying power. He reasoned that

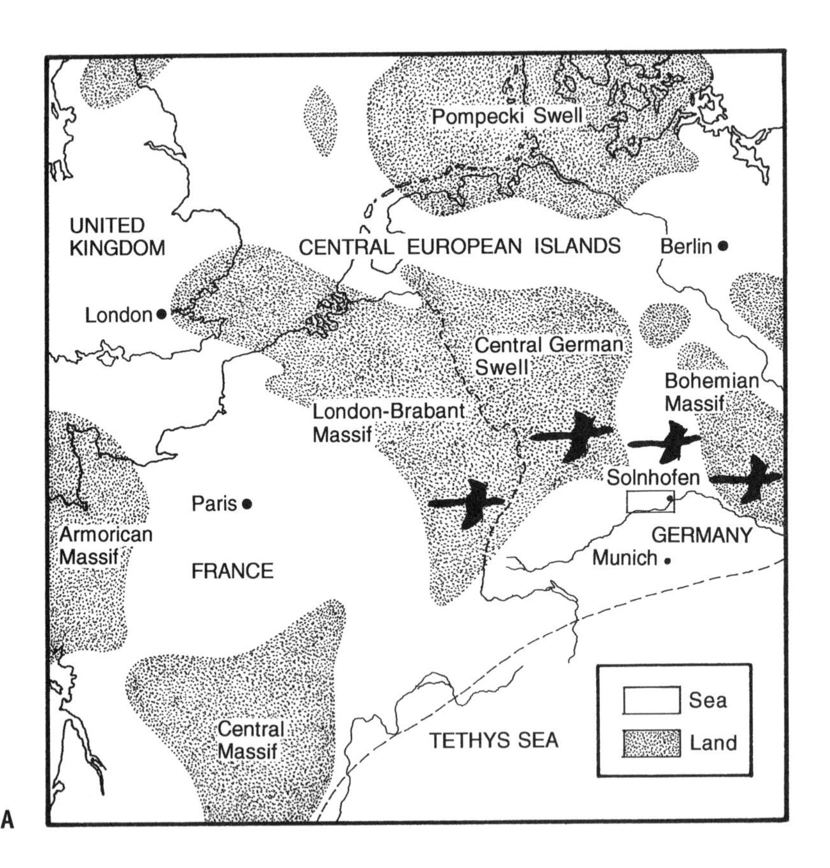

FIGURE 5.5

A, paleogeography of Central Europe during the Late Jurassic period. The *rectangle* marks the sedimentation area of the Solnhofen Lithographic Limestone. Much of Europe was covered by a shallow sea. *Archaeopteryx* probably lived in the Central European Islands, such as London-Brabant Massif, Central German Swell, and Bohemian Massif. The animals were probably downed in flight by occasional storms

it lacks two bony elements that seem to be essential for flight: the ossified, keeled sternum for the attachment of the powerful pectoralis muscle, which executes the downstroke, and a triosseal canal for the supracoracoideus pulley associated with the upstroke. Perhaps *Archaeopteryx* used its feathers to catch insects, Ostrom speculated. Critics such as Larry Martin (1983a) argued that, if the wings functioned as insect-catchers in early birds, natural selection would have acted strictly to improve them as fly swatters. In that case avian wings would not have been designed for flight.

Alan Feduccia (1980, 1996) and his colleagues have opposing

B

and blown into the Solnhofen lagoon (source: modified from Viohl 1985). *B,* paleoecology of *Archaeopteryx.* Its claw geometry suggests that it was primarily an arboreal bird and trunk-climber. *Archaeopteryx* was probably capable of undulating flight for a short distance.

views about the life habits and flight capability of *Archaeopteryx*. They pointed out that the shape and arrangement of the primary feathers of *Archaeopteryx* are similar to those of modern flying birds (Feduccia and Tordoff 1979). Each flight feather displays vane asymmetry, in which the smaller vane faces the leading edge. This is an important aerodynamic requirement for flight; flightless birds have symmetrical feathers. The sturdy furcula offers other evidence. It provided an expanded site for the attachment of the pectoral muscle, which is used in the downstroke (Olson and Feduccia 1979). The lack of the supracoracoideus pulley may simply indicate that *Archaeopteryx* probably could not take off from the ground. The highly curved claw geometry of both hand and feet and the reversed first toe indicate climbing and perching habits of *Archaeopteryx* (Feduccia 1993). It must have climbed trees and moved between treetops to pursue insects (figure 5.5B). The vegetation in these

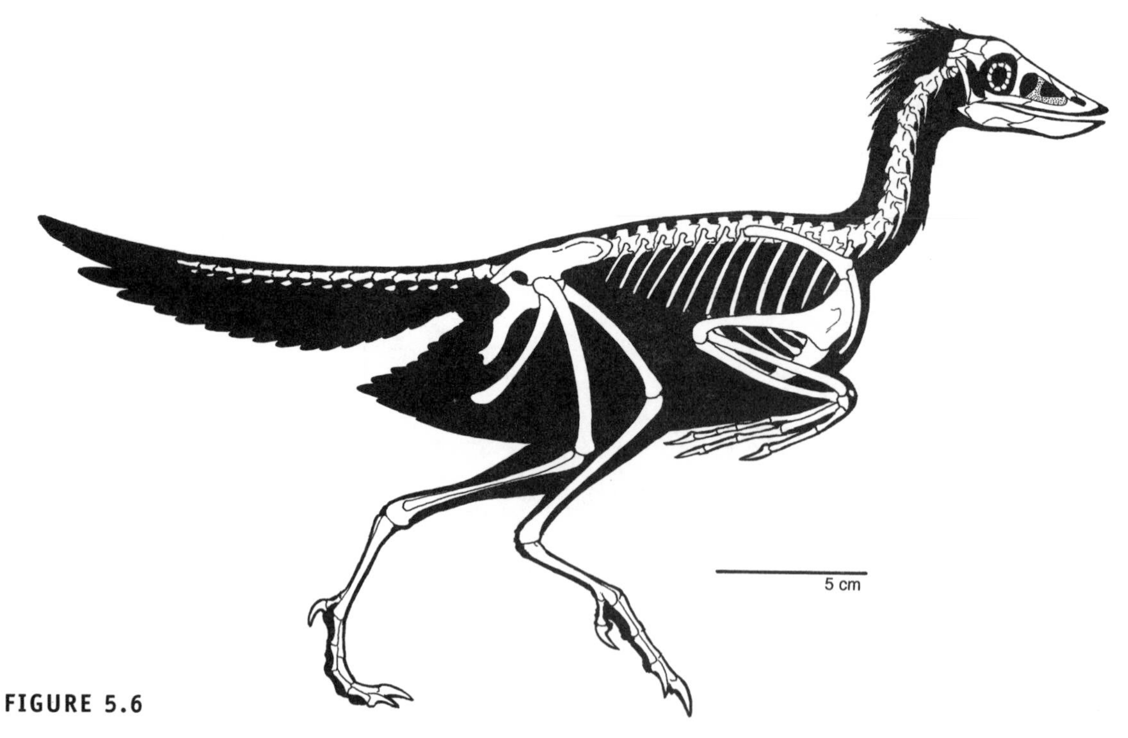

FIGURE 5.6

Skeletal restoration of the Jurassic bird *Confuciusornis* from China (source: modified from Hou et al. 1995).

islands was probably bushland of conifers, seed ferns, ginkgoes, and cycads. A variety of insect fossils, such as cicadas, dragonflies, and wood wasps, have been recorded in the Solnhofen.

Archaeopteryx might also have launched from cliff tops to glide above the beaches and shallow seas for food (Peters 1994). Their sharp, conical teeth, like those of the marine Cretaceous birds *Hesperornis* and *Ichthyornis* and of crocodiles (Martin 1985, 1991), could be used to catch slippery fish. The construction of the pectoral girdle, lack of an ossified sternum, and primitive wrist morphology suggest relatively weak specialization for flapping flight. A variety of birds have become secondarily flightless in island biogeography. Was *Archaeopteryx* on its way to losing its flying capability because of isolation? The unossified sternum in *A. lithographica* suggests this possibility. Although we may never know how well *Archaeopteryx* could fly, the consensus is that it could surely glide and could possibly fly in undulating fashion for a short distance at a cruising speed (see chapter 9).

Confuciusornis from China

Recently, Lianhai Hou and associates (1995) described a new pigeon-sized bird—*Confuciusornis sanctus* from the Yixian Formation (Jurassic-Cretaceous boundary) of Liaoning Province of northeastern China—that may be slightly younger than *Archaeopteryx.* Like *Archaeopteryx,* the Chinese species shows a patchwork of primitive and advanced features (fig. 5.6). The postcranial skeleton is primitively designed; the manus shows unfused carpal elements, long fingers, and large, curved claws. The shoulder girdle and vertebral column are unknown, but the pelvis and hindlimb are reminiscent of those of *Archaeopteryx,* with a retroverted pubis and a reversed hallux. Typical avian contour feathers are preserved on either side of the tibiotarsus. The presence of contour feathers in this Jurassic bird is significant because it may indicate endothermic physiology. Contour feathers might have been present originally in the Berlin *Archaeopteryx* specimen, as shown along the back and leg in one of the earliest drawings in 1879, but were all lost during extensive preparation (Ostrom 1985b). However, the skull of *Confuciusornis,* although highly crushed, seems to be more derived than that of *Archaeopteryx.* It has an edentulous beak, a nasofrontal hinge, a large premaxilla, and an enlarged external naris, as are seen in the Late Cretaceous bird *Gobipteryx. Confuciusornis* was apparently a good climber but a poor flier. It is currently grouped with enantiornithine birds (fig. 5.6).

Another contemporary Jurassic bird from Korea similar to *Confuciusornis* has recently been discovered, but the description is not yet available. The new discoveries of bird skeletons and the abundance of footprints suggest that by the Late Jurassic birds were numerous and evolving on many continents.

The Birds of the Early Cretaceous Period

The trees belong to birds; no one transgresses
these commands of this god Savitr.
—Rig-Veda, ca. 1200 B.C.

CHAPTER 6

The Cretaceous period (about 144–65 million years ago) marks the first global radiation of birds. During this time, Laurasia and Gondwana fragmented; shallow seas invaded North America and Europe. Dinosaurs became endemic because of the isolation of continents and appeared more varied and diversified. Forests thinned but the landscape grew prettier and more colorful with flowers. Flying creatures, ranging from the giant pterodactyls to modest-sized birds, were diversifying fast. Marine reptiles, such as plesiosaurs and mosasaurs, dominated the shallow seas, where the food was abundant.

We have known for a century about the Late Cretaceous toothed birds *Hesperornis* and *Ichthyornis* from Kansas, but the birds of the Early Cretaceous remained essentially unknown except for fragmentary remains of *Enaliornis* from England. *Enaliornis* is considered to be the earliest form of a hesperornithiform bird. Through the 1980s and into the 1990s, discovery after discovery revealed more wonder and variety of Cretaceous birds around the world (Chiappe 1995a, 1995b; Kurochkin 1995). The new discovery of several bird fossils from Mongolia, Spain, and China (fig. 6.1) has provided critical information about the early radiation of birds and the modification of flight in the post-*Archaeopteryx* period. Most of the specimens are incomplete and preserved in essentially two dimensions, however, making detailed description and interpretation of the bones difficult. Martin (1995b) has grouped most of these Early Cretaceous birds from Spain and China within Enantiornithes, or opposite birds. Evolving from a *Confuciusornis*-like ancestor, these enantiornithine birds became more cosmopolitan and widespread during the Cretaceous period.

FIGURE 6.1

Early Cretaceous continental configuration showing the locations of several genera of birds: *Enaliornis* (England), *Ambiortus* (Mongolia), *Sinornis* (China), *Cathayornis* (China), *Iberomesornis* (Spain), *Concornis* (Spain), and *Nanantius* (Australia).

During the Early Cretaceous, Pangea continued to break apart into the continental landmasses of Laurasia in the northern hemisphere and Gondwana in the southern hemisphere (fig. 6.1). Both of these supercontinents were separated by the Tethys Seaway. The sea level continued to rise, causing widespread flooding along the coastal areas. The overall global climate became increasingly warm. In this warm, equable environment, the radiation of birds left its first evidence in the fossil record. The broad coastal plain and lakes on the northern shore of the Tethys were inhabited by a wide spectrum of birds.

Ornithurine Birds

ENALIORNIS

Enaliornis is the earliest known genus of Hesperornithiformes, described by H. G. Seeley (1876) from the Early Cretaceous Greensand of Cambridge, England. The material consists of several partial skull fragments, the braincase, the heterocoelous cervical vertebrae, the femur, and the tarsometatarsus, all of which bear the signature of a foot-propelled diving bird.

AMBIORTUS

Evgeny Kurochkin (1985) described a partial associated skeleton of a crow-sized bird, *Ambiortus dementjevi*, along with feather impressions from the lacustrine deposits of Mongolia. The material consists of the left pectoral girdle and wing, including scapula, coracoid, proximal half of the humerus, carpometacarpus, partial sternum and furcula, and a series of thirteen presacral vertebrae. Unfortunately, like most other Early Cretaceous birds, the skull of *Ambiortus* is unknown. *Ambiortus* shows a refined level of flight capability, as seen in *Ichthyornis*. It has a keel on the anterior part of the sternum, a strutlike coracoid with a triosseal canal, a wide and spatulate sternal end of the coracoid, a U-shaped furcula, a humerus with a well-developed crest, a deltopectoral crest, and a carpometacarpus. Like other Cretaceous birds, *Ambiortus* displays a mosaic of primitive and advanced features. The lack of a bicipital crest on the humerus, the amphicoelous centra, the wide clavicular angle of the furcula, and the retention of the third phalanx in the second digit of the wing are some of the primitive characters.

CHAOYANGIA

In 1996, Lianhai Hou and his colleagues described an ornithurine toothed bird, *Chaoyangia*, from the Early Cretaceous of China. *Chaoyangia* is morphologically very similar to *Ambiortus*. It shows a deep, rounded scapular facet on the coracoid; its procoracoid process is long and straplike. The furcular hypocleidium is subdued, and the sternum is elongated with a prominent keel. This Chinese bird was probably adapted for wading, which is indicated by its long phalanges and elongated fourth toe.

Enantiornithes

In 1981, Cyril Walker described and identified a new subclass of birds, the Enantiornithes, from the Lecho Formation of Argentina. The Enantiornithes may represent the most dominant lineage in the radiation of Mesozoic birds. The outstanding feature of this group is the reverse articular arrangement between the scapula and the coracoid, where the coracoid forms a peg to fit into the socket of the scapula. Because of this reverse articulation, Walker named this group Enantiornithes, meaning "opposite birds." Enantiornithine birds were globally distributed during the Cretaceous period and occupied different niches. The early forms were mainly arboreal, perching birds. Some became aquatic, while others adapted the lifestyle of shore birds. Larry Martin (1995b) provided an extended classification of this group. He accommodated all of the Early Cretaceous Chinese and Spanish taxa, such as *Sinornis, Cathayornis, Iberomesornis, Concornis*, and *Noguerornis* within Enantiornithes.

SINORNIS

In 1992, Paul Sereno and Chenggang Rao described this sparrow-sized bird, *Sinornis santensis*, from the Early Cretaceous lake sediments of Liaoning Province of northern China (fig. 6.2A). *Sinornis* appeared about 10 million years later than *Ar-*

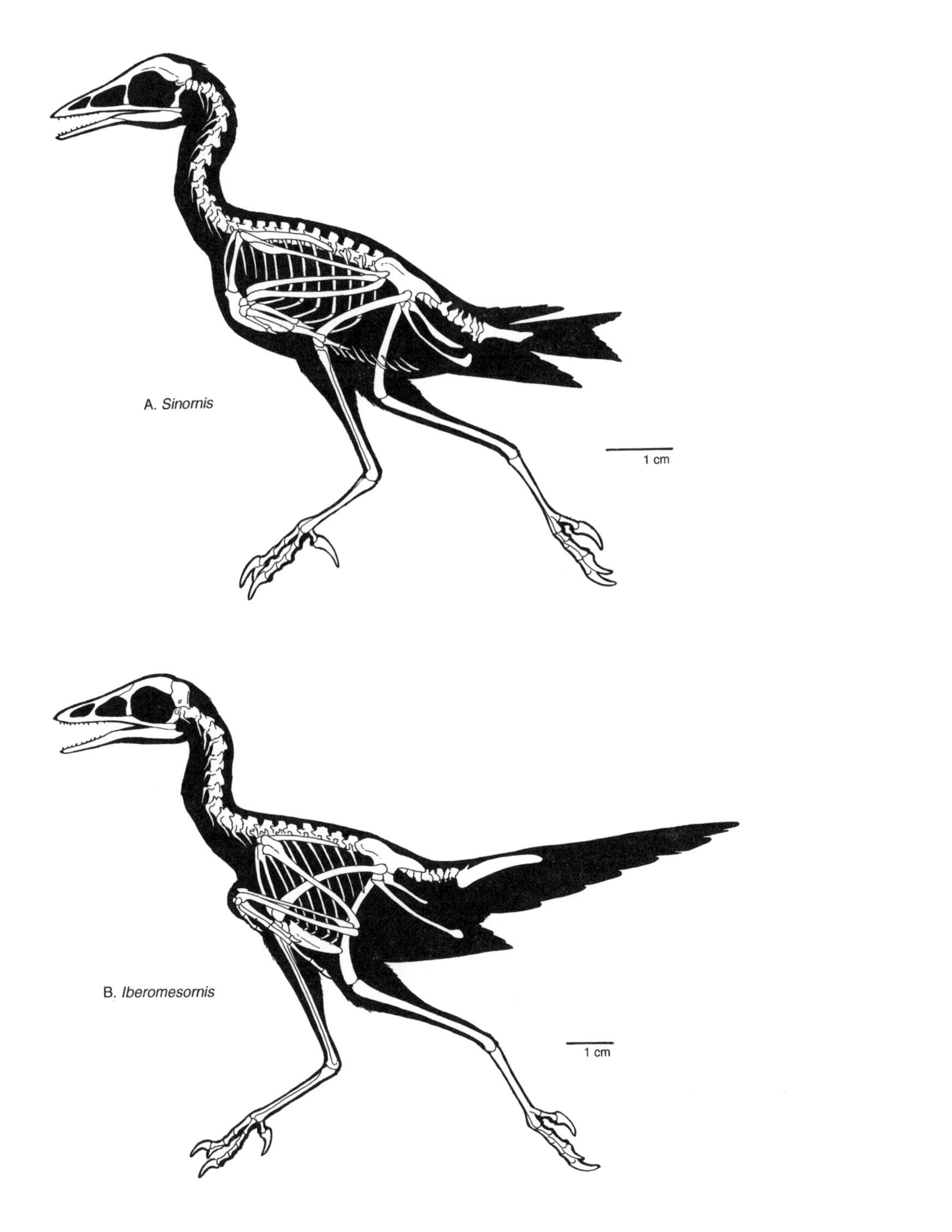

FIGURE 6.2

Early Cretaceous enantiornithine birds. *A,* skeletal reconstruction of *Sinornis santensis* from the Liaoning Province of northern China (source: modified from Sereno and Rao 1992). *B,* skeletal reconstruction of *Iberomesornis romeralli* from the Las Hoyas of Spain (source: modified from Sanz and Bonaparte 1992).

chaeopteryx but shows a great deal of modification in the flight apparatus. The skull is virtually unknown except for tooth-bearing jaws. The postcranial skeleton of *Sinornis,* like that of *Archaeopteryx,* shows several primitive features, such as unfused carpus and metacarpus, clawed manus, separate pelvic bones, footed pubis, separate metatarsals, gastralia, and limited skeletal fusion. However, the flight apparatus is more advanced than that of *Archaeopteryx. Sinornis* appears to have a strutlike coracoid, a furcula with elongate hypocleidium, and a broad, ossified sternum. The laterally facing glenoid permitted dorsoventral movement of the humerus during flight. The wrist joint is modified by the development of a large radiale and complex ulnare, which would interlock the short manus in place during the flight stroke. The finger bones are separate and the claws are reduced. The thorax is short, and the tail is truncated, with only eight free vertebrae and a large pygostyle. No doubt *Sinornis* was capable of powered flight. The pes shows a fully reversed hallux and highly recurved claws, indicating a perching adaptation and arboreal habitat. *Sinornis* is currently grouped with enantiornithine birds.

CATHAYORNIS

In 1992, soon after the discovery of *Sinornis,* three Chinese colleagues, Zhou Zhong, Jin Fan, and Zhang Jiang, reported another small bird, *Cathayornis yandica,* from the Early Cretaceous lake beds of the Jiufotang Formation of northern China. *Cathayornis* is represented by two well-preserved postcranial skeletons. Zhou (1995a) gave a more detailed description of *Cathayornis* in a subsequent paper. The skull is well preserved in the side view, showing a toothed beak. The premaxillae are probably fused along the midline. As in modern birds, the frontal bones are enlarged, the parietal is reduced, and the supraoccipital abuts against the parietal. The foramen magnum has been shifted ventrally. The quadrate seems to have an orbital recess. Like *Sinornis, Cathayornis* is a small perching bird with advanced flight adaptations, such as a strutlike coracoid, a

V-shaped furcula, a keeled sternum, a swivel wrist joint, and a pygostyle. The synsacrum has eight fused vertebrae. The carpometacarpus, pelvic bones, and metatarsals remain separate. The humerus lacks a bicipital crest, and the hindlimb is primitively built in the fashion of *Sinornis.* The pubis has a foot, as seen in *Sinornis.* Morphologically, *Cathayornis* is very similar to *Sinornis,* if not identical with it. The reverse scapulocoracoid articulation, broad furcular arms, anterodorsal ischial process, and proximal to distal metatarsal fusion indicate its affinity with enantiornithine birds.

IBEROMESORNIS

The celebrated bird from the Las Hoyas of Spain is another significant find linking *Archaeopteryx* to later birds. Although reported briefly in 1988, *Iberomesornis romeralli* was formally named by J. L. Sanz and José Bonaparte in 1992, when they analyzed its characters and compared it with other Mesozoic birds (fig. 6.2B). The specimen was found in a lacustrine or lagoonal deposit of lithographic limestone that has yielded several avian specimens, an isolated feather, and a diversified flora and fauna. *Iberomesornis* is based on a partial skeleton missing the skull, anterior cervical vertebrae, carpus, and manus. It is articulated but badly crushed. This sparrow-sized bird, about 8 cm long, is similar in many ways to *Sinornis* and shows a combination of primitive and advanced characters. The coracoid is typically avian, strutlike with an acrocoracoid process and an expanded distal end for the connection with the sternum. There is an ossified and probably keeled sternum. The furcula is springlike and narrow, ending with a large hypocleidium. The humerus has a stout deltopectoral crest and a well-defined ligamental groove. The tail is short, with the development of a pygostyle. *Iberomesornis* was obviously well adapted for powered flight. Unlike *Sinornis,* gastralia is not found in *Iberomesornis.* The cervical vertebrae show low neural spines but lack heterocoely and hypapophyses. The pelvis, sacrum, and hindlimb are primitively built in the fashion of *Archaeopteryx.* The foot is anisodactyl, and the hallux claw is very large. Martin (1995b) included *Iberomesornis* within enantiornithine birds.

CONCORNIS

J. L. Sanz and A. D. Buscalioni (1992) described a second bird, *Concornis lacustris*, from the same horizon of Las Hoyas of Spain. *Concornis* shows an asymmetrical feather impression, but the specimen lacks the skull. *Concornis* is twice the size of *Iberomesornis*, but its anatomy is poorly known. The flight apparatus is very similar to that of *Iberomesornis*, with a strutlike coracoid, a hypocleidium-bearing furcula, and a keeled sternum, all collectively indicating an advanced level of flight ability. *Concornis* seems to be more derived than *Iberomesornis* because it has a fused tibiotarsus and shows a trochlear surface on the distal end of the metatarsus. Larry Martin (1995b) has allied the Chinese and Spanish birds with Enantiornithes.

NOGUERORNIS

In 1989, A. Lasca-Ruiz described *Noguerornis gonzalezi* from the famous lithographic limestone quarry of Montsec Mountain, Spain. The material is poorly preserved except for the humerus, carpometacarpus, and furcula, which resemble those of *Iberomesornis.*

EOALULAVIS

Sanz et al. (1996) reported another new enantiornithine bird, *Eoalulavis hoyasi*, from the Lower Cretaceous Las Hoyas of Spain. The skeletal wingspan of this bird is 17 cm, about the size of a goldfinch. The outstanding feature of this skeleton is the preservation of the alula, or bastard wing, indicating that this bird could fly at slower speed and attained maneuvering flight 115 million years ago.

The Birds of the Late Cretaceous Period

The Lord did well when he put the loon and his music into this lonesome land.

—Aldo Leopold, *Round River,* 1953

CHAPTER 7

During the Late Cretaceous, between 84 and 65 million years ago, the continents continued to separate and began to assume the positions of the present day. Great warm seas covered much of the earth during this period, a time of mild climate and ice-free polar regions. The Western Interior Seaway, almost 1,500 km wide, at times bisected the North American continent. The western shore of the interior sea was inhabited by many species of diving and shore birds (fig. 7.1). Many terrestrial birds and birdlike forms thrived in South America and Mongolia.

In 1880, when Othniel Charles Marsh published his magisterial book on Odontornithes, the Cretaceous toothed birds from the Niobrara Chalk of Kansas, it became an instant favorite among evolutionary biologists. The acceptance of these fossils as true birds was universal. Marsh's monograph is the most detailed description, bone by bone, of five species of toothed birds: *Hesperornis regalis, H. crassipes, Ichthyornis dispar, I. victor*, and *Apatornis celer.* The text is accompanied by excellent drawings of the individual bones and skeletal restorations. This book still ranks as one of the outstanding contributions to avian paleontology. These toothed birds from Kansas represented a triumph for Darwin. Darwin did not cite *Archaeopteryx* as a long-sought evolutionary intermediate to bolster his theory. He wrote to Marsh on 31 August 1880: "Your work on these old birds . . . has afforded the best support to the theory of evolution, which has appeared in the last 20 years [i.e., since the publishing of the *Origin of Species*]."

FIGURE 7.1

Late Cretaceous continental configuration showing the locations of several genera of birds, including *Ichthyornis* (Kansas), *Hesperornis* (Kansas), *Enantiornis* (Argentina), *Patagopteryx* (Patagonia), and *Polarornis* (Antarctica).

Hesperornithiformes

The Hesperornithiformes are foot-propelled, flightless, diving birds that parallel the modern loons and grebes in many skeletal features. These birds are distinct from modern birds because they have true teeth in their jaws. They show evidence of great diversity and cosmopolitan distribution during the Cretaceous. The Niobrara Chalk and Pierre Shale formations of Kansas, South Dakota, and Wyoming have produced three genera: *Baptornis, Parahesperornis,* and *Hesperornis.* The skull of *Baptornis* is poorly known, but the postcranial material is well represented. *Baptornis* was a less specialized diving bird than its contemporary, *Hesperornis.*

Larry Martin (1984) described an almost complete skeleton of *Parahesperornis alexi.* Although the skull of *Parahesperornis* closely resembles that of *Hesperornis,* there are certain differences in the morphology of the lacrimal and quadrate bones, the frontoparietal suture, the tympanic recess, and the braincase between these two taxa. The presence of soft, hairlike plumaceous feathers indicates that in life these birds had a furry look.

Hesperornis is by far the best-known genus in this group and has been studied by many workers since Marsh. It was about 2 meters long and had completely lost the power of flight (fig. 7.2A). The skull is about 26 cm long, with an elongate beak presumably covered by a horny sheath. The top part of the bill overhangs the lower part. The external naris is holorhinal and posteriorly set. Teeth are retained in the maxilla and the dentary but are absent from the premaxilla. Sutures are present in the dermal bones. The premaxillae are elongate and contact the frontal bones, and the upper jaw moved prokinetically (Witmer and Martin 1987). The temporal configuration is like that of modern birds in that the orbit is confluent with the two temporal openings. The upper temporal opening is bordered by a deep recess. The postorbital and ectopterygoid bones are lost. The braincase is inflated and intimately fused with the skull roof. The quadrate is modern-looking, with a prominent orbital process. The dorsal head is single but expanded transversely so that it articulates with both the squamosal and the prootic. The two halves of the dentary are separate but connected to a small predentary bone. Along the roof of the orbits, there are well-marked depressions for salt glands, indicative of the aquatic habits of *Hesperornis.*

The postcranial skeleton was highly modified for foot-propelled diving habits and shows the evidence of flightlessness. The vertebrae are heterocoelous. The shoulder girdle is relatively primitive with an *Archaeopteryx*-like coracoid. It supported only a degenerate limb, the relic of a wing, the forearm and hand being lost. The sternum is flat and lacks a keel, and the furcula is not fused. The shoulder girdle and diminutive wings clearly indicate that *Hesperornis* could not fly. The pelvis

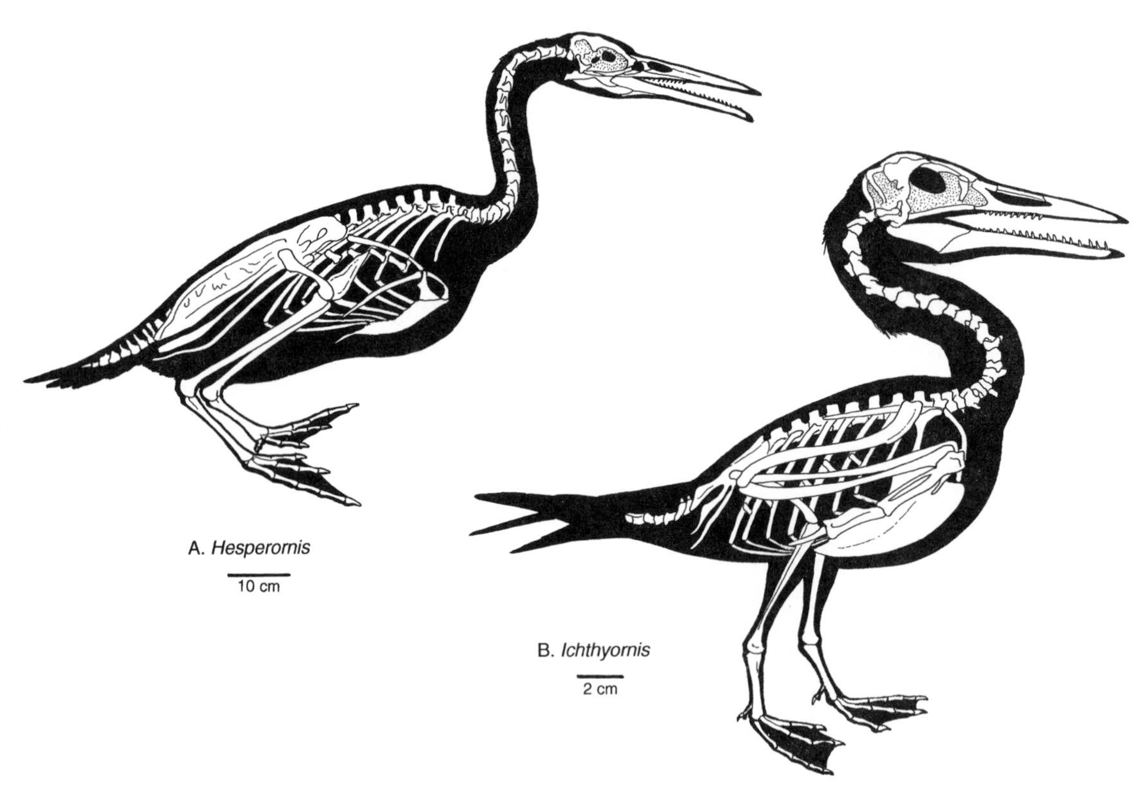

FIGURE 7.2

Late Cretaceous birds from Kansas. *A,* skeletal restoration of the foot-propelled seabird *Hesperornis* from Kansas; *B,* skeletal restoration of the highly volant bird *Ichthyornis* from Kansas. (Source: modified from Marsh 1880.)

and hindlimbs are typical of a foot-propelled diving bird, with relatively short, powerful legs placed far back on the body. The pelvis is narrow and compressed. The postacetabular process is much more elongated than the preacetabular blade. The pubis and ischium are horizontal and extend posteriorly at the level of the ilium. The ilium and ischium remain separate throughout their lengths, without any fusion. The acetabulum is partially closed internally by a bone, and there is a prominent antitrochanter. The femur is extremely short and stout, directed horizontally in a lateral direction to position the foot.

A special feature in a foot-propelled diving bird is the development of a large cnemial crest on the tibia; the extensive muscles necessary for the power strokes used in diving attach to this crest. The way in which the cnemial crest is developed among different diving birds may be an important phylogenetic character. In *Hesperornis,* the cnemial crest evolved from the development and expansion of the patella. In grebes, on the other hand, both patella and tibiotarsus contribute to the formation of the cnemial crest. The cnemial crest of loons is derived solely from the tibiotarsus, and the patella has been reduced to a tiny remnant of bone. The nonpneumatic nature of the limb bones may have been either a specialization for increasing their weight for more effective diving or a primitive retention.

Ichthyornithiformes

Marsh (1880) described two volant (capable of flying), toothed birds from Kansas, *Ichthyornis* and *Apatornis*; these birds superficially resemble gulls and terns, respectively. Of the two, *Ichthyornis* is better known and shows a wider geographic distribution. The premaxilla and maxilla are virtually unknown in *Ichthyornis,* but the posterior part of the skull, quadrate, and tooth-bearing lower jaw are well represented. The skull is large relative to the rest of the skeleton (fig. 7.2B). It has a salt gland depression along the roof of the orbits, as seen in *Hesperornis.* The quadrate and temporal configurations are similar to those of modern birds, and the braincase is expanded. The jaws of

Ichthyornis and *Hesperornis* show strong convergence because of a similar feeding adaptation. They probably relied heavily on fish. Both genera show similar tooth implantation. Teeth are set in a groove in young individuals but are fully socketed in the adult. The teeth are present in the maxilla and dentary. The lower jaw has a well-developed intramandibular joint for lateral spreading of the rami. This mechanism allows these birds to swallow a large amount of food. The two rami are separate but are probably united at the tip by a predentary bone.

The cervical vertebrae have large pleurocoels and are amphicoelous. A few primitive characters, such as teeth, amphicoelous vertebrae, and the lack of a bicipital crest in the humerus, persist in *Ichthyornis.* The rest of the skeleton is highly derived. The synsacrum contains ten vertebrae and there is a pygostyle. The morphology of the shoulder girdle and wing conform to the pattern seen in modern flying birds. The scapula is narrow, and the coracoid is strutlike, with the development of a triosseal canal; distally, it is expanded to fit into a horizontal groove of the sternum. The sternum is strongly built, with a prominent keel cranially. The furcula is partly preserved and appears to be U-shaped. The humerus is robust, with a prominent head and an enormous deltopectoral crest. Both dorsal and ventral tuberosities are well developed at the proximal end, with a distinct capital groove. Distally, the humeral condyles are modern-looking, with a prominent brachial depression. The ulna has a trochlea at the distal end, and the carpometacarpus is totally fused.

The pelvis and hindlimb of *Ichthyornis* are highly derived. The ilium has a long preacetabular but a short postacetabular process. The ilium, ischium, and pubis are fused around a large, perforated acetabulum. The acetabulum contains a prominent antitrochanter. The ischium remains parallel to the ilium, without any contact such as that seen in ratites. The pubis is rodlike and reverted. Ventrally, the pubis and ischium remain separate, with a large pelvic outlet. Dorsally, the two ilia form a large pelvic shield. The hindlimb elements have a modern appearance, with a large degree of fusion at the joints to form the tibiotarsus and the tarsometatarsus. The femur is relatively short and cylindrical, with a prominent greater trochanter, fibular groove, and popliteal space. The tibiotarsus has developed both cnemial crests but lacks a supratendinal bridge at the distal end.

Enantiornithes

In 1981, Cyril Walker described a new subclass of birds, the Enantiornithes. The type species, *Enantiornis leali*, was fairly large, about the size of a turkey vulture, with a wingspan of at least 1 meter. The skull is poorly known at present, except for a toothless jaw fragment. Walker recognized a great deal of morphological variation among the humeri and tarsometatarsi but a uniformity within the shoulder girdle. Reversed scapulocoracoid joints, a flat-headed humerus, a reduced outer metatarsal, and the lack of procoracoid and sternocoracoidal processes are the unique attributes of this group. The acrocoracoid abuts against the acromion of the scapula, so that the triosseal canal is essentially confined to the coracoid. Many advanced features in the skeleton indicate that *Enantiornis* was a volant bird (fig. 7.3A). The glenoid faces upward to allow wide excursion of the humerus. The sternum is ossified, with the development of a keel. There is some heterocoely in the cervical vertebrae, but the caudal vertebrae are reduced and terminate into a pygostyle. The humerus shows many avian features (bicipital crest, pneumatic fossa, small dorsal and ventral tuberosities, capital groove, ligamental furrow, and brachial depression), but its distal condyles are subdued. The distal carpals and metacarpals are fused to form a carpometacarpus. The pelvis is opisthopubic; the ilium and ischium are fused to enclose a large ilioischiadic fenestra. The femur is primitive and lacks the fibular groove. However, the head shows the ligamental pit, where the distal condyles are well developed and closely spaced. The tibia is fused with the astragalus and calcaneum to form the tibiotarsus. The metatarsals show some degree of fusion proximally but not distally.

Enantiornithine birds were globally distributed during the Cretaceous period and occupied different niches. The early forms were mainly arboreal, perching birds. Some became aquatic, while others adopted the lifestyle of shore birds. Recently, several other genera from Argentina, such as *Neuquenornis, Yungavolcuris, Lectavis,* and *Soroavisaurus,* were identified by Luis Chiappe (Chiappe 1995b). *Avisaurus* from North America, *Alexornis* from Mexico, *Nanantius* from Australia, and *Gobipteryx* from Mongolia are also included in this group. Larry Martin (1995b) provided an extended classification of this group. He accommodated all of the Chinese and Spanish taxa, such as *Sinornis, Cathayornis, Iberomesornis, Concornis,* and *Noguerornis,* within Enantiornithes. Many of these taxa are based on hindlimb elements. Cranial information is sparse, but partly available from *Gobipteryx.* The skull is crushed but shows a toothless beak formed by the fused premaxillae. The external naris has been shifted backward, and the mesethmoid is ossi-

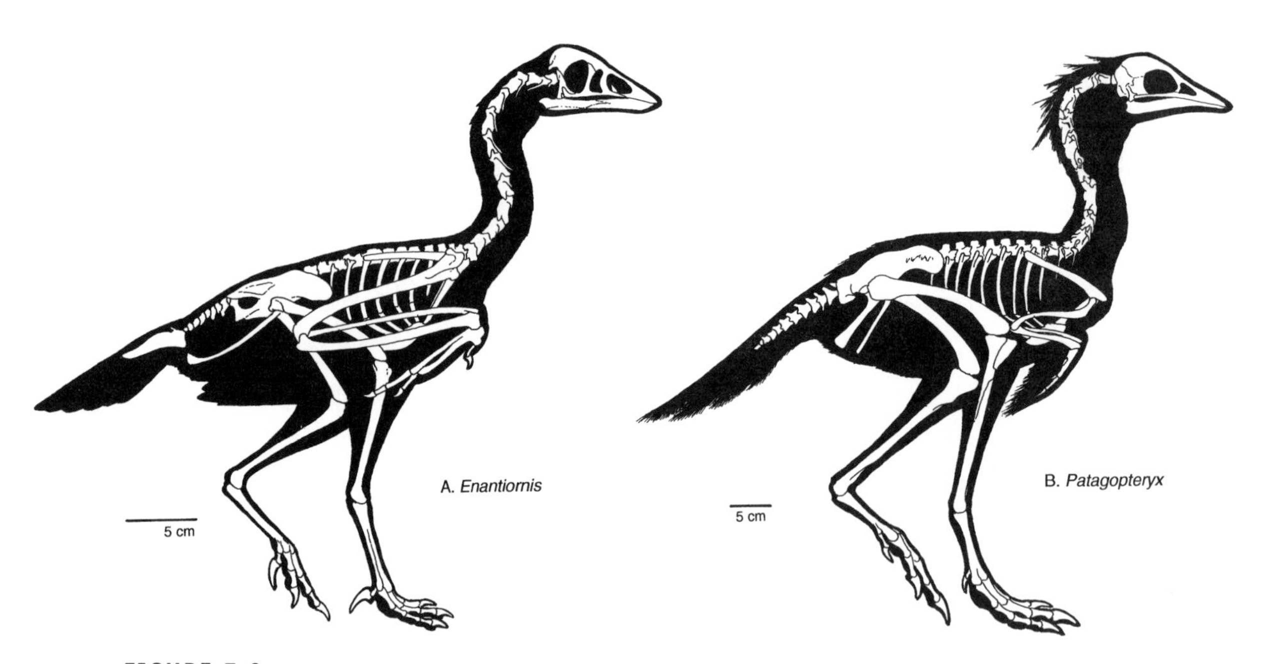

FIGURE 7.3

Late Cretaceous birds from South America. *A,* the vulture-sized enantiornithine (opposite) birds were widespread globally in the Cretaceous ecosystem and occupied different niches. In this reconstruction, the first-described species, *Enantiornis leali* from Argentina, is shown. *B,* reconstruction of the chicken-sized flightless bird *Patagopteryx deferrariisi* from Patagonia (source: modified from Chiappe 1996).

fied. The temporal configuration is unknown. In the palate the choana has been shifted backward and the pterygoid articulates with the vomer to form a median bar. The quadrate appears primitively built in *Archaeopteryx* fashion. The lower jaw is toothless, with a bony mandibular symphysis.

Patagopteryx

Herculano Alvarenga and José Bonaparte (1992) described a chicken-sized, flightless bird, *Patagopteryx deferrariisi*, from the Rio Colorado Formation of Patagonia (fig. 7.3B). *Patagopteryx* is known from several articulated specimens (Chiappe 1996). The skull is known only from the caudal part, showing a rounded braincase, large orbit, and zygomatic and orbital processes. The cervical and anterior dorsal vertebrae are heterocoelous. The wing is highly atrophied, a sign of secondary flightlessness. The humerus is reduced but retains a bicipital crest. The shoulder girdle is incomplete; the coracoid shows a depression on the posterior surface, as seen in enantiornithine birds. The clavicles are fragmentary. The pelvis and hindlimb are well preserved. The ilia form a pelvic shield with the synsacrum; the latter is dorsoventrally compressed and transversely wide. The pelvis is opisthopubic and resembles that of ratites, where the ischium and pubis are slender rods; the ilium and ischium separate posteriorly without enclosing the ilioischiadic fenestra. A large antitrochanter is present on the rim of the acetabulum. The hindlimb is well developed, as in all flightless birds. The femur is robust and shows a prominent popliteal space. The tibiotarsus has a lateral cnemial crest. The distal tarsals and metatarsals are fused to form the tarsometatarsus, and the latter develops a tendinal groove distally without a supratendinal bridge.

The Antarctic Loon

Loons are generally regarded as ancient birds; their fossil record extends as far back as the Late Eocene of England (Storer 1956). In 1989, I briefly reported the discovery of a partial, associated skeleton from the Late Cretaceous Lopez de Bertodano Formation of Seymour Island, Antarctica. The specimen, informally

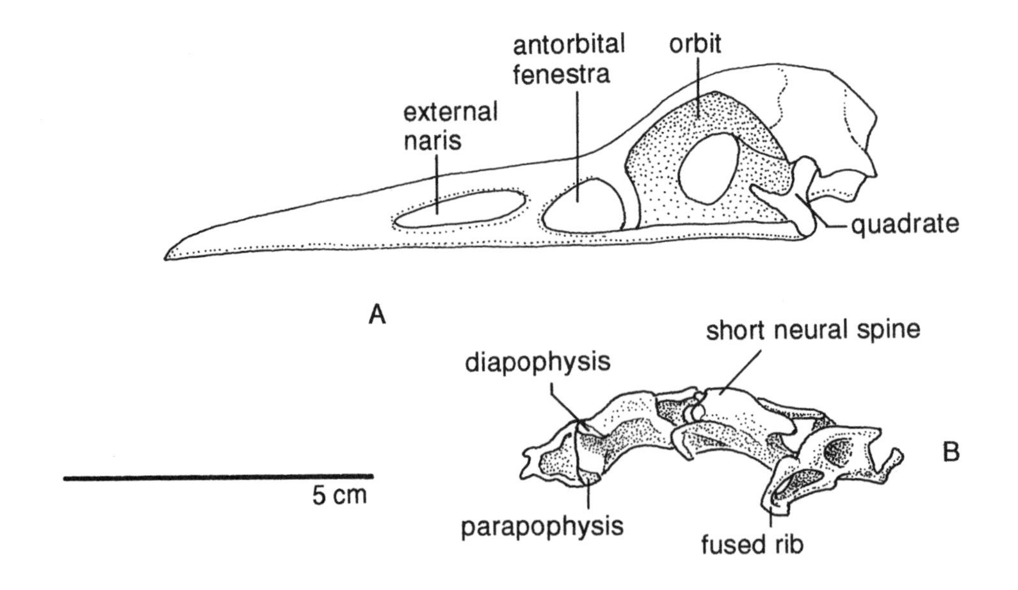

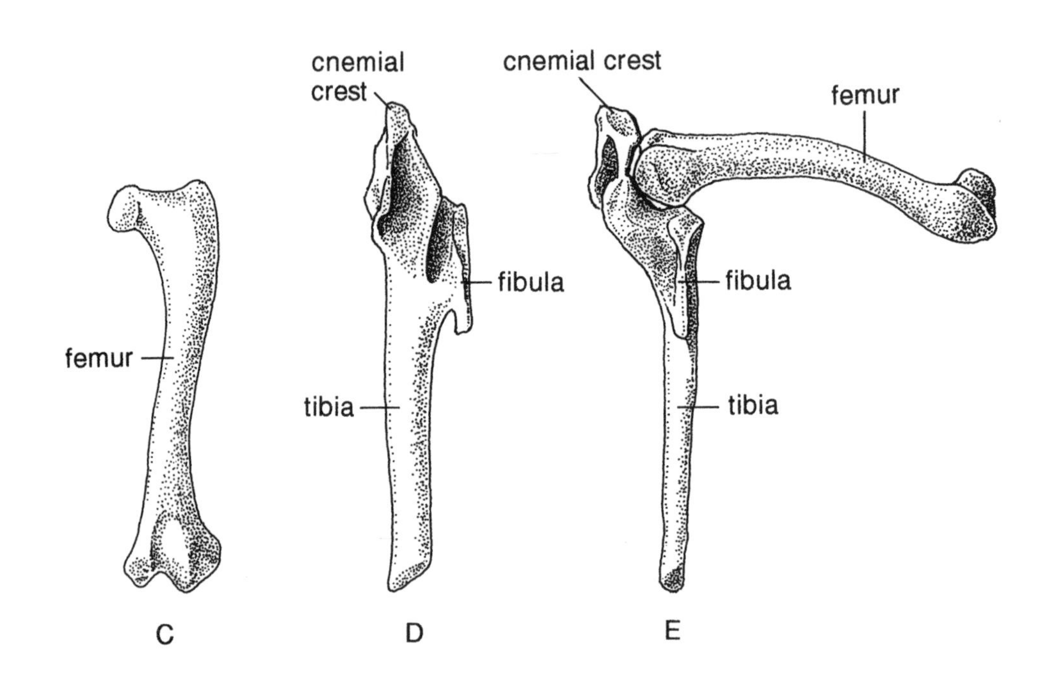

FIGURE 7.4

The Late Cretaceous loon, *"Polarornis,"* from Seymour Island, Antarctica. This is the earliest record of a loon from adequate material and is very similar to the modern loon in morphology. *A,* skull in the left lateral view; *B,* associated cervical vertebrae; *C,* left femur in the cranial view; *D,* proximal half of the left tibia and fibula in the cranial view. The prominent cnemial crest on the tibia is a hallmark for a foot-propelled diving bird. *E,* left hindlimb in articulation, lateral view. The femur and tibia are at right angles to one another. In a foot-propelled diving bird, the tibia is the main locomotor organ, and the femur remains passive.

named "*Polarornis*," was found in a shallow marine environment along with plesiosaurs, mosasaurs, and ammonites. It represents not only the world's oldest loon, but also the only Mesozoic bird that can be placed securely in the modern family of birds.

The skull of *Polarornis* is fairly intact, beautifully preserved, and entirely edentulous. It has a long and tapering snout and intimately fused dermal bones. In size and morphology, it closely resembles the modern red-throated loon, *Gavia stellata*. The external naris is elliptical and holorhinal and has been shifted considerably backward with the development of a typical beak (fig. 7.4A). The frontonasal hinge area is very similar to that of modern prokinetic birds. The orbits are large and forwardly placed. Along the roofs of the orbit are well-marked depressions for the salt glands. The orbital septum is fully ossified. The squamosal is reduced and forms an overhanging roof over the quadrate. The palate is of the neognathous type, with the development of palatine/pterygoid mobility. This is the earliest documentation of a neognathous bird (similar to modern flying birds) in the fossil record. The quadrate has a large orbital process, and its articular end shows two distinct heads for the squamosal and the prootic, respectively. Laterally, the quadrate has a deep socket for the reception of the quadratojugal.

Four articulated cervical vertebrae of *Polarornis* are preserved in the collection (fig. 7.4B). They are elongated and heterocoelous, with ribs fused to them. A strong hypapophysis on the centrum terminates cranially against a ventral depression. The wings and girdles are missing in the specimen. The femur is twice the size of that in the modern loon (fig. 7.4C). The tibiotarsus is long and its cnemial crest is strongly developed, indicating that the earliest loon had already acquired the foot-propelled diving adaptation (fig. 7.4D–E). As in all loons, the cnemial crest is formed solely by the tibiotarsus. The anatomy and relationships of the Antarctic loon will be described in a future paper.

Another loonlike bird, *Neogaeornis* from the Late Cretaceous

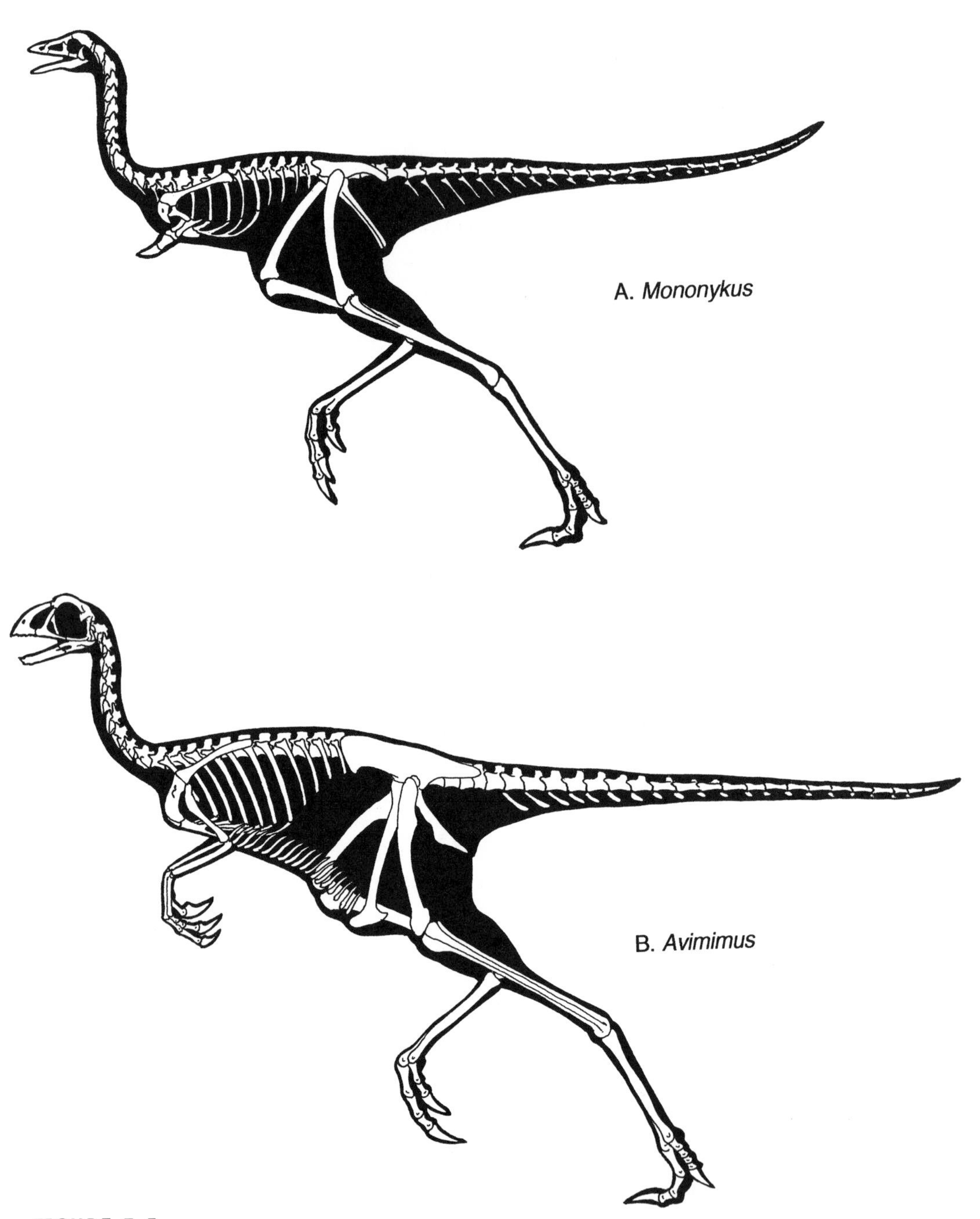

FIGURE 7.5

Dubious flightless birds from the Late Cretaceous Gobi Desert of Mongolia. Many experts believe that these two genera may be highly derived nonavian theropods, such as coelurosaurs or maniraptorans, with some avian attributes because of convergent evolution. *A, Mononykus olecranus* (modified from Perle et al. 1993); *B, Avimimus prorontus* (modified from Paul 1988). In both genera, the skull is extremely avian, but the postcranial evidence is equivocal.

of Chile, has been known from tarsometatarsi for more than fifty years. Recently, Storrs Olson (1992) evaluated this specimen and placed it within the loon family, Gaviidae. Olson believed that the loons and the penguins are sister groups. Both originated in the Southern Hemisphere, but they show different styles of adaptive radiation. The loons migrated to the Northern Hemisphere in the Early Tertiary (probably to avoid direct competition with the emerging penguins) and flourished there; the penguins stayed in the Southern Hemisphere and thrived in the cold water of the Antarctic.

Vorona

Recently, Forster et al. (1996) described the hindlimbs of a very primitive ornithothoracine bird, *Vorona berivotrensis*, from the Upper Cretaceous Maevarano Formation of Madagascar. Because of its fragmentary nature, the affinity of *Vorona* is difficult to assess. However, this bird certainly provides an important clue to the range and diversity of Mesozoic birds.

Dubious Flightless Birds

Two obligatory theropods, *Mononykus* and *Avimimus* from the Late Cretaceous Nemegt Formation of the Gobi Desert of Mongolia, have been allied to flightless birds in recent years, but many experts have questioned their avian identity.

MONONYKUS

In 1993, Altangerel Perle, Mark Norell, Luis Chiappe, and James Clark reported a bizarre theropod, *Mononykus olecranus*, from the recent expedition to the Gobi Desert. As the name suggests, the animal had an atrophied forelimb with a prominent olecranon process on the ulna and a single claw (fig. 7.5A). It has been proposed that this turkey-sized animal lived underground and dug holes with its shovel-like forefeet in a manner similar to that of the living golden mole. The jaws are probably edentulous except for the tips. The braincase and quadrate resemble those of dromaeosaurs. The scapulocoracoid is primitively built and lacks the biceps tubercle. The furcula is unknown. Several features in the skeleton, such as a keeled sternum, an antitrochanter on the ilium, the absence of a pubic foot, a trochanteric crest on the femur, a reduced fibula, and a cranial cnemial crest on the tibia, were cited by Perle and colleagues as avian attributes. More likely, these characteristics evolved convergently with the fossorial adaptation (adaptation to digging). For example, the keeled sternum, cited by these authors as a diagnostic character of the bird, is also present in moles. No bird has evident structural adaptations for digging. Most paleontologists agree that *Mononykus* was not a bird and never had a flying ancestor (Zhou 1995b). However, a recently discovered skull of *Mononykus* shows some avian attributes.

AVIMIMUS

The Russian paleontologist S. M. Kurzanov described the enigmatic theropod *Avimimus* from the Gobi Desert in 1985 and 1987. He documented in this turkey-sized animal a suite of avian characteristics that are not present in *Archaeopteryx*. The skull of *Avimimus* is remarkably avian where the orbit and lower temporal opening are confluent. The jugal bar is a slender rod, as is seen in modern birds. The squamosal-quadratojugal bar is lost in front of the quadrate, making the quadrate free. However, the quadrate is primitive and fused secondarily with the squamosal to lose its streptostyly. The upper temporal arch is intact and retains the postorbital. The palate retains the ectopterygoid bone. The braincase is inflated and modern-looking, whereas the occipital condyle is reduced. The skull is about 85 mm long, with a large orbit and edentulous jaws, but the premaxilla has toothlike denticles (fig. 7.5B).

The cervical vertebrae of *Avimimus* show pleurocoels and hypapophyses. The shoulder girdle is unknown. The humerus has asymmetrical distal ends, and the carpometacarpus is fused like that of modern birds. The pelvis and the hindlimbs, however, are primitively built. The ilium has a prominent antitrochanter and brevis shelf. The two pubes are fused distally, with an expanded foot. The orientation of the pubis is forward and downward, as seen in primitive nonavian theropods. The femur has a spherical head and a well-developed trochanteric crest; the lesser trochanter rises to the level of the greater trochanter. In addition, there is a posterior trochanter, as is seen in some maniraptorans. However, the femoral shaft retains a fourth trochanter, which is a primitive feature. The tibia, fibula, and proximal tarsals are fused distally. The metatarsals are also fused proximally; a splintlike fifth metatarsal is present. *Avimimus* is clearly an unusual theropod showing a curious mixture of primitive and advanced features. Because of these conflicting characteristics, some paleontologists believe that *Avimimus* may indicate the mixture of two species. Certainly, the skull, cervical vertebrae, and limbs show many avian attributes. The lack of a shoulder girdle makes it difficult to assess its avian level of organization. *Avimimus* may represent a flightless bird.

Eggs and Embryos, Feathers and Footprints

CHAPTER 8

The music of the moon

Sleeps in the plain eggs of the nightingale.

—Alfred, Lord Tennyson, "Aylmer's Field," 1864

And, departing, leave behind us

Footprints on the sands of time . . .

—Henry Wadsworth Longfellow, "A Psalm of Life," 1838

Ontogeny, the development of the individual from egg to adult, is an important component of evolutionary study. Avian hatchlings display remarkable differences in their skeletal morphology, external appearance, activity, and behavior. Some birds can fly shortly after hatching and are independent of parents. Others are born prematurely and need constant parental care and feeding until they become self-sufficient. The study of fossil eggs, nests, and embryos provides insights into the early developmental history of birds from growth to family life.

Fossil Eggs

The avian egg is a marvel of architectural design. Light and strong, it is a self-contained life support system for the developing embryo. The eggshell acts as a protective covering for the embryo as well as an important source of calcium for the chick. It is strong enough to hold the embryo securely during development, yet weak enough for the hatching chick to break. It is porous enough to permit the respiration of the embryo and yet closed enough to prevent the entry of dirt and microorganisms.

Avian eggs vary greatly in color, size, and shape. Egg size is related to bird size, but the relationship is much more complex. Hummingbirds (3 g) lay the smallest eggs (0.3 g), and ostriches (100 kg) lay the largest (1 kg). If we compare these two extreme birds, there is a 3,000-fold range of increment of egg weight, but the body weight increases 30,000-fold. Thus, larger birds lay proportionately smaller eggs relative to their body size. The largest known egg is that of the extinct flightless elephant bird, *Aepyornis* of Madagascar (500 kg). Its egg measured 34 by 24 cm and weighed about 10 kg. It could hold the contents of 7 ostrich eggs, 180 hen eggs, or 50,000 hummingbird eggs. It becomes apparent that the relationship between body weight and egg weight is exponential in a fashion similar to that between body weight and brain weight; body weight increases more rapidly than egg weight. H. Rahn and his colleagues plotted bird egg weight against body weight on logarithmic coordinates and deduced the following allometric equation (Rahn, Paganelli, and Ar 1975):

$$\text{egg weight} = 0.277\ (\text{body weight})^{0.77}$$

As a rule, large eggs have thick shells and small eggs have thin shells. The eggshell is largely made of polycrystalline calcite minerals held in a matrix of collagen fibers. The microscopic structure of the shell in radial section reveals two main layers—an inner mammillary layer and an outer spongy or palisade layer. The mammillary layer is composed of numerous roughly conical knobs or mammillae, which are oval to circular in cross

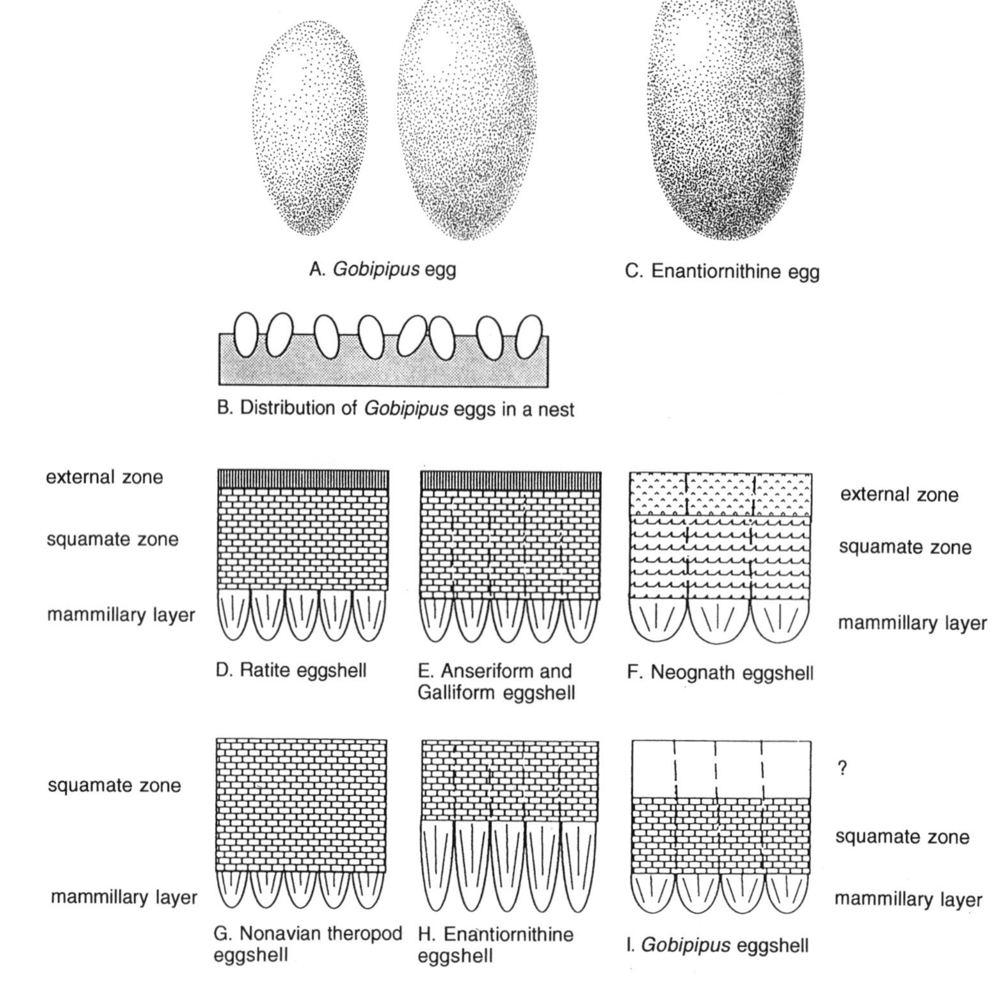

FIGURE 8.1

Eggshell and microstructure. *A,* two varieties of asymmetrical (ovoid) eggs are attributed to *Gobipipus;* the smaller variety is about 30-46 mm long, whereas the larger one is about 53-70 mm long. *B,* distribution of *Gobipipus* eggs in a single stratigraphic layer and their subvertical position in the substrate. *C,* the symmetrical (ellipsoidal) egg of *Laevisoolithus* probably belonged to an enantiornithine bird. *D–I,* general variants of ornithoid eggshell structure in radial section, based on data from scanning electron microscopy (SEM). Three structural strata of eggshell are shown diagrammatically: a basal mammillary layer, a middle squamate zone, and an outer external layer; vertical dashed lines indicate the expressiveness of columns in the continuous or spongy layer (source: simplified from Mikhailov 1992). *D–F,* modern bird eggshells: *D,* ratite eggshell; *E,* anseriform and galliform eggshell; *F,* typical neognath eggshell. *G–I:* fossil theropod eggshells: *G,* nonavian theropod eggshell (Oofamily Elongatoolithidae); *H,* enantiornithine eggshell (Oofamily Laevisoolithidae); *I, Gobipipus* eggshell (*?,* diagenetically altered zone) (source: simplified from Chatterjee, Kurochkin, and Mikhailov n.d.).

section with an organic core. They are tightly compressed side by side in a single stratum. The spongy layer is thicker, with calcite crystals arranged in the protein matrix as vertical palisades separated by minute pore canals. The chicken egg contains about 10,000 pores, which permit gas exchange (Rahn, Ar, and Paganelli 1979). The outer surface of this calcite layer is overlain with a protein cuticle that gives a smooth, shiny appearance.

Avian fossil eggs from the Mesozoic sediments are extremely rare, known thus far only from the Late Cretaceous Barun Goyot Formation of the Gobi Desert of Mongolia. The Russian paleontologist Konstantine Mikhailov (1992) has worked on this Gobi eggshell material. He recognized two kinds of avian eggs associated with embryos. The first type of egg is about 40 mm long and 20 mm wide. It is a small, ovoid egg belonging to a neornithine bird, *Gobipipus.* The eggs are found with their narrower end down, either vertically or obliquely in the sediment (fig. 8.1A–B). The second type of egg is relatively large, about 70 mm long and 40 mm wide. It has a symmetrical ellipsoidal shape and is attributed to an enantiornithine bird (fig. 8.1C).

Mikhailov studied the microstructure of dinosaur and avian eggshells under the scanning electron microscope. He recognized three horizontal strata or zones of microstructure in avian eggs from the inner to the outer surface: (1) a basal mammillary layer with radial organization of biocrystalline material, (2) a middle continuous or spongy layer with polycrystalline squamatic ultrastructure, and (3) an outer external zone (fig. 8.1D–I). Mikhailov proposed two morphotypes for the avian eggshells, emphasizing the microstructure of the spongy layer: Ratitae and Neognathae. In the ratite morphotype (as seen in ratites, galliforms, and anseriforms), the squamate layer does not exhibit distinct columns and the external zone is a dense crystalline layer (fig. 8.1D–E). In the neognathous morphotype, which is the most prevalent among living birds, the squamate layer exhibits distinct columns and prisms (fig. 8.1F). Interest-

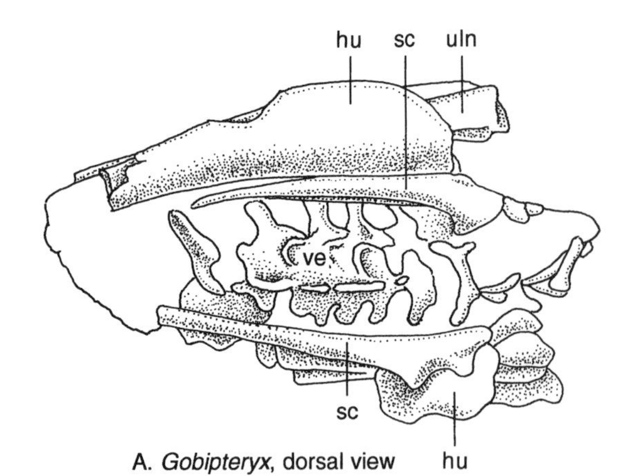

A. *Gobipteryx*, dorsal view

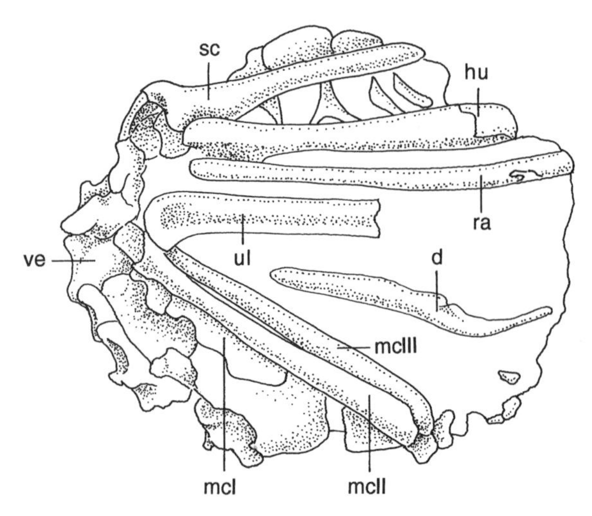

B. *Gobipteryx*, left lateral view

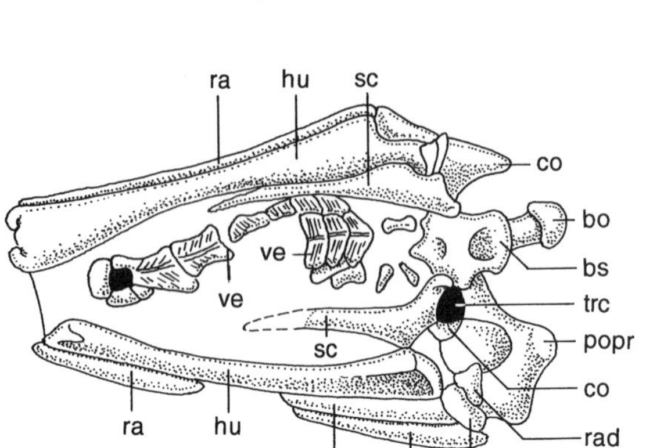

C. *Gobipipus*, ventral view

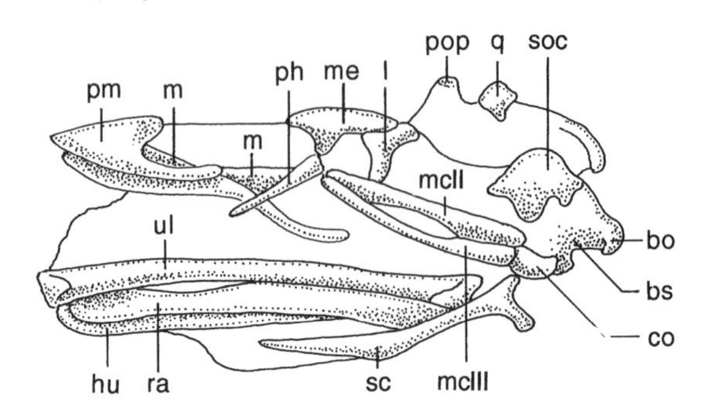

D. *Gobipipus*, left lateral view

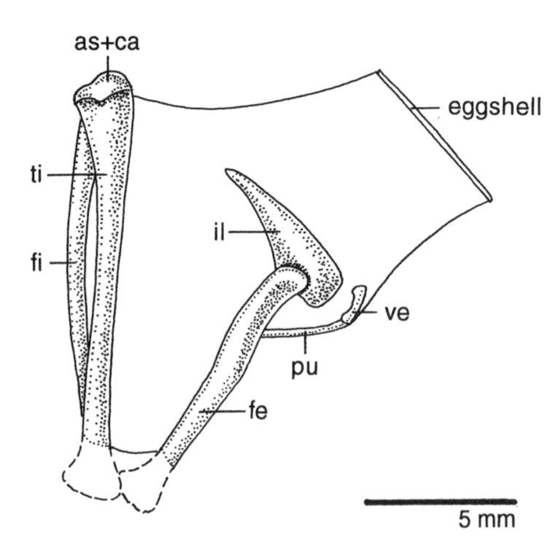

E. *Gobipipus*, lateral view of right ilium and hindlimb

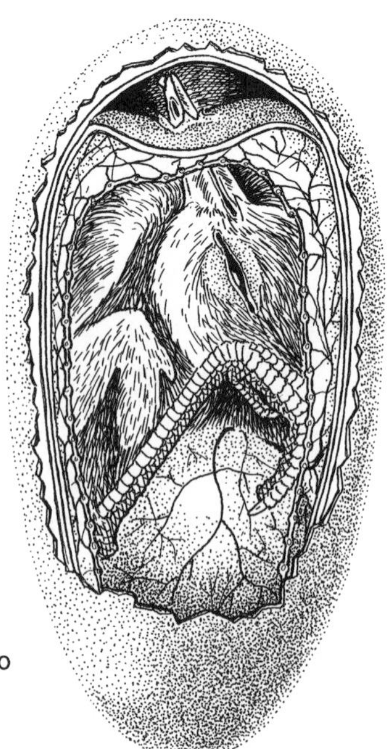

F. Reconstruction of *Gobipipus* embryo within the egg just before hatching

ingly, both types of Cretaceous eggshells from the Gobi Desert show a microstructure of the ratite morphotype (fig. 8.1H–I).

Fossil Embryos

Embryonic skeletons of birds are certainly among the rarest of fossils. In 1971, a joint Polish-Mongolian team discovered several avian nesting sites in the Upper Cretaceous Barun Goyot Formation of Khermeen Tsav in the southern Gobi Desert in Mongolia, along with a rich trove of nonavian dinosaurs and mammals. Many of these avian eggs contain precocial embryonic skeletons, which were described by Anjay Elzanowski (1981). Precocial birds are capable of a high degree of independent activity from birth. Various specimens show partial skull, wings, shoulder girdle, vertebrae, and part of the hindlimbs. Elzanowski concluded that these embryonic specimens belong to an enantiornithine bird, *Gobipteryx minuta,* which he described earlier from an adjacent area (fig. 8.2A–B).

In 1997, in collaboration with my Russian colleagues Evgeny Kurochkin and Konstantine Mikhailov, I described another new embryonic bird from the Gobi Desert on the basis of two specimens (Chatterjee, Kurochkin, and Mikhailov n.d.). The specimens were found by Valeri Reshetov from the eastern side of Khermeen Tsav in 1977. In the first specimen the skeleton, preserved in exquisite detail, is in a fetal position revealing skull, shoulder girdle, wings, and a series of neck vertebrae (fig. 8.2C–D). The second specimen shows a partial hip girdle and hindlimb (fig. 8.2E). Surprisingly, most of the bones are well ossified, indicating that the embryos were close to hatching (fig. 8.2F). We have named this bird *Gobipipus reshetovi. Gobipipus* is

FIGURE 8.2

Embryos of Late Cretaceous birds from the Gobi Desert, Mongolia. *A–B, Gobipteryx minuta,* dorsal and left lateral views (source: simplified from Elzanowski 1981). *C–D,* ventral and left lateral views of *Gobipipus reshetovi;* the delicate skeleton in a fetal position is about 18 mm long, showing a partial skull, vertebrae, the shoulder girdle, and wing elements. *E, Gobipipus reshetovi,* lateral view of the right ilium, femur, tibiotarsus, and fibula. *F,* this reconstruction shows a 35-mm-long *Gobipipus* embryo just before hatching. (*C–F,* source: after Chatterjee, Kurochkin, and Mikhailov n.d.) For abbreviations, see figures 4.1–4.5.

more advanced than *Gobipteryx* and is included within Neornithes, or modern birds.

The importance of understanding how animals grow and develop and the functions of parental care in the survival of birds has been recognized by early evolutionary biologists for over a century. In regard to their maturity at hatching, there are two broadly different types of birds—altricial and precocial. Altricial chicks are born helpless, blind, naked, and feeble, depending completely on parental care (fig. 8.3A). At the other extreme are precocial chicks, born fully clothed, alert, bright-eyed, and able to run after their parents and feed themselves independently (fig. 8.3B). The fundamental functions of parental care are, of course, protection and provision of nourishment. By

FIGURE 8.3

Altricial versus precocial mode of development among recent birds. *A,* the one-hour-old chick of an altricial bird, the Java sparrow (*Lonchura oryzivora*), is blind, naked, and utterly helpless. Most tree-nesting and cavity-nesting birds have altricial chicks, which require long periods of feeding before they are strong enough to leave the nest on their own. *B,* the one-hour-old chick of a precocial bird, the killdeer (*Charadrius vociferus*), by contrast, has open eyes and a thick coat of natal down and is able to fend for itself. Most shore birds, game birds, and ducks have precocial chicks, which can walk away from the ground nest within hours of hatching.

staining bone and cartilage, J. Mathias Starck (1993) has studied in detail the patterns and sequence of ossification of a wide range of embryonic birds and has found two different styles. The skeletons of precocial birds, which move actively but grow slowly, show a high proportion of bone and a low percentage of cartilage. Conversely, the skeletons of altricial birds, which grow quickly, show a high proportion of cartilage and a low percentage of bone. Interestingly, both *Gobipteryx* and *Gobipipus* show a high degree of ossification in the embryonic stage, suggesting a precocial mode of development (fig. 8.2F). It is likely that the hatchlings of both genera were able to fly as soon as they emerged from their eggs, as is evident from their ossified wing skeletons. Starck concluded from his ontogenetic analysis that the precocial mode of development is primitive for birds, whereas the altricial mode is more advanced. Cretaceous embryos from the Gobi Desert support this contention. In these birds, the bones are so well ossified that the hatchlings probably were self-sufficient and steady on their feet from the beginning and fed themselves. They could even fly when they were quite small, similar to modern gallinacious birds. This is no doubt a useful safety device for terrestrial birds. These birds probably buried their eggs in the ground and gave little parental care to the chicks. Altriciality is a more sophisticated mode of development, which might have evolved later in the evolution of birds, when eggs were placed in an arboreal nest with associated parental care.

Trace Fossils

Although skeletal remains are the most important fossils for tracing the evolution of Mesozoic birds, feathers and footprints provide additional evidence for their existence and activities in the Mesozoic ecosystem. These trace fossils are of little taxonomic importance except to demonstrate the birds' size, diversity, and paleoecology. Feathers and footprints reveal the natural history of ancient birds, their environments and distributions in time and space.

Feathers

Feathers, the unique evolutionary novelty of birds developed from the reptilian scales, are also their most characteristic possession. Feathers are the most complex form of body covering to be found in any vertebrates, varying from long flight feathers to mere puffs of down. They reveal a spectacular range of patterns and colors. Keratin, the same type of protein that gives strength to hair and nails, makes feathers light, strong, and flexible. Although flight evolved many times during the history of life, birds are the only animals that possess feathers. Feathers provide the aerodynamic power necessary for flight, insulation for maintaining body temperature, and coloration for courtship and camouflage. For such soft, keratinous tissues as feathers (or rather their impressions) to be preserved in the fossil record is extremely rare.

FEATHERS IN MODERN BIRDS

There are six major types of feathers in modern flying birds: contour feathers, down feathers, semiplumes, filoplumes, powder feathers, and bristles (fig. 8.4). Contour feathers are the basic vaned feathers of the body and wings; they coat the body, giving it a streamlined shape. A few of the contour feathers on the wing (remiges) and tail (rectrices) have become large and highly specialized for flight (or display). Both remiges and rectrices show asymmetrical vanes for aerodynamic function. On the other hand, the smaller contour feathers that cover the body have symmetrical vanes. The distal vane area of the body contour feather is firm and pennaceous, but its proximal region is soft and plumaceous because its barbules lack hooks. Contour feathers grow only in the feather tracts (pterylae). The coverts are short feathers that cover part of the remiges and rectrices and give them additional rigidity. The other kinds of feathers are primarily used for insulation, protection, grooming, waterproofing, and sensation.

The key to successful flight is the structure of the feather. A typical contour feather is composed of a long, tapering central shaft (rachis) with a broad, flexible vane on either side. Vanes may be symmetrical or asymmetrical. Vanes are asymmetrical in flight feathers, where the leading edge is narrower and stronger than the trailing edge. This is a requirement for powered flight because air pressure is greater along each feather's front edge. This asymmetrical design gives the feather a curved,

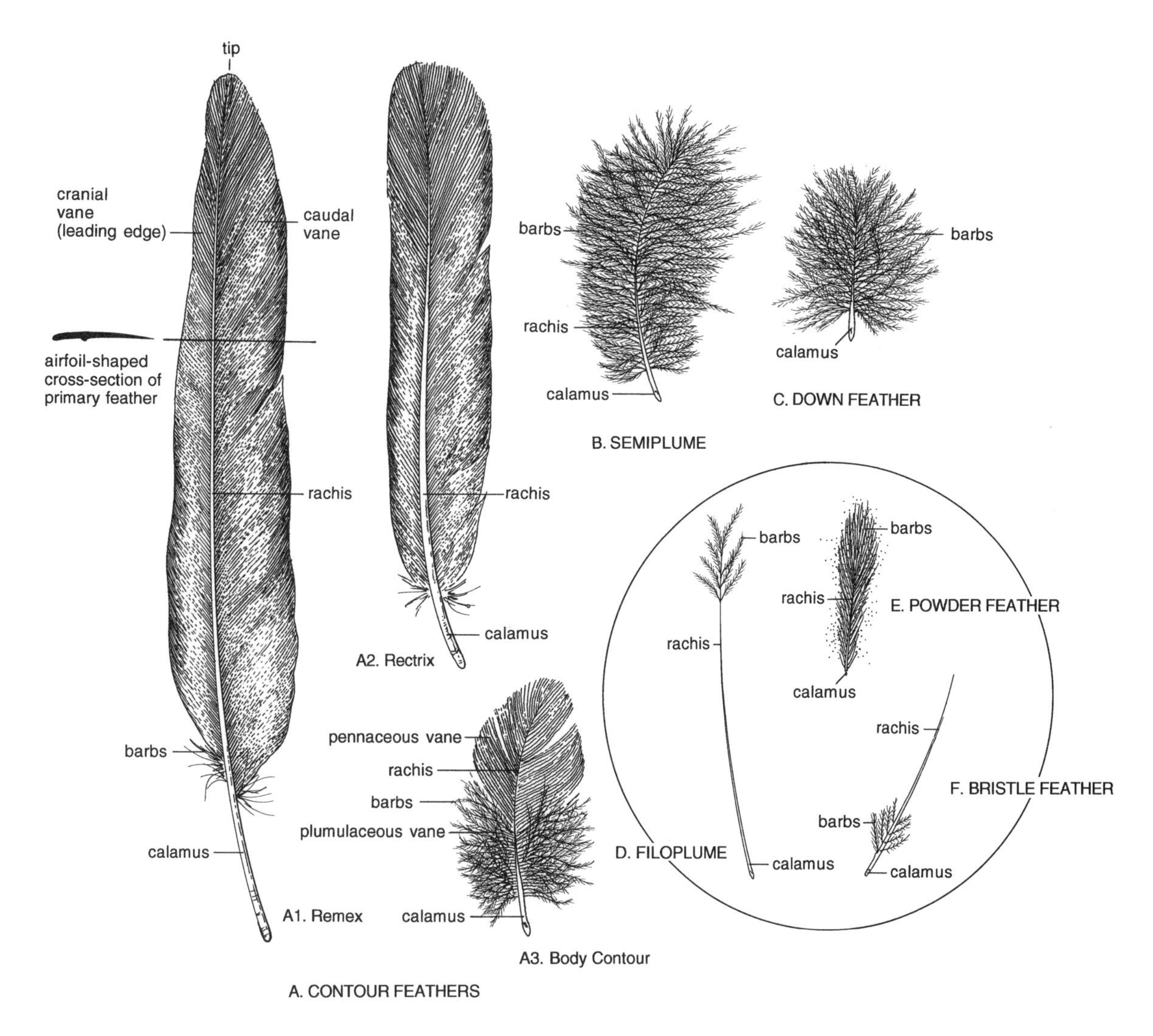

FIGURE 8.4

The major types of feathers in modern birds. *A,* contour feathers are the basic vaned feathers, including the large flight feathers of the wing (*A1,* remex) and tail (*A2,* rectrix) and the smaller body feathers (*A3,* body contour); *B,* semiplume; *C,* down feather; *D,* filoplume; *E,* powder feather; *F,* bristle.

airfoil cross section for aerodynamic function. In flightless birds, the vanes are symmetrical.

The vanes are composed of many barbs, which in turn bear many still smaller interlocking barbules; the barbules form a strong, light, flexible surface, one that can be restored by preening after it is broken apart. Feathers need constant care (preening, oiling, powdering, shaking, and stretching) to keep them in good shape. However, feathers are dead, horny structures and deteriorate with time; once a year birds must renew their contour feathers, usually after the breeding season.

THE FOSSIL RECORD

Fossil feathers are sparse in the paleontological record. Since feathers decay easily, the preservation of this soft, delicate structure requires an unusual setting, such as the calm waters of a lacustrine or lagoonal environment, and rapid burial without any turbulence. Recently, Davis and Briggs (1995) discussed the mode of fossilization of feathers. According to them, the feathers of Mesozoic birds were preserved in three forms: imprints, carbonized traces, and amber inclusions (fig. 8.5). From studies with scanning electron microscopes, they concluded that a bacterial glycolax played an important role in the fossilization of feathers preserved as imprints and carbonized traces. Because of its antibiotic properties, however, amber preserves the most perfect fossil feathers. The reported Mesozoic feathers are contours, remiges, rectrices, semiplumes, and downs, often isolated. That down feathers might be the external covering of the body is shown by the Jurassic bird *Confuciusornis.*

IMPRINTATION The earliest record of a fossil feather comes from *Archaeopteryx* from the Upper Jurassic of Bavaria (fig. 8.5A). The feathers of *Archaeopteryx* are preserved mainly by imprinting after early lithification of Solnhofen Limestone by bacterial decay. Alan Feduccia (1980) has pointed out that the presence of vane asymmetry in the wing feathers of *Archaeopteryx* indicates their flight capability. There are about twelve primaries and fourteen secondaries arranged in the same basic avian design. Surprisingly, this earliest fossil feather is identical with the structure of that of a modern volant bird. Even dark brown pigment is preserved in the isolated feather, as in a thrush. Imprinted down feather from the leg region of the Jurassic bird *Confuciusornis* has also been reported (Hou et al. 1995).

CARBONIZED TRACES Carbonized traces are the most common type of preserved feathers found in the Cretaceous sediments. The Early Cretaceous records include an isolated contour feather from Montsec, Lerida Province of Spain (Ferrer-Condal 1954), asymmetrical feathers associated with *Concornis* from the Las Hoyas Province of Spain (Sanz and Buscalioni 1992), colored down feathers from Mongolia associated with *Ambiortus* (Kurochkin 1985), three isolated contour feathers from Australia (Waldman 1970), and a semiplume (Martil and Filgueira 1994) and a down feather (Kellner, Maisey, and Campos 1994) from Brazil. The solitary record of carbonaceous fossil feathers, the plumaceous type in *Parahesperornis*, comes from the Late Cretaceous deposits of Kansas (L. D. Martin 1984). Fossil feathers become more abundant in Tertiary freshwater deposits and are often associated with insect fossils.

AMBER Amber often contains beautiful fossil feathers as inclusions, such as the semiplume specimens from the Lower Cretaceous of Lebanon (Schlee 1973). This is the oldest record of a feather in amber. The structure of this feather is reminiscent of certain species of present-day waterfowl. The second oldest amber specimen is known from the Late Cretaceous of New Jersey (Grimaldi and Case 1995). It is also a semiplume feather and probably belonged to a terrestrial bird. Feathers are common in the Late Cretaceous amber deposits of Alberta, Canada (Martil and Filgueira 1994).

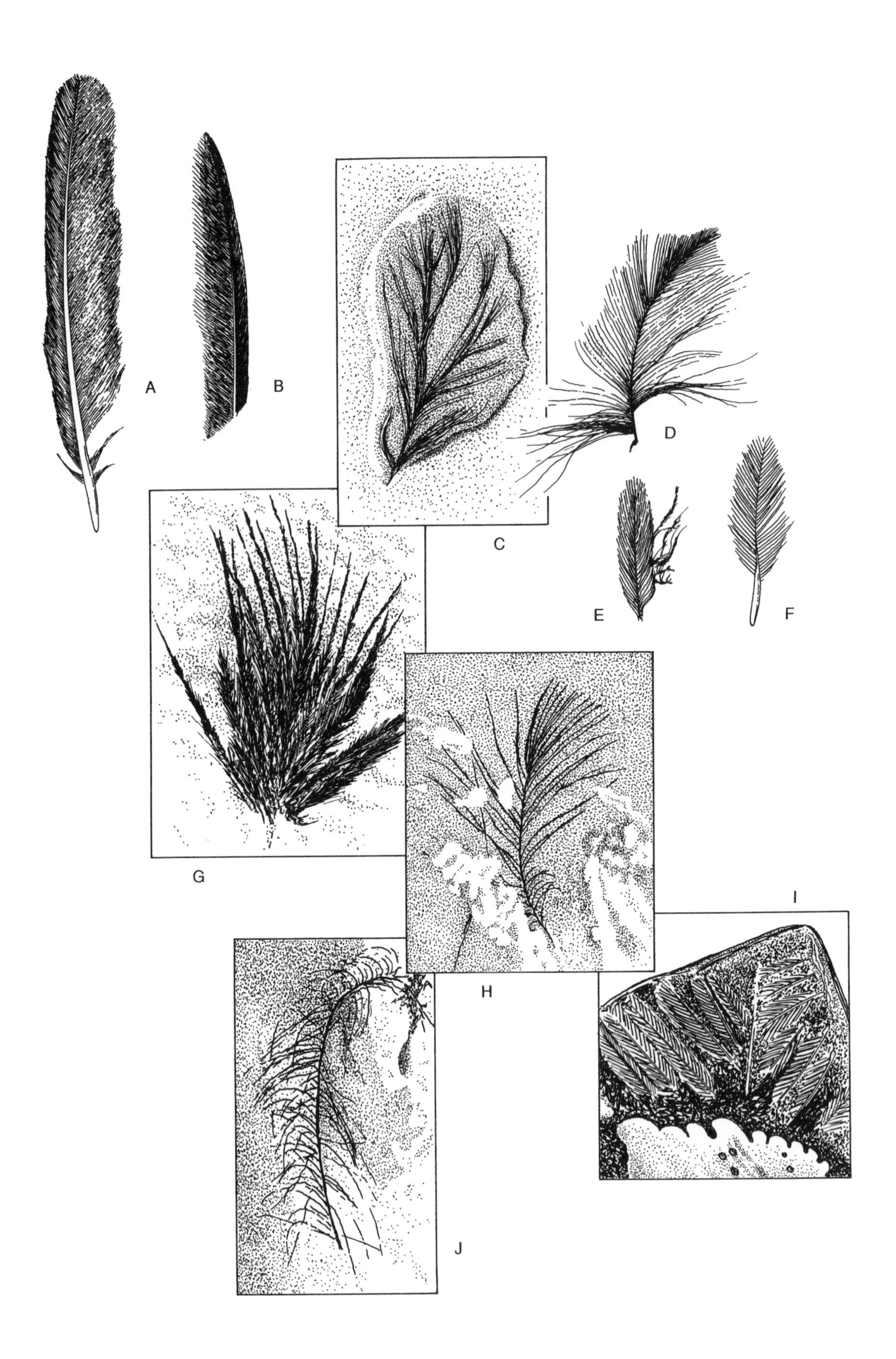
A
B
C
D
E
F
G
H
I
J

Footprints

Bird tracks are very distinctive because of their small size and tridactyl or tetradactyl configuration. Footprints reveal much of the anatomy, ecology, speed of locomotion, habits, and behavior of birds. The size and shape of the footprints, the number of toes, their spreading angle, and their arrangement serve to identify the birds and to provide us with clues to their mode of life. Prints that show partially webbed feet indicate an aquatic existence of their markers. Perching birds generally hop on the ground and leave their prints paired. Game birds, on the other hand, generally walk or run; thus, their track pattern consists of a series of alternate prints.

Mesozoic bird tracks are relatively poorly known compared to the tracks of nondinosaurs. In recent years there has been an increasing number of reports of footprints of Mesozoic birds from East Asia, Africa, and North America. These tracks extend the geographic distribution of birds and suggest their diversity. Martin Lockley and associates (1992) recently reviewed the classification and distribution of the ichnofauna of Mesozoic birds. These avian tracks are small, about 2 to 6 cm long, and are identical to those of modern birds. They are generally tridactyl or tetradactyl with slender digit impressions. There is a

FIGURE 8.5

Mesozoic fossil feathers were preserved in three ways: imprintation (*A*), carbonized traces (*B–H*), and amber inclusions (*I–J*). *A,* the solitary remex of *Archaeopteryx* reported by Herman von Meyer in 1861 from the Late Jurassic of Germany; asymmetrical vanes indicate aerodynamic function. *B,* primary feather that probably belonged to *Ambiortus* from the Early Cretaceous of Mongolia; asymmetical vanes suggest that *Ambiortus* was a volant bird (source: Kurochkin 1985). *C,* isolated down feather associated with *Iberomesornis* from the Early Cretaceous Las Hoyas of Spain (source: Sanz et al. 1996). *D–F,* a variety of feathers recovered from the Early Cretaceous sediments of southern Victoria, Australia (source: Waldman 1970). *G,* down feather from the Early Cretaceous Santana Formation of Brazil (source: Kellner, Maisey, and Campos 1994). *H,* semiplume feather from the Early Cretaceous Crato Formation of Brazil (source: Martill and Filgueira 1994). *I,* semiplume feather in amber inclusion from the Early Cretaceous of Lebanon (source: Schlee 1973). *J,* semiplume feather in amber inclusion from the Late Cretaceous of New Jersey (source: Grimaldi and Case 1995).

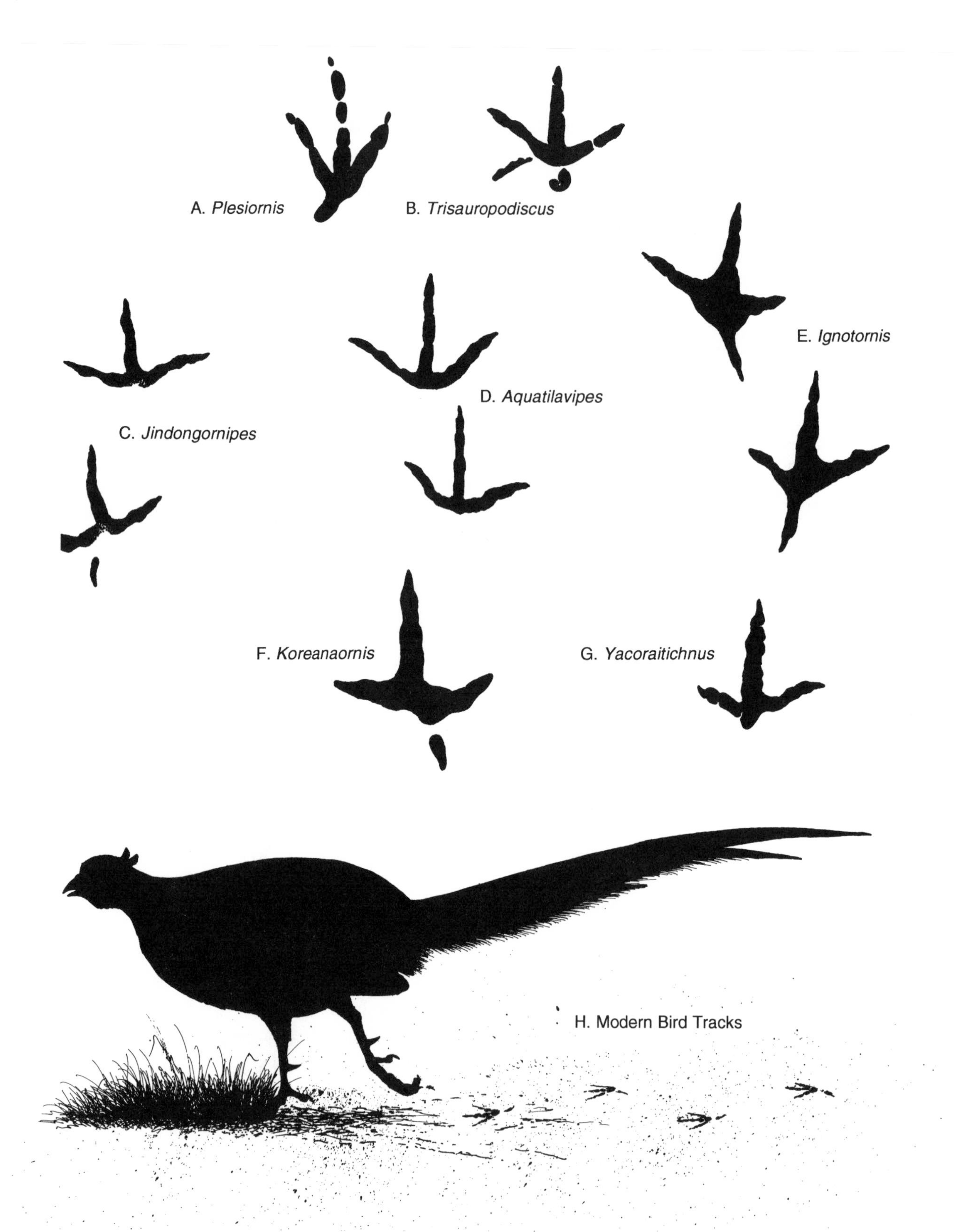
A. *Plesiornis*
B. *Trisauropodiscus*
E. *Ignotornis*
D. *Aquatilavipes*
C. *Jindongornipes*
F. *Koreanaornis*
G. *Yacoraitichnus*
H. Modern Bird Tracks

wide divarication angle between digits II and IV, the hallux is caudally directed, and the claws are slim (fig. 8.6).

The earliest record of bird tracks comes from the Late Triassic Manassas Sandstone in the Culpepper Basin of Virginia, which is coeval with the Dockum Group. Robert Weems and Peter Kimmel (1993) described these small tridactyl bird tracks as *Plesiornis pilulatus* (fig. 8.6A). The prints are small, about 3 cm long, with narrow toes and indistinct foot pads; the divarication angle between digits II and IV is 70°. Weems and Kimmel concluded that these tracks were made by a *Protoavis*-like bird. *Plesiornis* was originally described from the Early Jurassic Portland Formation of Massachusetts (Lull 1953).

The next oldest record of bird tracks is reported from the Early Jurassic rocks of Africa and North America. Ellenberger (1972) described several varieties of *Trisauropodiscus* (fig. 8.6B) from the Stormberg Beds of South Africa. These slender, tridactyl or tetradactyl toes range in size from 2 to 7 cm, with a large divarication angle. Similar tracks are known from the contemporary beds of North America and Morocco. The Upper Jurassic bird tracks are poorly known except for the dubious footprints of Ribadesella, Spain. The existence of these Late Triassic and Early Jurassic bird trackways is quite exciting. It suggests that birds were numerous and widespread 50 to 75 million

FIGURE 8.6

Mesozoic bird tracks. *A, Plesiornis* from the Late Triassic Manassas Formation of Virginia. This earliest bird track was probably made by a *Protoavis*-like animal; similar tracks are also known from the Early Jurassic Portland Formation of Massachusetts (source: Weems and Kimmel 1993; Lull 1953). *B, Trisauropodiscus* tracks are known from Early Jurassic rocks of Africa, Morocco, and North America (source: Ellenberger 1972). *C, Jindongornipes* tracks from the Early Cretaceous of South Korea (source: Lockley et al. 1992). *D, Aquatilavipes* track from the Early Cretaceous of Canada (source: Currie 1981). *E, Ignotornis* tracks from the Early Cretaceous of Colorado (source: Lockley et al. 1992). *F, Koreanaornis* track from the Early Cretaceous of South Korea (source: Lockley et al. 1992). *G, Yacoraitichnus* track from the Late Cretaceous of Argentina (source: Alonso and Marquillas 1986). *H,* bird tracks produced by a ring-necked pheasant (*Phasianus colchicus*).

years before the appearance of *Archaeopteryx*, thus filling important gaps at a crucial time in the history of bird evolution.

Bird tracks are abundant and dense in the Early Cretaceous deposits. Many of these footprints were made by the shorebirds and waders. Interesting ichnotaxa include *Jindongornipes* (fig. 8.6C) from the Jindong Formation and *Koreanaornis* (fig. 8.6F) from the Haman Formation of South Korea, *Ignotornis* (fig. 8.6E) from the Dakota Group of Colorado, and *Aquatilavipes* (fig. 8.6D) from the Gething Formation of Canada (Lockley et al. 1992). These tracks are generally tetradactyl with a caudally directed hallux, and the second and fourth digits show a high divarication angle. The footprints average 4 to 5 cm wide and 3 to 6 cm long. Bird tracks similar to those of *Aquatilavipes* are also known from Japan and China. The abundant track records of Early Cretaceous birds indicate their adaptive radiation and diversity in lacustrine and shoreline environments.

Late Cretaceous bird tracks are poorly documented. The earliest recorded webbed foot tracks are from the Uhangri Formation of South Korea and were probably made by a flamingo-like bird. These tracks were not named or described in detail. Probable bird tracks are also known from Utah and Morocco, but detailed description is not available. Alonso and Marquillas (1986) reported bird tracks from the Late Cretaceous of Argentina that display clear digital pad impressions (fig. 8.6G). Some bird tracks from the Dunvegan Formation of Alberta may be attributable to hesperornithiform birds (Lockley et al. 1992).

The Origin of Flight

Some of these diminutive Dinosaurs were perhaps arboreal in habit, and the difference between them and the birds that lived with them may have been at first mainly one of feathers.
—O. C. Marsh, "Jurassic Birds and Their Allies," 1881

CHAPTER 9

Birds are the best flyers among vertebrates because of their streamlined body and feathered wings. The evolution of flight in birds is recognized as the key adaptive breakthrough that contributed to the biological success of this group. Since ancient times people have watched and envied the enchanting flight of birds, even being inspired to try to conquer the air themselves. The human body was not, however, designed for flight. Undaunted by structural deficiencies, early inventors dreamed up ideas for flying machines. The flight of fancy was realized after the trials and tribulations of many workers. During the last few centuries men like Leonardo da Vinci, Sir George Caley, Otto Lilienthal, and the Wright brothers kept careful notes and diaries on their observations of birds during their pursuit of artificial flight. Manned flight, like avian flight,

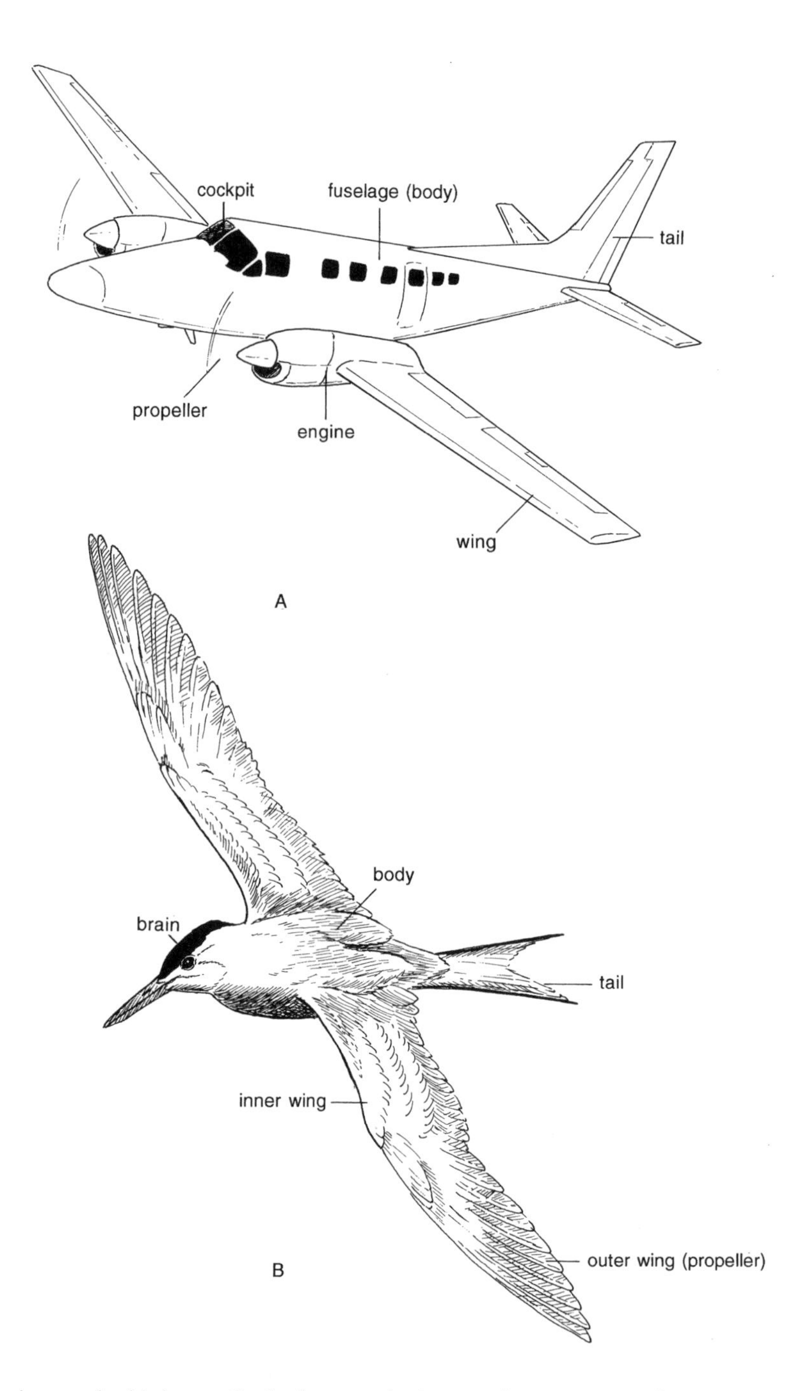

FIGURE 9.1

Comparison between a propeller airplane and a bird. *A,* in an airplane, the wing provides the lift; the propeller, powered by the engine, gives the forward thrust; and the cockpit is the controlling system. *B,* in a bird, the wing functions as both a lifting surface and a propeller; the inner wing supplies most of the lift, the outer wing serves as a propeller, and the brain controls the complex neuromuscular movement (proprioception) of the wing. In both cases, the tail provides stability and has an important function in controlling flight, steering and braking. Bodies are streamlined and landing gears are tucked close to the body during flight to reduce drag but are lowered during landing.

evolved through successive stages, first in hot air balloons, then in gliders, and finally in true flying machines. The beginning of the twentieth century heralded the age of aviation when the Wright brothers invented and built the first successful airplane. On 17 December 1903, Orville Wright, assisted by his brother Wilbur, made the first historic flights over Kitty Hawk, North Carolina, with their "Flyer" biplane and proved once and for all that humans could fly like birds across the mountains and oceans. Paradoxically, it was the human's own inventions—first the glider and then the propeller airplane—that were the keys to recognizing the secrets of avian flight.

How do birds fly so gracefully? Birds fly in air by flapping their wings, steering mainly with their tails. Air, like water, has weight and exerts pressure. By taking advantage of these properties, both birds and airplanes can maintain themselves in the air. Flight is a combination of lift and propulsion. Like airplanes, all flying animals are heavier than air and require a lift force to keep them aloft. In an airplane, the wing provides the lift, while the propeller, powered by the engine, gives the forward thrust. Unlike the fixed wings of airplanes, avian wings are not rigid but flexible, and their geometry can be changed in response to flight. The adjustable wing can be shortened or lengthened by flexion, whereas the feathers of the tip can be spread or closed. Bird flight is a complex performance of wing movements, aided by an equally complex set of moment-by-moment body and tail adjustments to provide the fine balance and control required. In birds, each wing functions as a combined wing and propeller. The inner wing supplies most of the lift, while the outer wing serves as propeller. The power of propulsion is provided by the flight muscles (fig. 9.1). In both airplanes and birds, the body is streamlined to reduce drag forces. In airplanes, the cockpit is the command post for flight; in birds, the brain controls the complex neuromuscular movement.

The Mechanism of Flight

Flight is accomplished by use of the forelimbs, which are specialized as wings. Most birds fly, and the few that do not had flying ancestors. Gliding and powered flight are the basic types of avian flight. Flight is arduous, which is why many species of birds have discovered how to stay airborne without flapping their wings. The trick is to make use of speed and altitude either previously gained by wing beats or available from wind currents. Gliders stretch their motionless wings to form one lifting surface and passively descend through the air by the aid of gravity without any flapping. They can glide a great distance without any muscular effort. Gliding is always accompanied by loss in height with an angle of descent of less than 45°. Soaring flight is gliding in circles. In soaring, a large bird obtains energy from the upward currents of air in which it flies. The only way a gliding or soaring bird can maintain or gain altitude is to find updrafts that exceed the rate at which it is sinking. Speed is maintained by extracting energy from the wind.

Powered flight requires much more energy than does gliding (Brown 1951). In powered flight, most flying is done by active beats of the wings. Powered flight requires lift to keep the bird airborne and horizontal thrust to move forward; lift is achieved by the airfoil action of the wing, and thrust by flapping the wings up and down in a complex manner.

When a bird is in level flight in still air with no flapping of the wings, four forces—lift, weight, thrust, and drag—are equal and opposite (fig. 9.2A). Weight acting downward under the pull of gravity is balanced by lift generated by the wing. Similarly, the forward thrust provided by the wing tip is equalized by the resisting force of drag. Gliding is passive flight without wing strokes; it costs a minimum of energy. In gliding, lift balances weight but drag remains uncompensated (fig. 9.2C). As a result the animal must descend. In flapping flight, the thrust generated by wing strokes balances the drag (fig. 9.2B). Flapping or powered flight is much cheaper locomotion than is running (Rayner 1991). Changes in lift or thrust cause birds to change

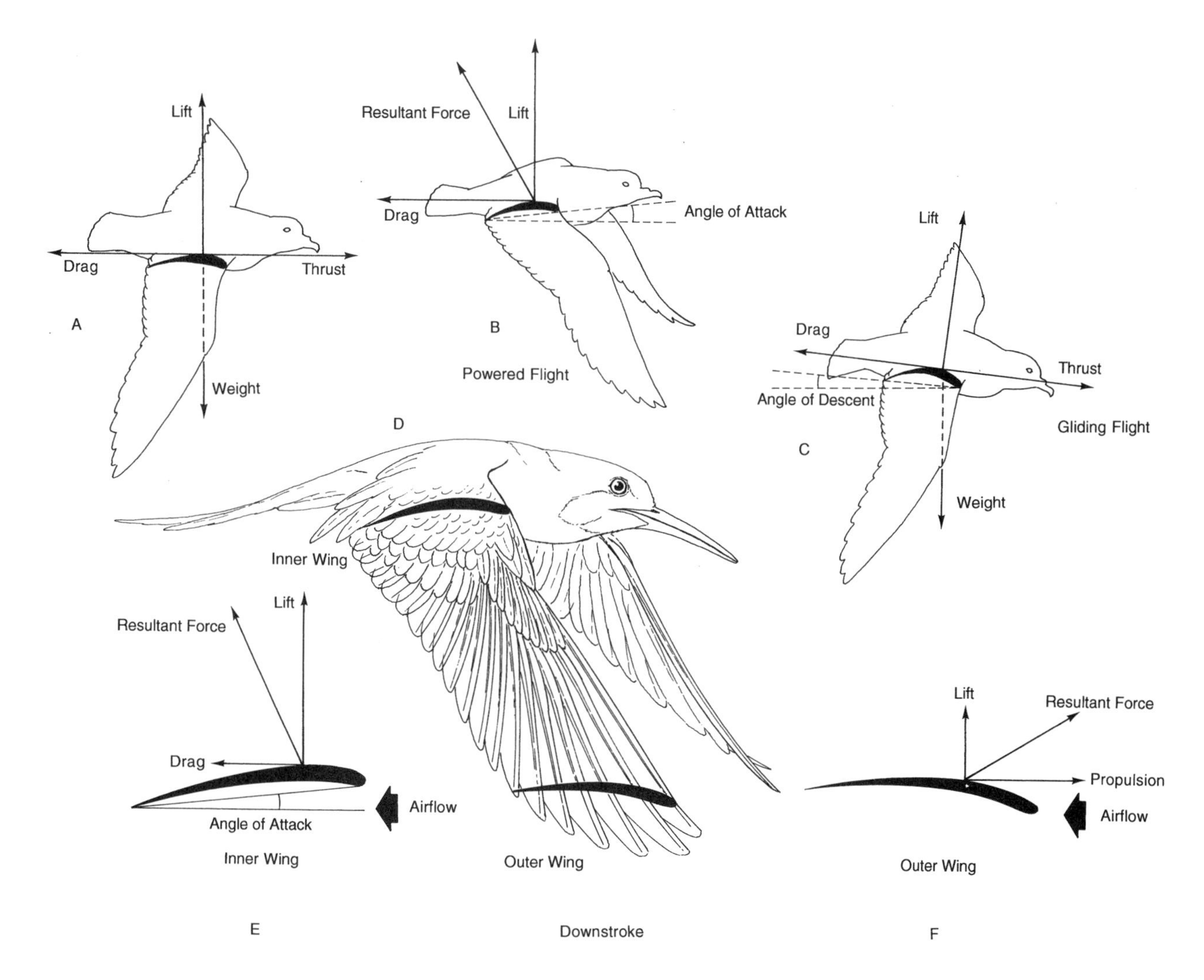

FIGURE 9.2

The biomechanics of flight. *A,* when a bird is in level flight, four forces (lift, weight, thrust, and drag) are equal and opposite. Changes in lift or thrust cause the bird to change altitude or speed. *B,* the bird's wings act as an airfoil in powered flight. The air flow over an airfoil generates forces of lift and drag. The flying wing maximizes lift and minimizes drag. The angle between the airfoil and the air flow is called the *angle of attack.* As the angle of attack increases, lift becomes greater than drag, and the resultant flight is more vertical than horizontal. *C,* in gliding flight, the airfoil of the wing provides the lift, while the pull of gravity is used as the thrust. Gliding is always accompanied by loss in height with an angle of descent of less than 45°. *D,* airfoil cross sections of the inner and outer wings of a bird during the downstroke in relation to the direction of airflow. *E,* the inner wing provides the lift as the angle of attack is about 5°; *F,* the outer wing is twisted and tilted downward in relation to the direction of the airflow, so that the airfoil creates a forward thrust or propulsion.

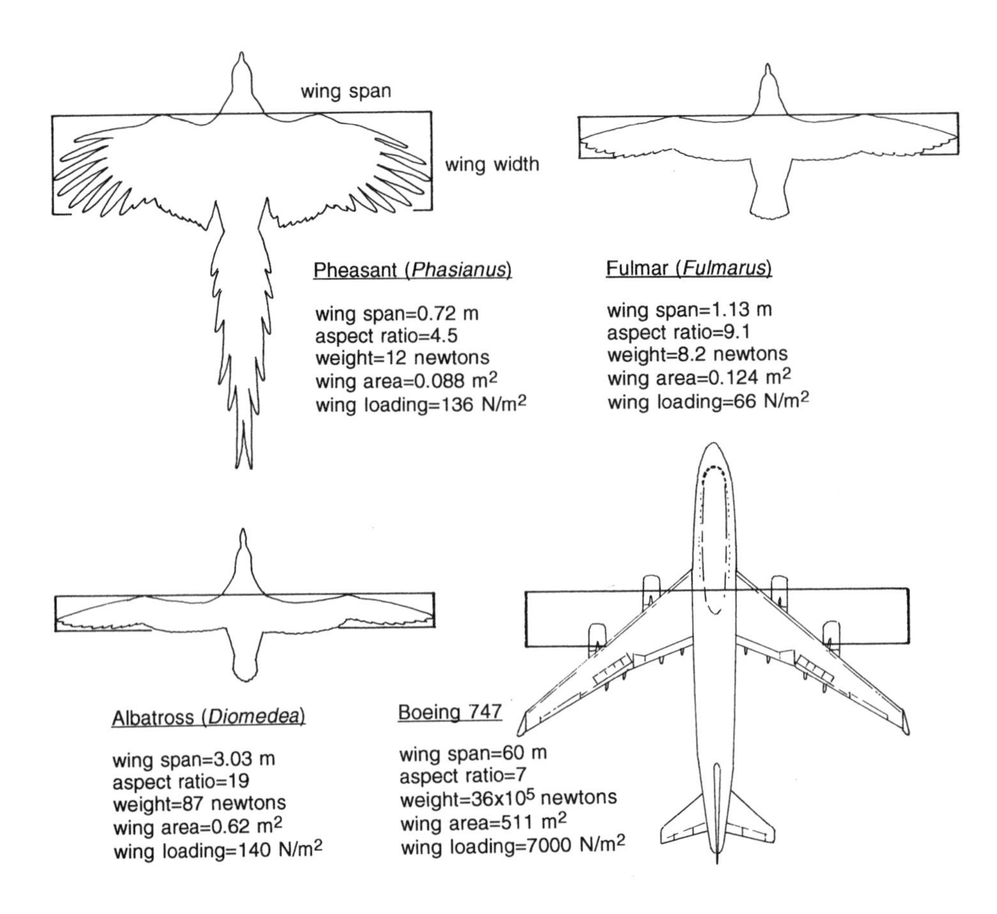

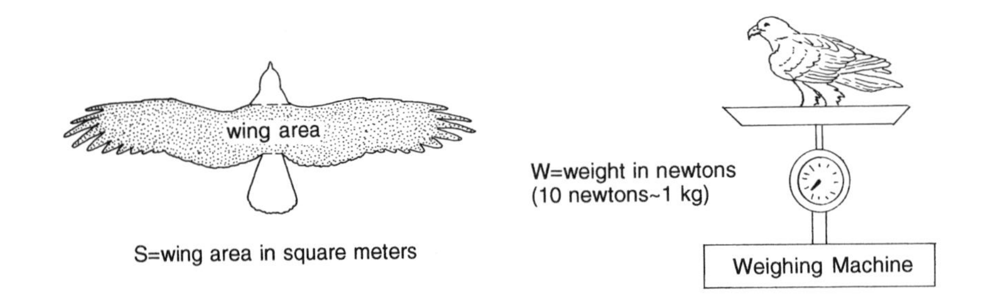

FIGURE 9.3

Wing shape, wing size, body weight, and flight style of birds compared with those of a Boeing 747. The wing shape of birds reflects the way in which they live, particularly the ratio of wing span to wing width, known as the *aspect ratio*. *A,* comparisons of aspect ratios of different birds and the Boeing 747. *B,* wing loading is the ratio of total weight to wing area. (Data from Tennekes 1996.)

speed and altitude. A flying animal must overcome the force of gravity by generating vertical lift and producing horizontal thrust to counter drag. Lift is produced by the inner part of the wing, where the wing area is largest, whereas thrust is produced by twisting the outer part of the wing.

The shape and size of wings determine the flight pattern and lifestyle of birds. Two features of wings are linked to flight performance, the aspect ratio and wing loading (fig. 9.3). To fly economically, the wings must be slender. The slenderness of the wings can be measured by the *aspect ratio*—the ratio of the wing span to wing width (fig. 9.3A). Birds with long, narrow wings, such as gulls and albatrosses, have a high aspect ratio and a high lift/drag ratio. The long wing is structurally weak but is aerodynamically efficient for gliding and soaring. Long wings are, therefore, most suitable to birds that live in open airspace. Flight is cheap for these long-winged birds, as it requires little muscular effort, but takeoff is difficult. Birds of prey also have developed wings of high aspect ratio; their wings taper to a point and tend to be swept backward. Wings of harriers and vultures are relatively broad, with slotting of the primary feathers, which are adapted for soaring and for precise control in the pursuit of the prey.

Birds living in woodland habitats would find long wings a handicap and thus evolved elliptical wings. Short, broad wings with a low aspect ratio provide rapid, powered lift and maneuverability through the forest canopy. Most fast flyers designed for sophisticated acrobatics, such as swifts and pigeons, possess short, wide wings with a low aspect ratio and a low lift/drag ratio. As a result, such birds have to flap the wings continuously to remain aloft. Flight is expensive for these groups of birds.

Weight is another important factor in flying performance and is linked to *wing loading*—the ratio of the bird's weight to the wing area (fig. 9.3B). Small birds with a low wing loading are more maneuverable than are large birds with a high wing loading; the latter have to fly faster to generate the lift to stay airborne. A Boeing 747 flies much faster than a sparrow because of its high wing loading (Tennekes 1996).

There are four basic requirements for flight.

A LIGHTWEIGHT BODY

Flight necessarily imposes limitations on body weight. Modern flying birds are very light for swift flying; the mean weight ranges from 10 g to 1 kg, dwindling sharply above and below these limits (Pennycuick 1986). Beyond a certain size, sustained flight becomes prohibitive because of energy requirements and the physical limitations of bone and muscle. Pennycuick (1972) empirically estimated that the largest bird capable of flapping flight would be close to 12 kg, about the size of a mute swan (*Cygnus olor*). Larger birds (10–15 kg) like vultures, storks, swans, cranes, bustards, and albatrosses spend most of their time soaring, which extracts energy from the air with very little muscular effort (Pennycuick 1986). The largest known flying bird was *Argentavis magnificens* from the Late Miocene of Argentina (Campbell and Tonni 1983). It was a vulture-like soaring bird with a wing span of 7.5 meters, weighing as much as 77.5 kg. At the opposite end of the scale is the tiny bee hummingbird (*Melisuga helenae*) of Cuba, the smallest bird. When fully grown, it measures about 5 cm and weighs about 3 g. Birds became lightweight by reducing their size and evolving thin, hollow, pneumatic bones. Strength is maintained by support struts inside the bone and the fusion of many bones where motion is not required. Birds must have a shock-resistant skeleton to withstand the impact of landing and to provide firm attachment surfaces for powerful flight muscles. There must be landing gear, supplied by modification of the hindlimbs. A light and strong skeleton is a basic requirement for a flying bird.

THE PRODUCTION OF LIFT

Aerial vertebrates must have wings and a flight surface developed by transformation of the forelimbs. The wing of a bird is an airfoil of high camber, presenting a curved, streamlined surface that is thicker at the leading edge but gradually thins toward the trailing edge. In a flapping wing, each section performs a separate task; the inner wing, with secondary feathers, provides the lift, while the outer wing, with large primaries, provides the thrust (fig. 9.4A). Both lift and thrust depend on the angle of wings with respect to the direction of airflow, the *angle of attack.* Birds change the angle of attack of their wings to fit the circumstances. When the angle of attack is low, the airflow over the wing's airfoil generates forces of lift and drag. The interaction of these two forces, the lift/drag ratio, produces a resultant force that varies with the angle of attack. As the angle of attack rises to about 5°, the lift/drag ratio is maximum. Birds generate lift by tilting up the inner wing against the air current to an appropriate angle (fig. 9.2B). Since the top surface of the wing is curved more than the bottom, the upper air stream moves faster than the lower air stream to reach the trailing edge. The airfoil design of the wing generates unequal pressures, less above and more below, to produce lift. According to Bernoulli's principle, the accelerated airflow over the top surface of wing exerts less pressure than does the airflow across the bottom, thus creating an upward force on the wing. It is the continuing pressure differential over the upper and lower surfaces that creates and sustains lift. If the angle of attack becomes too great (>15°), the air stream cannot follow the upper surface and the air flow breaks away, resulting in a turbulent wake with less negative pressure and a loss of lift. The angle at which this occurs is called the *stalling angle.* The alula on the first digit of the wing reduces the chance of stalling. By moving that digit, a bird can separate the feathers of the alula from the rest of the wing, creating a slot that helps channel air over the flight feathers to prevent eddies.

POWER OR THRUST

Once lift is achieved, power or thrust must be generated for forward motion. This power comes from the flight muscles that can move the wings up and down and back and forth, relative

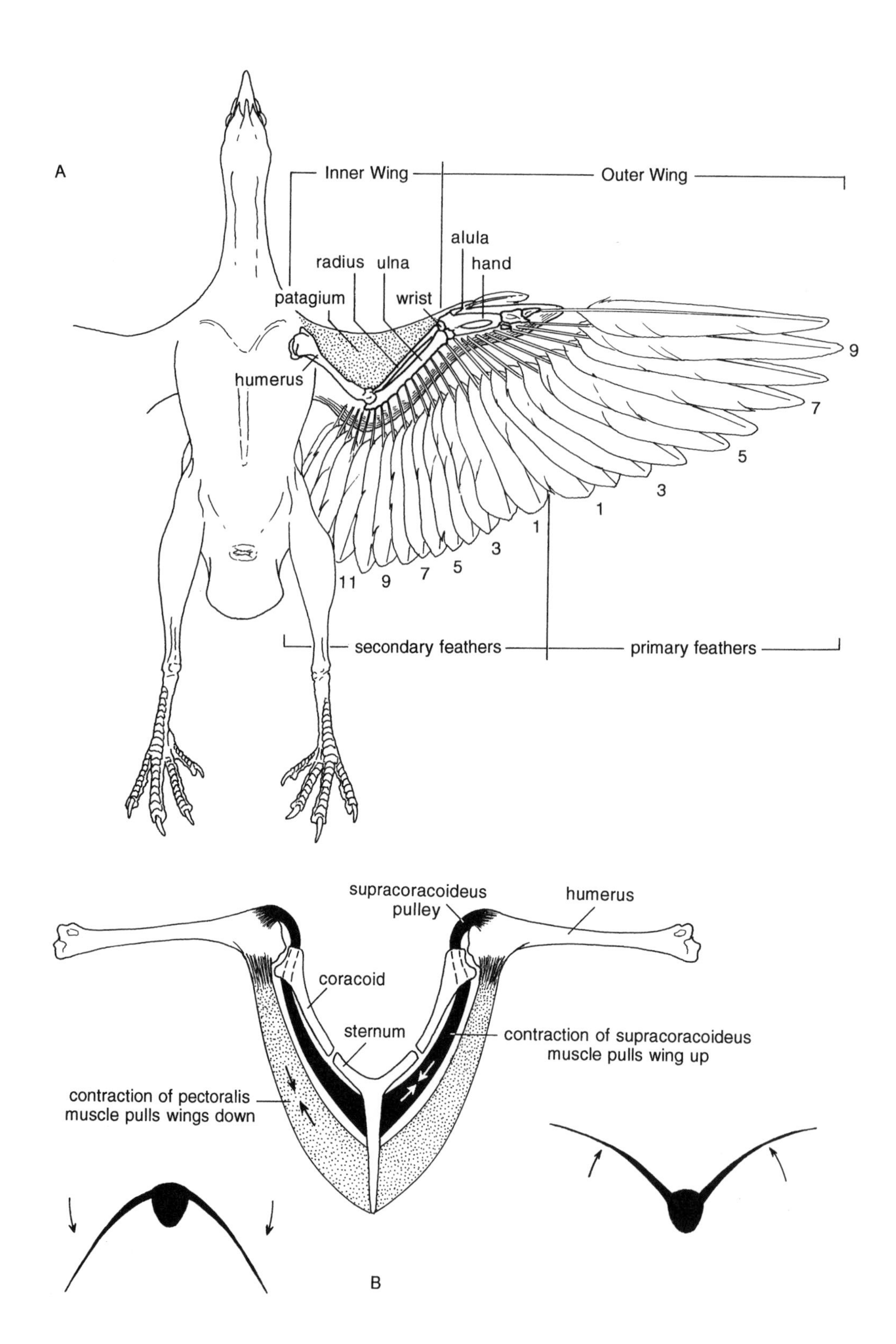
A
Inner Wing
Outer Wing
alula
radius
ulna
hand
patagium
wrist
humerus
9
7
5
3
1
1
3
5
7
9
11
secondary feathers
primary feathers
supracoracoideus pulley
humerus
coracoid
sternum
contraction of supracoracoideus muscle pulls wing up
contraction of pectoralis muscle pulls wings down
B

to the glenoid of the shoulder girdle and the longitudinal axis of the body (fig. 9.4B). The tips of the wings not only flap up and down but also twist forward on the downstroke to propel the bird forward. Flapping flight entails a downstroke, when the wing moves down and forward to provide thrust, and an upstroke, when the wing moves up and back in medium flight speed (Rayner 1981). With a complete wing beat cycle, the wings create a lazy figure 8. The primary feathers, attached to the trailing edge of the hand in such a fashion that they push against the air during the downstroke of the wing, propel the animal forward (fig. 9.2E–F); this, in turn, creates vortex wakes that are forced backward as the bird moves (Rayner 1988). The twisting of the wing toward the tip to reduce the angle of attack is the most important element in the downstroke. The upstroke serves to position the wing for the subsequent downstroke. The outer wing is more flexible than the inner wing and does most of the propelling at the tips of the primary feathers. While the inner wings tilt up to provide lift, the outer wings tilt down so that their airfoils create forward rather than upward thrust (fig. 9.2B). As the inner wings descend, their thrust pulls them forward, and they in turn pull the outer wings. Birds maneuver in flight by controlling the patterns of lift and thrust.

NEUROSENSORY CONTROL FOR MANEUVERING

Flight requires fine coordination between muscles and senses. The brain sends signals to the nerves, which operate muscles. The neural control of the fine movements of flight is still poorly

FIGURE 9.4

Wing structure and flight muscles in birds. *A,* ventral view of the left wing of the pigeon (*Columba*) showing the inner and outer wings, as well as the number, location, and attachments of major flight feathers. The primary feathers attached to the hand propel the bird forward, whereas the secondary feathers attached to the ulna provide most of the lift. Note the strong asymmetry of the primary feathers; each feather acts as an individual airfoil. *B,* diagrammatic view of the shoulder girdle bones of a bird to show how the flight muscles operate during flapping. The flight muscles are anchored to the keel of the sternum and pull the humerus of the wings up and down.

understood, but the flying animal demands acute vision to guide its flight and a large cerebellum for balance and coordination. This is evident from the enlargement of its brain in the regions of the cerebellum and optic lobe (Jerison 1973). The flying animal also needs the transmission of information from the body to the brain about the continual adjustment of the wings in response to airflow over them.

The Origin of Flight

The fossil record shows that true powered flight evolved independently three times in the history of vertebrates: in pterosaurs and birds in the Late Triassic and in bats in the Early Tertiary. True flyers have more control over their flight path; they can ascend, descend, or move horizontally. What makes avian flight unique is that birds are the only flyers that have used feathers as airfoil and propeller. The origin of avian flight is one

FIGURE 9.5

The origin of flight. The cursorial (ground up) theory is based on the idea that the proavian was essentially a small running theropod that leapt up into the air and became an active flyer without any gliding stage.

of the greatest challenges to evolutionary biology and has been debated for more than a century. It is often equated with the evolutionary origin of birds. One of the prime reasons for this controversy is the lack of intermediate fossils linking flyers to nonflying ancestors. Once the initial stage of flight was achieved, as documented by *Archaeopteryx,* the successive stages of powered flight can be reconstructed from the Mesozoic birds.

There are two major conflicting theories concerning the origin of flight. (1) The cursorial (ground up) theory maintains that flight evolved in running bipeds through a series of short jumps; as these jumps became extended, the wings were used for balance and propulsion, and the animals began to fly without a gliding intermediate. (2) The arboreal (trees down) theory states that flight originated in tree-living animals that leaped from branch to branch or tree to tree, steadying themselves with outstretched wings; they began to glide and then fly downward from heights. Each theory relies heavily upon speculation about the paleoecology and functional adaptation of the proavian (the hypothetical bird ancestor) and *Archaeopteryx.*

THE CURSORIAL THEORY

The first cursorial theory was briefly outlined by Samuel Williston (1879). He believed that flight evolved through a series of steps: running, leaping, jumping from heights, and finally flying. This is how the forelimbs of dinosaurs might have transformed into wings. Surprisingly, he gave no other details to support his theory. In 1907 and 1923, Franz von Nopsca elaborated on Williston's idea by suggesting that wings would help to increase the speed of an animal as it ran along the ground. He concluded that birds evolved directly from terrestrial cursorial theropods without any gliding stage, and that bipedality preceded the development of flight (fig. 9.5). However, Nopsca acknowledged that *Archaeopteryx* had already achieved some level of arboreal capacity. Nopsca's ideas on the origin of flight were severely criticized by other workers because the use of wings to increase running speed has no living analogs and because outstretched wings would increase drag (Heilmann 1926; Ostrom 1974; Bock 1986). As soon as the animal left the ground, its legs could no longer propel it forward and it would immediately fall back down.

The cursorial theory was revived by John Ostrom (1974, 1979, 1985a, 1986) with a new vigor after almost fifty years. He believed that the hindlimbs and bony claws of *Archaeopteryx* represent the lifestyle of a typical cursorial terrestrial biped and are very similar to those of dromaeosaurs such as *Deinonychus*. Ostrom argued that the bipedalism of *Archaeopteryx* would al-

FIGURE 9.6

The origin of flight. Modified version of the cursorial (ground up) theory. *A,* the insect-net theory of Ostrom (1979) states that proavians used proto-wings to catch insects. As the feathers were enlarged for batting insects, the wing motion would turn from swatting into actual flapping flight. *B,* Caple, Balda, and Willis (1983) proposed that proavians leapt into the air in pursuit of insect prey. Feathers and wings assisted and extended their leaps until flapping flight evolved.

low free movement of the forearms, making them available for catching insects. In this view, wings evolved from arms used to capture insects; feathers were first present as insulators and later became elongated fly swatters. The feathers would enlarge through time, making them better tools for batting insects. In fact, the entire forelimb would become a large, lightweight insect net. These protobirds would catch insects by chasing and swatting them with their wings. Eventually, as the forelimb evolved into a better tool for catching insects, the motion would turn from swatting into actual flapping flight (fig. 9.6A).

Ostrom's "insect net" theory was heavily criticized on several grounds. His model has four major problems that are difficult to explain (Feduccia 1980, 1995b, 1996; Martin 1983b; Caple, Balda, and Willis 1983; Bock 1983): (1) The motion used to catch insects is very different from the motion used in a flight stroke. Therefore, it is unlikely that such an insect net would evolve into a structure used in active flight. (2) *Archaeopteryx* would damage its wings and feathers for flights if it used them to bat down prey. (3) To catch an insect, *Archaeopteryx*'s wings would need air holes to let air pass, as in a fly swatter. *Archaeopteryx*'s wings did not have air holes; therefore, it would be unable to catch insects in this way. (4) The insect nets would have generated several instabilities and loss of balance. All of these problems show that the insect net theory cannot possibly work as a step toward flight evolution. If wings really did develop first as insect-catchers, as Ostrom suggests, natural selection would have acted strictly to improve them as flyswatters, not transform their function to flying. Because of these difficulties, Ostrom (1986) later rejected his hypothesis.

Although Ostrom's model has many flaws, it did encourage other groups to revive the cursorial theory, partly because of the phylogenetic relationship of *Archaeopteryx* with dromaeosaurs (Gauthier and Padian 1985; Padian 1985; Sereno and Rao 1992; Ruben 1991). As an alternative, Caple, Balda, and Willis (1983) proposed a "fluttering model" (fig. 9.6B). According to these authors, the proavians might have used their jaws to catch prey but employed their wings as bilateral stabilizers during a jump into the air. Caple and colleagues maintained that the rudimentary wings of proavians were effective for balance while running, jumping, and turning, until they were able to take off at high speed. They speculated that, when the proavians extended their forelimbs, minute increments of lift made it easier to jump further and capture more prey. The motion of the forelimbs for stabilization, according to them, would mimic the flight stroke of a bird. Eventually, the proavians evolved larger airfoils that enabled them to obtain even greater lift. Also, as lift increased it aided in landings. Therefore, the proavians could slow down and direct their landings. As a result, power flight evolved (fig. 9.6B).

Taking off requires more energy than does level flight because the bird must accelerate and climb. Whenever birds take off from the ground, they need speed to become airborne. It is the speed with respect to the air that matters, not the speed with respect to the ground. The bird must beat its wings more vigorously to obtain the extra lift it needs. Taking off from the ground is hard work that requires four times as much power as ordinary flight (Tennekes 1996). For this reason, most birds prefer to take off from elevated objects such as a tree, a tele-

FIGURE 9.7

Kangaroos (*Megaleia*) share with theropods several functional traits, such as obligate bipedality, cursorial posture, short forelimbs, and a long tail; they offer a modern analog for testing of the cursorial theory. Successive jumping stages of the kangaroo show a great range of flexion and extension of the hindlimbs. The forelimbs, however, are held close to the body in a parasagittal plane without any activity. It is unlikely that jumping theropods would extend their forelimbs to become flyers because they would thus create drag forces. The origin of flight from the ground up is biomechanically untenable.

phone pole, a window sill, a roof ledge, or a cliff to secure the needed lift. As soon as the proavians jumped into the air from the ground, they would loose airspeed and altitude and fall back. There was no lift or thrust to keep them aloft. The cursorial theory, even in its modified form, is biomechanically untenable.

The strongest criticisms against the cursorial theory are based on ecological, adaptive, and mechanical grounds (Jerison 1973; Rayner 1979, 1985, 1988, 1989, 1991; Bock 1983, 1985, 1986; Martin 1983b; Feduccia 1980; Norberg 1985, 1990; Pennycuick 1986). There are no contemporary analogs of cursorial bipeds using forelimbs for stability; outstretched wings would increase drag and slow down locomotion. In contrast, modern cursorial mammals such as kangaroos and kangaroo rats have forelimbs that play a passive role during jumping. To minimize the drag force, they are kept in a folded position in a strictly sagittal plane during takeoff, midway through the leap, and during landing (fig. 9.7). Thus, the use of forelimbs for jumping is minimal in these animals and does not mimic a rudimentary flight stroke. As a consequence, selection has favored relatively shorter forelimbs in these animals, quite opposite the trend proposed by Caple and associates. If their model were correct, we would expect to see the emergence of flying kangaroos from their ground-bound ancestors.

It is not clear why selection should favor flapping wings for foraging in a terrestrial animal. In a fluttering model (Caple, Balda, and Willis 1983), as soon as the proavians jumped into the air (fig. 9.6B), they would lose airspeed and fall down. Flight at slow speed is aerodynamically more sophisticated and complex than is flight at high speed. It is unlikely that early birds acquired this sophisticated slow speed flight at the beginning. Foraging in flight requires very fine coordination between the senses and muscles, implying instability and a great degree of maneuverability. It is unlikely that such a fine control developed in jumping proavians with a long stabilizing tail.

The cursorial theory fails to explain fully why the primary

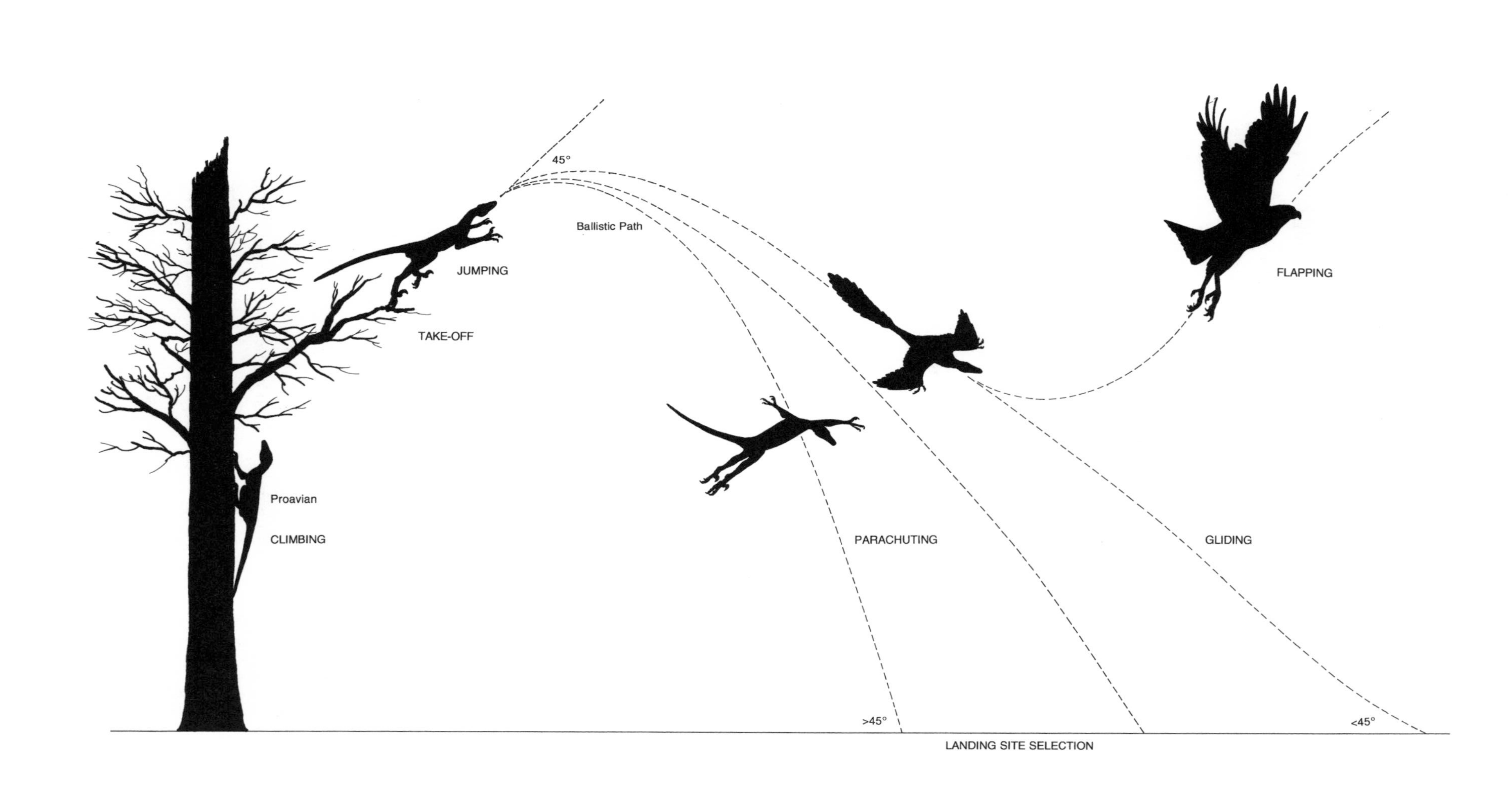
45°
Ballistic Path
JUMPING
TAKE-OFF
Proavian
CLIMBING
PARACHUTING
GLIDING
FLAPPING
>45°
<45°
LANDING SITE SELECTION

vanes of *Archaeopteryx* are so asymmetrical and complex, a condition seen only in modern volant birds. If *Archaeopteryx* were a ground-dwelling bird, as Ostrom depicted, it would have had hairlike feathers, like those of ostriches and rheas. The cursorial theory works against gravity and is energetically more expensive. The effects of gravity would create additional stress on proavians during takeoff. To overcome this added stress, the supracoracoideus pulley system would be required during takeoff; its lack in *Archaeopteryx* indicates that *Archaeopteryx* took off from trees to become airborne, not from the ground up.

The cursorial theory does not address the necessary transitional form between the preflight stage and the active, flapping flight stage; flight would thus have evolved rapidly, from jumping to active flying, almost by saltation without any intermediate gliding stage. This theory does not explain adequately the origin of feathers, the origin of endothermy, or the origin of brain enlargement and three-dimensional perceptual control. These are important characteristics of birds that must be accounted for in any evolutionary and adaptive model of flight. In the cursorial model, landing was perfected first, followed by improvement of the flight apparatus. In the fossil record of Mesozoic birds, we see the opposite trend: flight evolved first followed by the development of sophisticated landing (Chiappe 1995a).

THE ARBOREAL THEORY

The arboreal theory begins with a climbing proavian that started to glide between trees and then gradually to fly with wing strokes.

FIGURE 9.8

The origin of flight. The arboreal (tree down) theory begins with a climbing proavian that started to parachute from tree tops and eventually began to glide as the angle of descent decreased from 45°. Gliding would increase maneuverability and slow landing. Once gliding was perfected, flapping would begin to prolong flight. In the arboreal theory, gravity is used as a source of power to convert potential energy into kinetic energy.

The central theme of the arboreal theory is the use of gravity as a source of power by proavians to convert potential energy to kinetic energy. Charles Darwin (1859) was probably the first to propose the arboreal theory for the origin of bats, and Othniel Charles Marsh (1880) extended this idea for birds. Marsh observed that a large number of Holocene birds live in trees or bushes and that there are many modern parachuting and gliding tetrapods, all of which are arboreal. Marsh believed that flight evolved from arboreal proavians whose scales developed into rudimentary feathers. These feathers were used as parachutes, slowing the descent of the animals as they leaped from branch to branch and controlling their jumps and falls.

In 1926 Gerhard Heilmann argued strongly in favor of the arboreal theory. Like Nopsca, he also believed that the proavians were bipedal animals. Heilmann developed an adaptive model where ground-dwelling runners became tree climbers. Once in the tree, the proavians would leap from limb to limb.

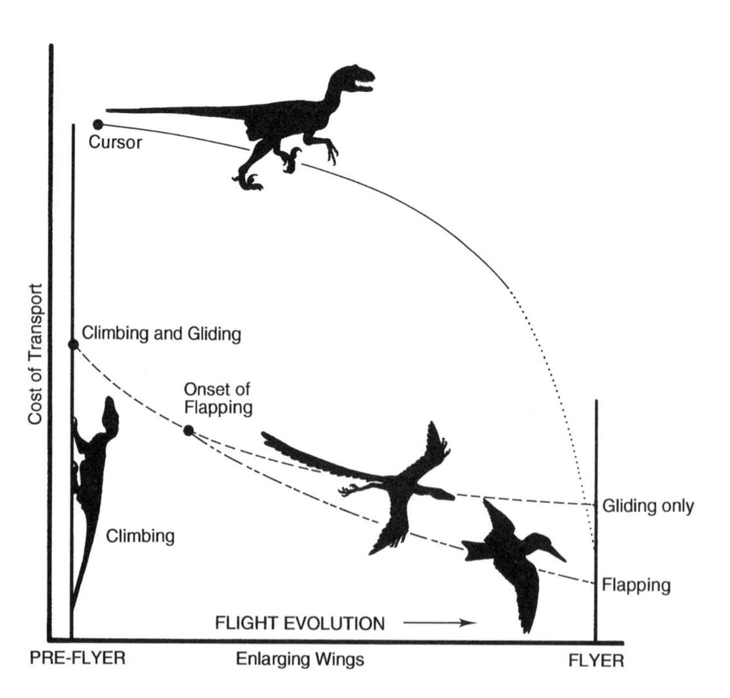

FIGURE 9.9

An energy-saving mode of locomotion. It costs less for a proto-flyer to climb a tree and then glide or flap its wings than to run (source: modified from Rayner 1991).

They would start leaping from low branches to the ground and eventually begin to glide. As gliding became specialized, active flight would evolve. Heilmann thus set the stage for the arboreal theory that was later refined by other workers.

More recently, Walter Bock (1965, 1969, 1983, 1985, 1986) has presented a more detailed version of the arboreal theory in the context of the historical narrative, which means that earlier events affect later ones. Bock reconstructed the life history of proavians from recent examples of birds. He believed that proavians spent some time in ground feeding and the rest of the time sleeping, hiding, and nesting in trees. He argued that proavians climbed trees with the aid of manual and pedal claws. Once they became more confident with climbing, they would jump from limb to limb and make short jumps to the ground to conserve energy and time. The next step would be to make longer jumps or to parachute, which was an unspecialized glide with a steep angle of descent. As the angle of descent decreased to less than 45°, gliding would commence (fig. 9.8). Gliding is a more efficient way to travel over horizontal distances than is walking or running. Gliding would also increase maneuverability and slow down landing. These three actions—climbing, parachuting, and gliding—were important because they allowed flight muscles to become powerful. Once gliding became specialized, flapping could begin. The initial gliding flight would obtain energy from gravity. Bock suggested that flapping began when proavians were landing. Flapping would add additional power, thus prolonging flight and reducing the energy cost of transport (fig. 9.9).

In recent times, the arboreal theory has been criticized by proponents of the cursorial theory on biomechanical grounds (Padian 1982; Caple, Balda, and Willis 1983; Balda, Caple, and Willis 1985; Ruben 1991). These authors claimed that gliders were evolutionary dead ends and could not give rise to flapping flight. Moreover, they argued that flight feathers would easily be damaged during tree climbing. However, these objections are flawed and have been refuted by other workers, who showed

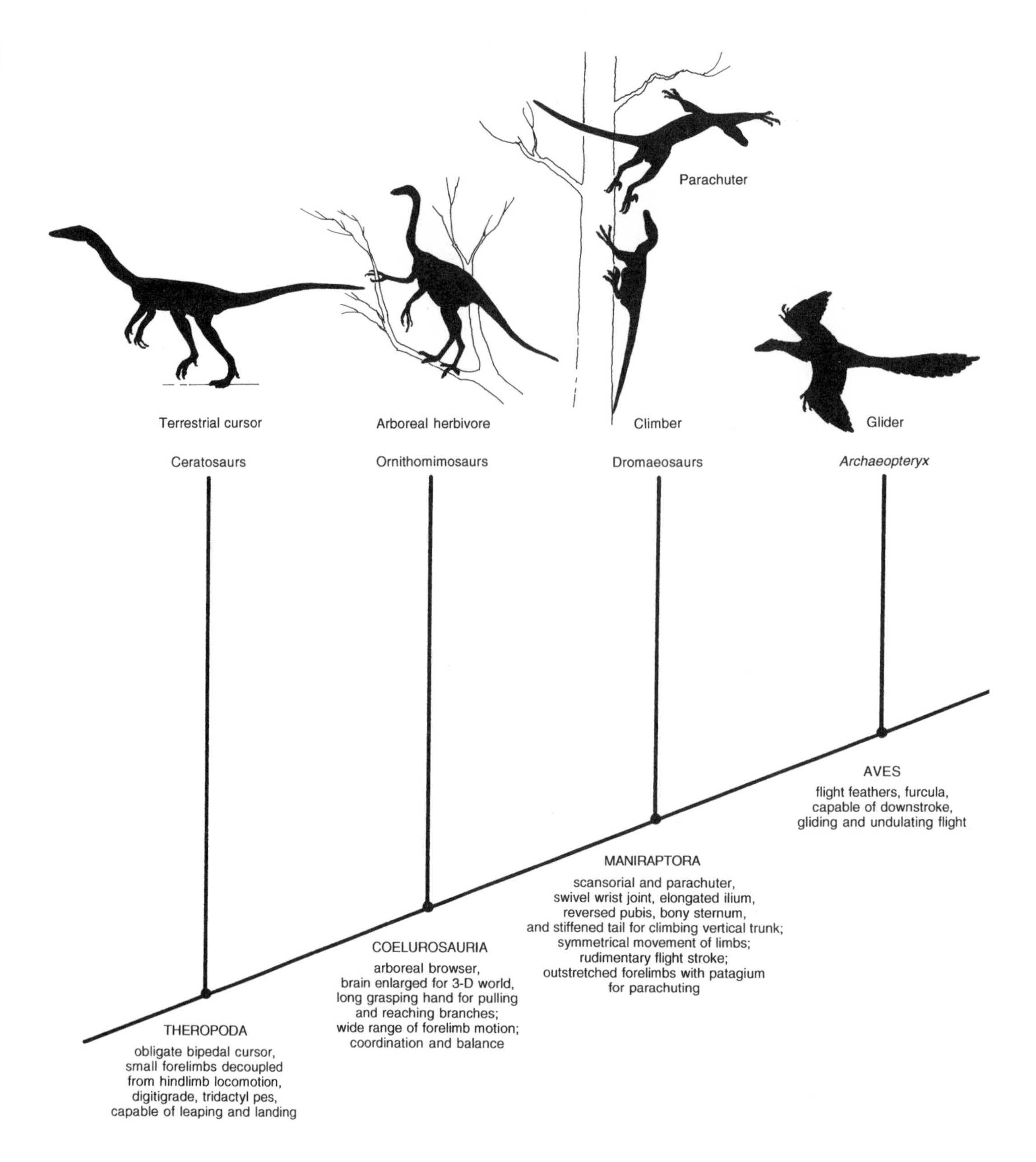

FIGURE 9.10

Cladogram showing the paleoecological evolution of avian flight. This sequence of events, fully adaptive at each stage, led not only from bipedal terrestrial cursor to long-distance flyer, but also from ceratosaurs to modern birds. The emphasis is on characteristics related to evolutionary changes leading to avian flight. Many of the flight-related characters were exaptated in protodromaeosaurs (proavians) during climbing and parachuting.

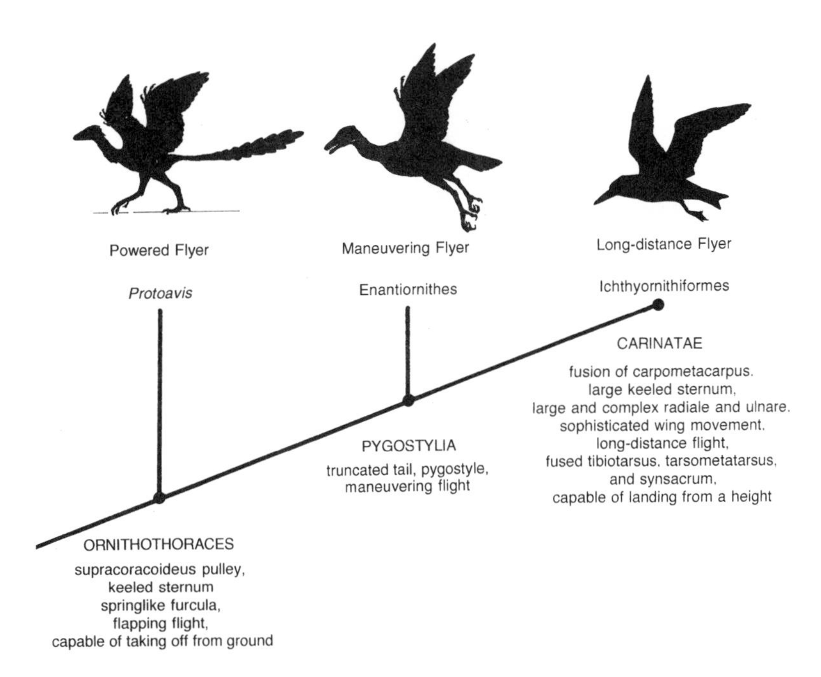

how a transition from gliding to active flight is mechanically feasible with intermediate adaptive stages (Norberg 1985, 1990; Bock 1986; Rayner 1981, 1985, 1988, 1989; Pennycuick 1986). Thus, according to these authors, gliding was an essential exaptation (preadaptation) for powered flight. The second criticism is shown to be invalid by the modern hoatzin chick, which can climb trees without damaging its wings (fig. 9.11B). The curvature and sharply pointed tip of the manual and pedal claws of *Archaeopteryx*, as well as the swivel wrist joint, indicate strongly that these primitive birds climbed trees (Yalden 1985; Bakker 1986; Feduccia 1993). If *Archaeopteryx* were a ground cursor, its pedal claws would have been blunt, like those of modern ground birds.

The Evolution of Avian Flight: Phylogenetic, Functional, and Aerodynamic Analyses

Although the arboreal theory is intuitively more attractive because it makes use of gravity to power flight, a combination of phylogenetic, paleoecological, adaptive, functional, and aerodynamic evidence is presented here to resolve the long-standing question on the origin of flight. I propose a modified version of the arboreal model in the context of recent theropod phylogeny.

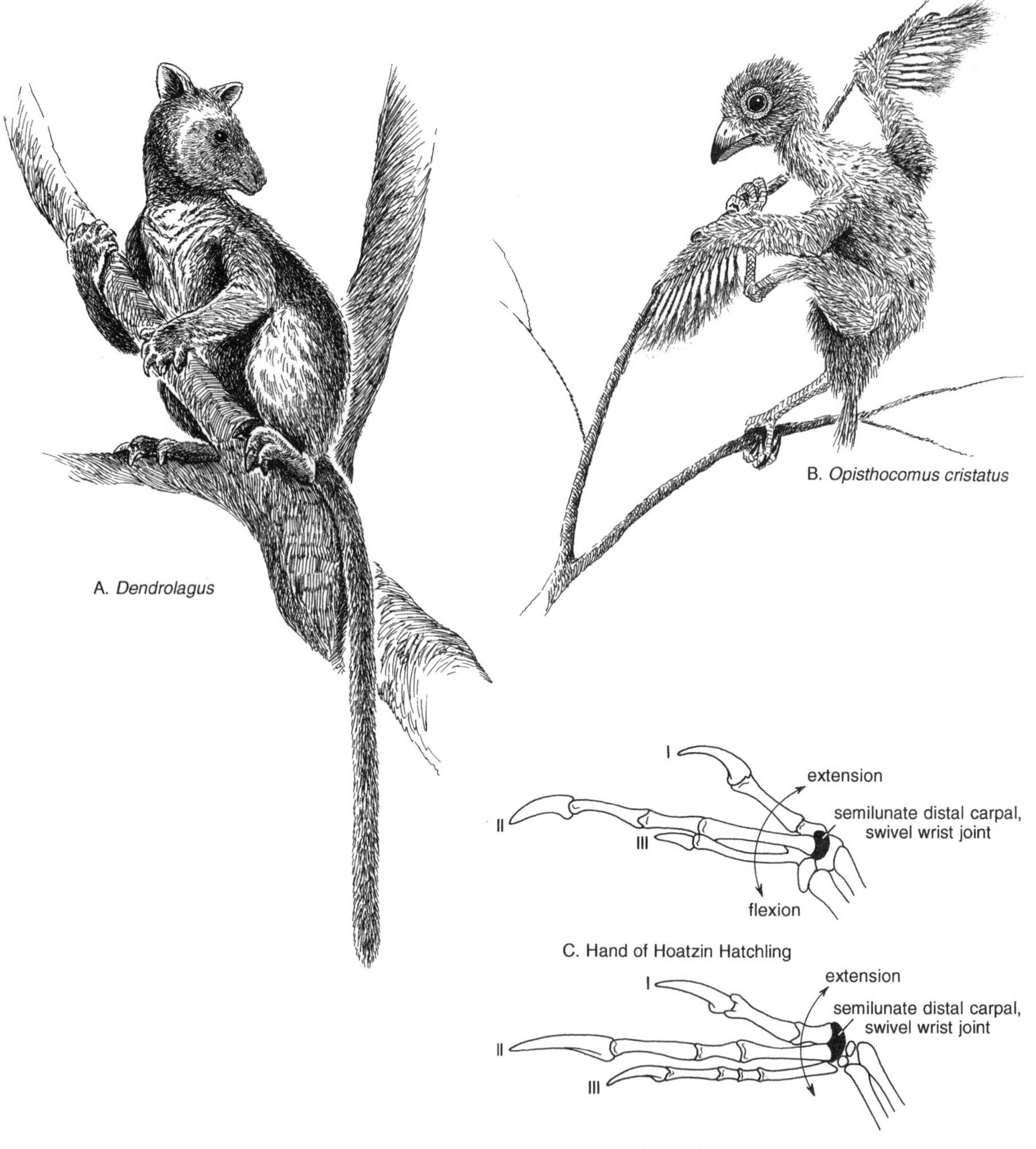

FIGURE 9.11

Climbing and arboreal habits of kangaroo and bird. *A,* an arboreal biped, such as the tree kangaroo (*Dendrolagus*) in New Guinea and Queensland, is adept at climbing with its recurved claws. *B,* a hoatzin (*Opisthocomus*) chick climbing trees with its wing claws. When alarmed, it jumps into water. *C–D,* wrist joint of a young hoatzin (*C*) compared with that of dromaeosaur (*D*); the swivel wrist joint with a semilunate carpal is an adaptation for climbing trees and is co-opted for the folding and unfolding of a wing.

Many anatomical features of proavians that are often equated with flight adaptation seem to have arisen during climbing and parachuting and to have later been modified for flight.

ECOLOGICAL PHYLOGENY

Phylogenetic analyses indicate that dromaeosaurs such as *Deinonychus* and *Velociraptor* were the sister-group of birds (Gauthier 1986; Gauthier and Padian 1985). It is now widely accepted that birds are the descendants of a small, unknown dromaeosaur. Gerhard Heilmann (1926) coined a neutral name, proavians, for these hypothetical ancestors of birds. The phylogeny of theropods (Gauthier 1986; Chatterjee 1995) offers a framework with which to trace the sequence of acquisition of avian characters with changing adaptations. Birds are here today because of countless historical accidents. If ceratosaurs had not stood up bipedally with digitigrade pes, if the brain of ornithomimosaurs had not grown so rapidly, if the wrist of dromaeosaurs had not developed the swivel joint for climbing trees, if the proavians had not attempted parachuting and gliding from tree branches, it is likely that birds might not be here. Several flight-related characteristics of birds appeared in their nonavian ancestors (fig. 9.10). I present these novelties in the phylogenetic order in which they evolved. Many of the characters will be discussed in detail in the following chapter.

With ceratosaurs, the bipedal erect posture was established in a lightweight skeleton with hollow bones. With freeing of the front limbs, the hindlimbs had to bear the entire weight of the body. Supporting additional weight, the ilium became elongated and was strengthened by five coalesced sacral vertebrae; the neck and head were counterbalanced by a long and slender tail. The pes became tridactyl and digitigrade for cursorial adaptation. These primitive theropods, such as ceratosaurs and allosaurs, were essentially terrestrial cursors living in a two-dimensional world, like modern dogs.

The next stage of theropod evolution is represented by the coelurosaurs. They were lightly built, agile, and highly adapted for three-dimensional woodland habitats, like modern cats. Among coelurosaurs, we see a distinct departure from a carnivore lifestyle in ornithomimosaurs. In this group, we see the first sign of brain enlargement of the avian type; this may indicate their invasion into a new arboreal habitat, partly for safety and partly for exploitation of new food resources, such as cones and seeds, well above the ground (fig. 9.10). The hands, especially the thumbs, became highly grasping organs, probably to grab fronds of ferns and cycads and pull them to the mouth (Paul 1988). Moving through horizontal branches required a sequence of actions of limbs and the ability to coordinate series of movements in a three-dimensional world. As in modern cats, this activity might enhance brain enlargement (see fig. 10.6).

In dromaeosaurs, with the development of long hands, the swivel wrist joint (fig. 9.11D), a stiffened tail, an ossified sternum, the biceps tubercle, and a streamlined body (opisthopubic pelvis), the scansorial, or climbing, adaptation on a vertical tree trunk is perfected. The synchronous forelimb motion used during climbing would mimic a rudimentary flight stroke. The animals could possibly parachute and glide from treetops with the development of the patagium in the proximal part of the forelimb (fig. 9.12).

In *Archaeopteryx,* with flight feathers and furcula, gliding and bounding flights were achieved. In *Protoavis,* the development of the supracoracoideus pulley and keeled sternum allowed flapping flight for a short burst in a straight, horizontal course, as well as taking off from the ground level. In enantiornithine birds, the tail was truncated, with the development of a pygostyle that allowed complex, maneuvering flight for a short distance. In *Ichthyornis* and other birds, fusion of the carpometacarpus, the complex radiale and ulnare, and the large and keeled sternum allowed long-distance travel, while the development of the tarsometatarsus and synsacrum indicate that landing from a height was perfected at this stage.

THE PALEOECOLOGY OF PROAVIAN DROMAEOSAURS: SCANSORIAL ADAPTATION

A miniature version of protodromaeosaurs, about the size of *Protoavis,* is taken as a structural model for the hypothetical proavians. The lifestyle of proavians is crucial in understanding how flight might have started (fig. 9.12).

All modern versions of the cursorial theory depend entirely on the assumption that dromaeosaurs and *Archaeopteryx* were exclusively bipedal cursors and thus could not climb trees (Ostrom 1974, 1979; Gauthier and Padian 1985; Padian 1982, 1987; Caple, Balda, and Willis 1983; Balda, Caple, and Willis 1985). Ostrom (1979) found it difficult to accept that bipedal proavians could climb trees. He argued that all flyers (including gliders and parachuters), both extinct and extant, are essentially quadrupedal, except for birds. According to him, birds became secondarily arboreal after active flight had evolved. However, his arguments are not entirely convincing from an ecological or anatomical point of view. First, many species of living birds, such as woodpeckers (Picidae), woodhewers (Dendrocolaptidae), and creepers (Certhiidae), actually climb trees using their tails as a supporting prop against the tree trunk. Obviously, they adapted arboreal habitat in the preflight stage because feathers *per se* do not help in climbing. Young hoatzins climb trees using wing claws (fig. 9.11B). Second, we know that arboreal bipeds do exist in nature, contrary to Ostrom's assertion. For example, tree kangaroos (*Dendrolagus*) in New Guinea and Queensland are an excellent modern analog of arboreal bipeds (fig. 9.11A). They have short, broad hind feet with strong recurved claws and are proficient climbers. Normally resting and sleeping in trees, they are capable of long leaps between trees or from tree to ground and can move comfortably along tree branches. The arboreal adaptation is not apparent in the skeletal features of these animals, however, except for the recurved pedal claws.

If bipedal tree kangaroos can be adept in arboreal habitats, there is no reason why small proavians could not climb trees. Most likely the proavians were both cursorial and arboreal, like many living species of birds. Most dromaeosaurs are recorded from the Cretaceous sediments. By that time, they had become large, heavy, and specialized for predatory habitats. Yet their body plan provides a clue to the climbing heritage of their Triassic ancestors, the proavians. I speculate that these proavians were small and acquired dual locomotory habits: they were bipedal cursors on the ground and along horizontal branches but were quadrupedal climbers on tree trunks. It is quite possible that these proavians fed on ground or water and roosted and nested in the trees, as do some gallinaceous and marsh birds. They could use tree branches as a vantage point for stealth attacks on larger terrestrial animals.

The proavians and early birds share three structural adaptations that may indicate their scansorial capability: the elongated forelimbs, the swivel wrist joint, and clawed digits. It has been suggested that the forelimbs of dromaeosaurs such as *Deinonychus* were used for subduing and killing prey (Ostrom 1979; Gauthier and Padian 1985). However, it is difficult to visualize how a manus with restricted wrist movement in the plane of the forearm could be an effective raptorial structure, since its claws, if used for stabbing and slashing, would be pointing at right angles to the motion of the hands (Chatterjee 1988). The swivel wrist joint permitted movement of the hand only in the plane of the forearm (fig. 9.11D). Moreover, these predatory movements of the forelimbs would be forwardly directed to grasp prey, whereas the movement of flight is essentially dorsoventral. It is inconceivable that this forward movement could transform into a flight stroke, as Ostrom has claimed. The hands of dromaeosaurs were doing something other than catching prey. They could be folded on the side of the body when not in use, as in birds. If so, what was the main function of these hands equipped with sharp claws? The elongated forelimbs and swivel wrist joint in proavian dromaeosaurs can best be interpreted as features evolved for the climbing of vertical tree trunks in a style similar to that of hoatzins (fig. 9.12), which use recurved claws for clinging (Bakker 1986).

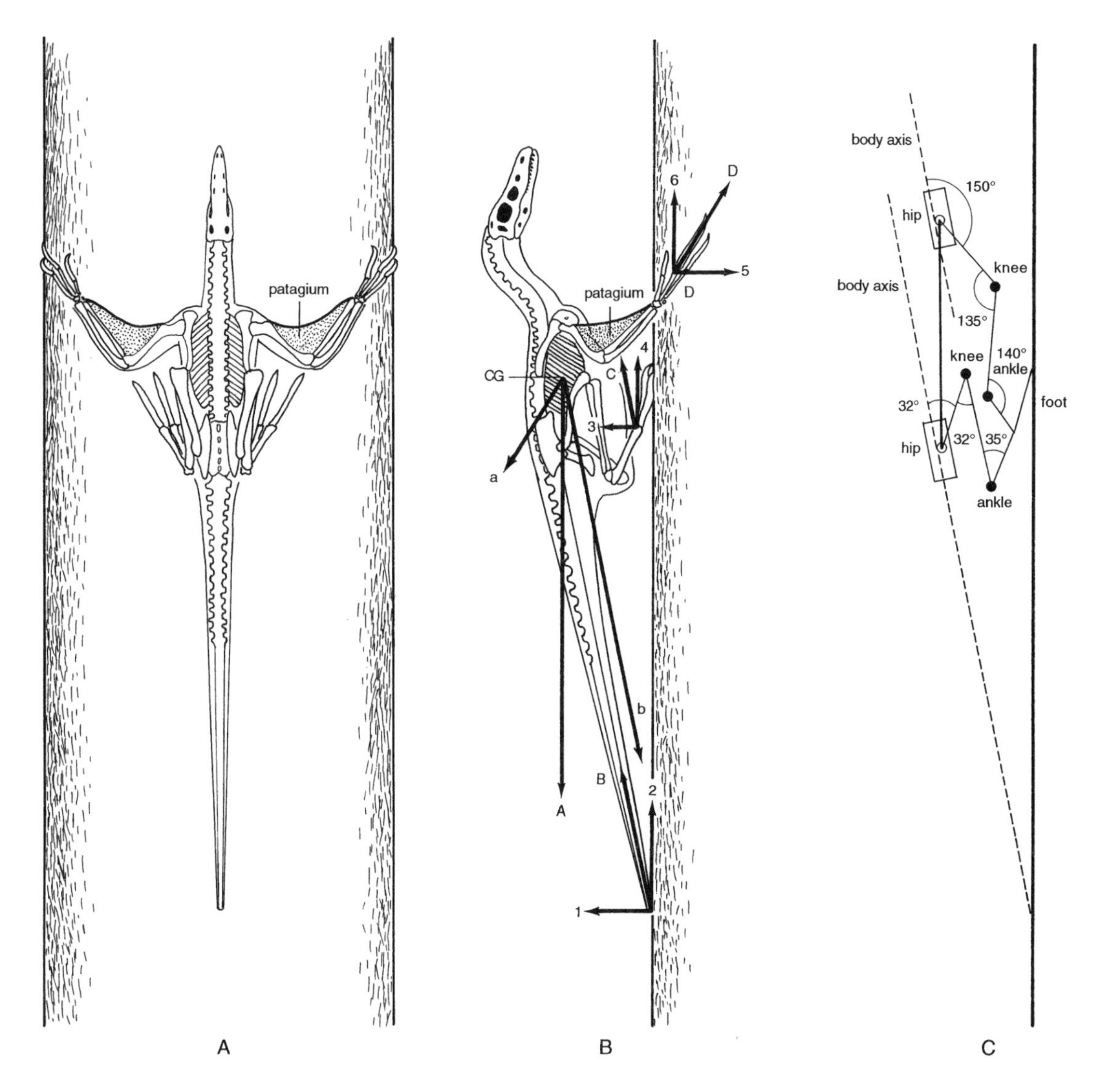

FIGURE 9.12

Scansorial adaptation of a hypothetical proavian, such as protodromaeosaur. *A,* dorsal view of the climbing protodromaeosaur, clinging quadrupedally to a vertical trunk; the patagium might have developed in front of the elbow as an elastic strap to keep the body close to the trunk; *B,* free-body diagram of the climbing protodromaeosaur in equilibrium in the right lateral view. The force of gravity **A** is acting on the animal along the center of gravity (*CG*) and is counterbalanced by forces at three points: **B** (tail), **C** (foot), and **D** (hand). **1** to **6** are components of these forces. When the system is at rest, the sum of all translation forces must be zero (vertical: **A** = **2** + **4** + **6**; horizontal: **1** + **3** = **5**). *C,* stick diagram of protodromaeosaur in the right lateral view showing the climbing sequence; the positions of only the hip and hindlimb segments are shown. As the hindlimbs are unfolded, feet provide the thrust that propels the body upward. In this position, the femur retracts considerably, about 150° from the body axis, and the knee and ankle joints are extended.

Climbing on vertical substrates is a difficult endeavor, and the proavians required two abilities to overcome this problem (Hildebrand 1982). They needed mechanisms to avoid falling from trees, and they had to propel themselves on tree branches, a discontinuous, three-dimensional substrate with variable width and orientation.

The proavians acquired various structural adaptations for clinging, hooking, and bracing vertical substrate to avoid falling. They climbed with wings, feet, and tail. They had shorter bodies and longer limbs, with grasping hands. The claws of both hands and feet were highly recurved, similar to those of tree-climbing birds, so that they could be dug securely into the bark. In climbing a vertical trunk, the outstretched forelimbs were pressed against the lateral side of the trunk, while recurved claws provided a hooking grip. The body was held in a vertical position without much undulation. The sharp, recurved claws allowed adduction force to keep them imbedded in the bark. They helped to secure the grip during climbing and supported part of the body weight (fig. 9.12). Presumably proavians used their limbs alternately in unison, first the paired forelimbs and then the paired hindlimbs. Modern flying squirrels climb rapidly in this manner.

Numerous trunk climbers use their tails as braces, struts, or props. The proavian dromaeosaurs had a rigid stiffened tail, the function of which is poorly understood. Ostrom (1969, 1990) believed that dromaeosaurs used their rigid tail as a dynamic stabilizer during cursorial progression. I believe that their stiffened tail could also have served as a prop during climbing, as in woodpeckers, to support part of the body weight (fig. 9.12).

The climbing movement of birds (Stolpe 1932) and primates (Cartmill 1985) on a vertical trunk has been studied in detail. Using these modern analogs as a guide, we can speculate about the scansorial ability of proavians. In climbing a tree, a proavian would work against gravity, and the role of friction is important in the prevention of slipping. A free-body diagram shows the action of forces on a climbing proavian clinging quadrupedally to a vertical support (fig. 9.12B). The force of gravity, **A**, was acting on the animal along the center of gravity (*CG*) and was counterbalanced by forces at three points, **B** (tail), **C** (foot), and **D** (hand). Forces **a** and **b** were two components of those opposing forces. The force **a** was directed downward and inward parallel to the axis of the tail and would be counterbalanced by a combined action of the forces **B** (provided by the tail) and **C** (by the pes). The outward component of **A**, the force **a**, tended to topple the animal away from its support and was counterbalanced by force **D** at the hand. When the proavian was at equilibrium, the sum of all translation forces would be zero (vertical: **A** = **2** + **4** + **6**; horizontal: **1** + **3** = **5**).

The sequence of limb kinematics of proavians during the propulsive and recovery phases of climbing is interesting. The forelimbs and hindlimbs moved in a reciprocal manner. In the recovery phase, the forelimbs were outstretched and swung upward and the manus was extended to grasp the cylindrical trunk; the weight of the body was carried out by the highly flexed hindlimbs (fig. 9.12C). In the next propulsive phase, the hands and the forelimbs were gradually flexed to propel while the hindlimbs were extended to move the animal upward. The highly flexed hindlimbs provided much of the propulsive thrust. The swivel wrist joint and lengthening of the forelimbs in proavians were obvious adaptations that facilitated the climbing of vertical trunks.

EXAPTATIONS FOR GLIDING FLIGHT

When a species shifts to a new niche, a structure that was useful in one way can be co-opted for another purpose. Such evolutionary change of function for a given structure with minimal change of form is called *preadaptation* or, more accurately, *exaptation* (Gould and Vrba 1982). Exaptation is a shortcut method of evolving new structure.

Surprisingly, the arboreal proavians had already developed numerous structural features that were exaptated for gliding flight. A gradual modification of the bone-muscle system in the locomotor apparatus of proavians toward the avian condition can be linked to the climbing adaptation. These flight antecedents were later finely tuned for achieving lift, producing thrust, and controlling stability. Figure 9.12 illustrates the acquisition of avian characters by the climbing proavians.

THE PATAGIUM Since the center of gravity (*CG*) of the proavian was not coplanar with the points of support (hands, feet, and tail), there would be a tendency for the animal to topple away from its support (see fig. 9.12A–B). To prevent this disaster, the long hind legs would be held in a flexed-knee posture, as seen in many climbing primates, to keep the body's center of gravity close to the substrate (Cartmill 1985). The force of gravity on proavians on a vertical trunk must be counteracted chiefly by the flexor muscles (Stolpe 1932). It is likely that a flap of skin, the patagium, evolved along the leading edge of the forelimbs, between the shoulder and carpal joints, initially as an elastic strap to keep the body close to the trunk during climbing and later was co-opted for lift in an aerial environment (Pennycuick 1986). All modern parachuters possess patagia.

TRANSFORMATION OF THE FLIGHT APPARATUS In the climbing mode, the humerus extended laterally and horizontally from the body, so that its proper ventral surface now faced outward. The elbows projected sideways so that hands could be pressed firmly on the trunk by the powerful adductor pectoralis muscle (fig. 9.13A). The development of a bony sternum suggests the presence of large pectoral muscles in proavians. The climbing ability was considerably improved as the elbow was brought not only lateral to the glenoid, but at the same time on a horizontal level with it. Because of the reorientation and axial twist of the humerus and the forward and upward movement of the elbow, the glenoid fossa gradually shifted laterally and somewhat dorsally from the posterior position. In consequence, the site of the origin for the biceps brachii muscle, the main flexor of the elbow joint, changed, which is reflected by the upward migration of the biceps tubercle on the coracoid. As the scansorial adaptation was more refined, the action of the biceps brachii muscle was elevated to keep the elbow movement at the glenoid level. Both Walker (1972) and Ostrom (1976b) suggested that the biceps tubercle of dromaeosaurs and *Archaeopteryx* was a precursor to the avian acrocoracoid process. This structural transformation is one of the most intriguing features in avian flight and is related to the evolution of the supracoracoideus pulley for the upstroke. The modification of this structure can now be explained in the context of the scansorial adaptation of proavians.

Because the biceps brachii muscle provided the propelling thrust of the body during upward climbing, the forward and upward migration of the biceps tubercle improved the effectiveness of the action of this muscle and thus enhanced climbing capability. As the biceps tubercle moved forward and upward to meet the furcula, it also changed the course and action of the supracoracoideus muscle. In proavians, the biceps tubercle is small and lies just below the glenoid, around which the short supracoracoideus muscle forms a loop. The supracoracoideus muscle must have pulled the humerus forward, as is apparent in figure 9.13A, and could have assisted in climbing. In *Archaeopteryx* (fig. 9.13, B and E), the biceps tubercle is considerably forward of the glenoid and is somewhat enlarged so that the humerus could be protracted subhorizontally by the combined actions of brachialis brachii and supracoracoideus muscle. Because the biceps tubercle in *Archaeopteryx* still lay below the glenoid, the supracoracoideus muscle could function as a wing protractor. As gliding was perfected, the biceps tubercle moved upward above the level of the glenoid, and the coracoid became more deep and strutlike, as seen in *Protoavis* and modern birds. The supracoracoideus tendon would pass through the triosseal canal and form a pulley to elevate the wings (fig. 9.13, C and F). Thus, the evolution of the avian coracoid and flight muscles can be traced through the scansorial adaptations of proavians.

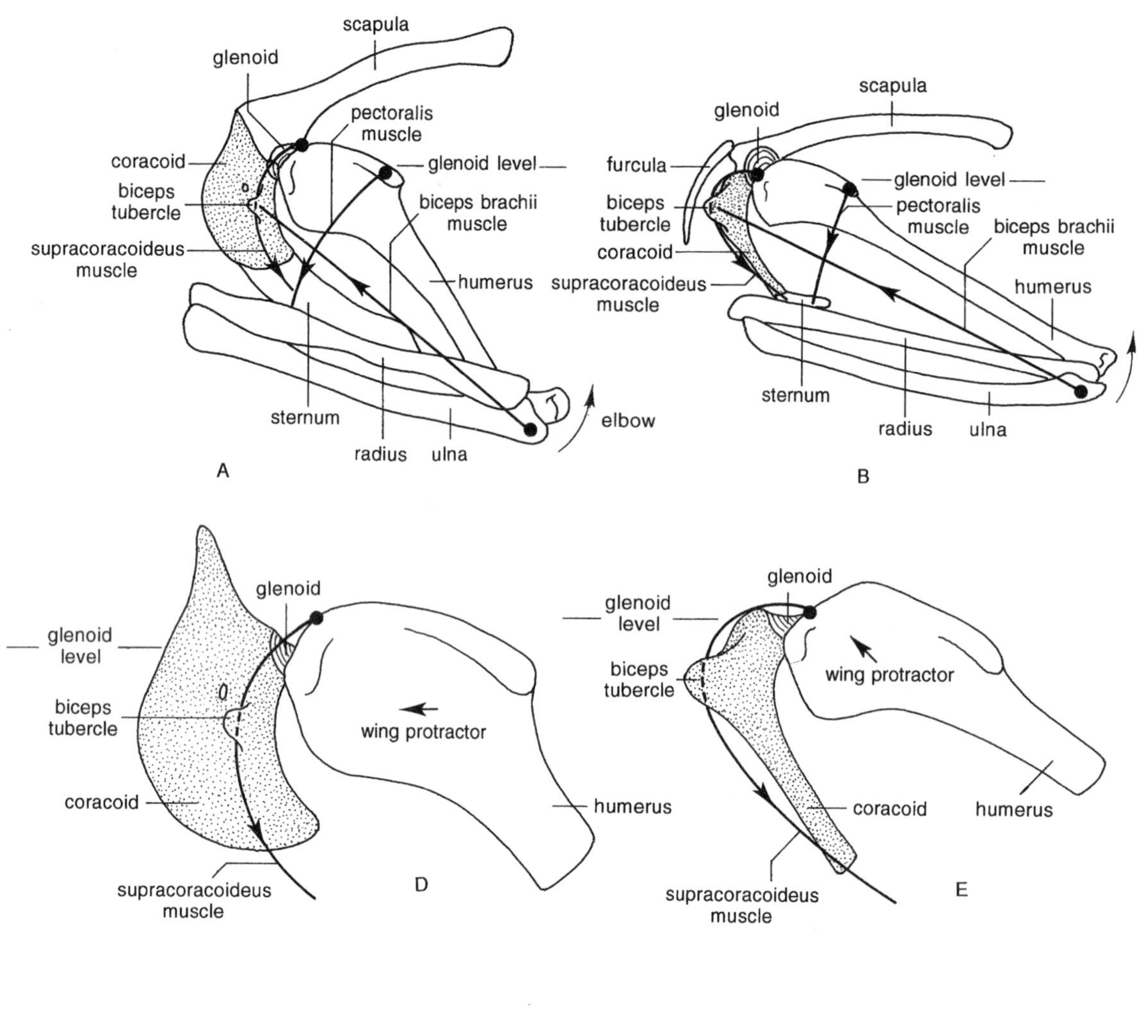

FIGURE 9.13

Evolution of the avian coracoid (*stippled*) shows the progressive transformation of the origin of the biceps brachii and supracoracoideus muscles. The biceps tubercle in the proavian, such as the protodromaeosaur, moves progressively forward in *Archaeopteryx* and then farther upward in *Protoavis* to form the acrocoracoid process and triosseal canal. With modification of the biceps tubercle to the acrocoracoid process, the action of the supracoracoideus muscles changes from wing protractor to wing elevator. The remodeling of the coracoid from proavian to *Archaeopteryx* might have taken place during climbing and gliding, whereas the development of the acrocoracoid process and triosseal canal from *Archaeopteryx* to *Protoavis* might have occurred with changes from gliding to flapping flight. Each representation is the left lateral view of the shoulder girdle and forelimb.

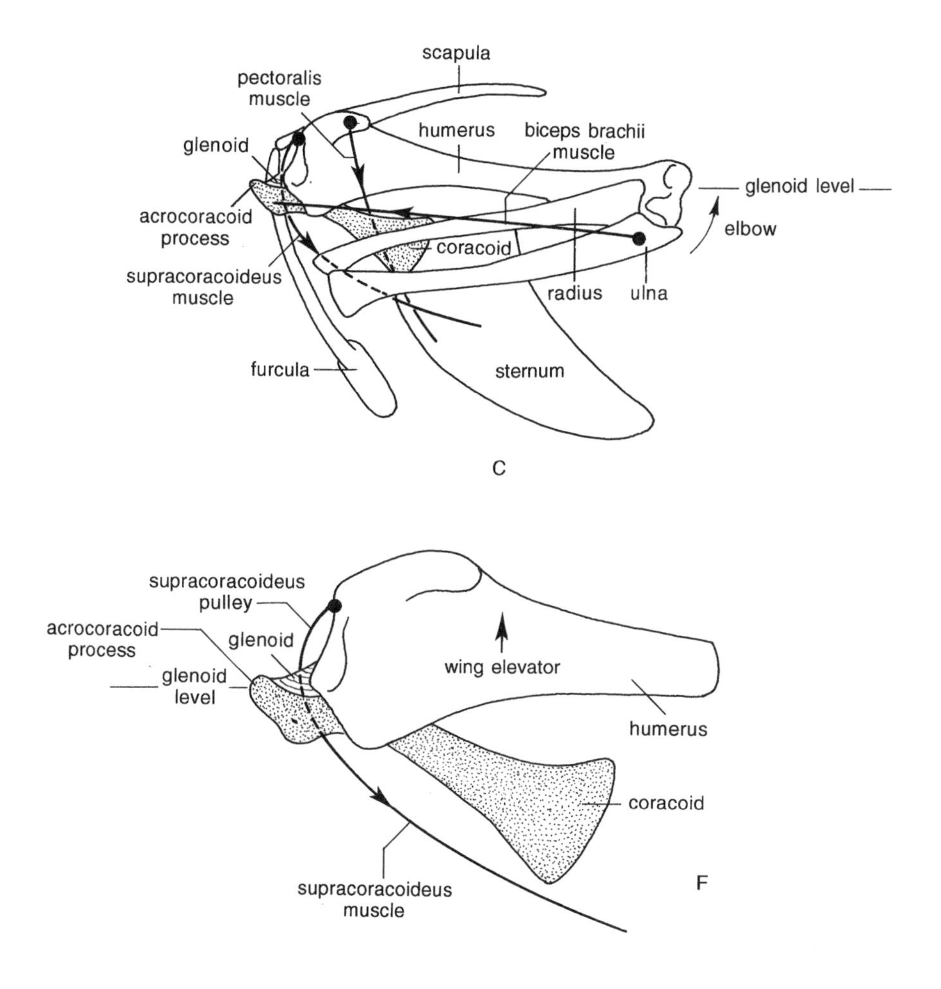

Synchronous motions of the forelimbs for climbing would resemble a rudimentary flight stroke. The outstretched forelimbs of proavians were useful for controlling the body orientation in air, especially roll and yaw. Forelimbs, which evolved during climbing as rigid, integrated structures with a swivel wrist joint, would be appropriate to withstand distortion of the wing against the air currents but would permit sweeping of the wing tip during upstroke to reduce drag (Rayner 1988).

THE REMODELING OF THE PELVIS AND HINDLIMB In terrestrial theropods, both pubis and ischium project considerably ventrally, but in birds both elements are rotated backward to reduce drag. The modification of the pelvic girdle and hindlimb in proavians toward the avian condition can be attributed to climbing adaptation (see fig. 9.12C). The magnitude of femoral swing in the climbing mode was strikingly much greater than that required in cursorial locomotion. In the recovery phase, the hindlimbs were folded to carry most of the body weight. The femur protracted considerably and made an angle of about 32° with the body axis. As the hindlimbs were unfolded during the beginning of the propulsive stroke, the feet provided the thrust that drove the body upward. In this position, the femur retracted considerably, about 150° from the body axis, and the knee and ankle joints extended (fig. 9.12C). The resulting considerable hip displacement propelled the animal upward. In the climbing cycle, the femur circumscribed as arc of 118° (fig. 9.12C). Since the degree of protraction was far greater in the climbing mode than in cursorial locomotion, the cranial process of the ilium became elongated to provide the site of the femoral protractor, the iliotibialis cranialis. Thus, the evolution of a subhorizontal femur and an elongated cranial iliac process in birds may be linked to their scansorial adaptation (fig. 9.12B). Since the center of gravity of the body lay along the vertical axis with the hip joint and the latter carried much of the body weight, the backward rotation of the pubis would be advantageous for postural support, keeping the body close to the trunk. As the pubis turned backward, the puboischiofemoralis internus, which protracted the femur in primitive theropods, was transformed to a retractor—the obturator medialis (Galton 1970). Thus, the pelvis and hindlimb of proavians were strongly streamlined for body support during climbing and were exaptated for gliding.

THE TAIL Because the tail was used for bracing in climbing proavians, the function of the caudofemoralis longus muscle changed from femoral retraction to tail depression. The dorsoventral movement of the rigid, stiffened tail, used as a prop in climbing, would act as an aerodynamic stabilizer during gliding to change the pitch axis and extend the gliding path.

THE EVOLUTION OF POWERED FLIGHT

The primary stimulus for aerial locomotion was to search rapidly and efficiently for food and shelter, extend the range of foraging, and avoid confrontations with large, terrestrial predators. Also, proavians would choose to move in such a way as to minimize the cost of transport and maximize the foraging range. They would spend a portion of the time in trees sleeping, hiding, nesting, and parachuting. There was a progressive decrease in the angle of descent during the evolution of avian flight. Initially, as the proavians would parachute from tree branches, the line of takeoff to landing would be steeper than 45°. With the development of some maneuvering in the air during the gliding stage, the angle of descent would be less than 45°. Finally, in the flying stage proavians would be capable of sustaining themselves in the air without any fall. In the initial stage of flight evolution, an important constraint was the need to launch from a height, where potential energy could be converted cheaply into kinetic energy. The ability to parachute also implies the ability to land without damaging the body. In the early stage, when proavians lacked a large lifting surface in the forelimb to retard a fall, the safest landing site would be water. A modified version of the arboreal theory is proposed here to trace the crucial stages of avian flight. The proavian forelimbs, which were primarily used for climbing, evolved successively for use in parachuting, diving, gliding, and active flapping. A sequence of seven phylogenetic and adaptive changes leading to avian flight among Mesozoic birds is presented here, with each step setting the stage for the next (fig. 9.14).

STAGE 1: PARACHUTING My model begins with small, active scansorial proavians, a miniature version of dromaeosaurs, about the size of *Protoavis* (figs. 9.12 and 9.14). These proavians would climb trees in the Triassic riverine forests and parachute regularly from tree branches. Like any other parachuters, they had possibly developed a small lifting surface in front of the elbow, in the form of a patagium, to increase the drag and retard

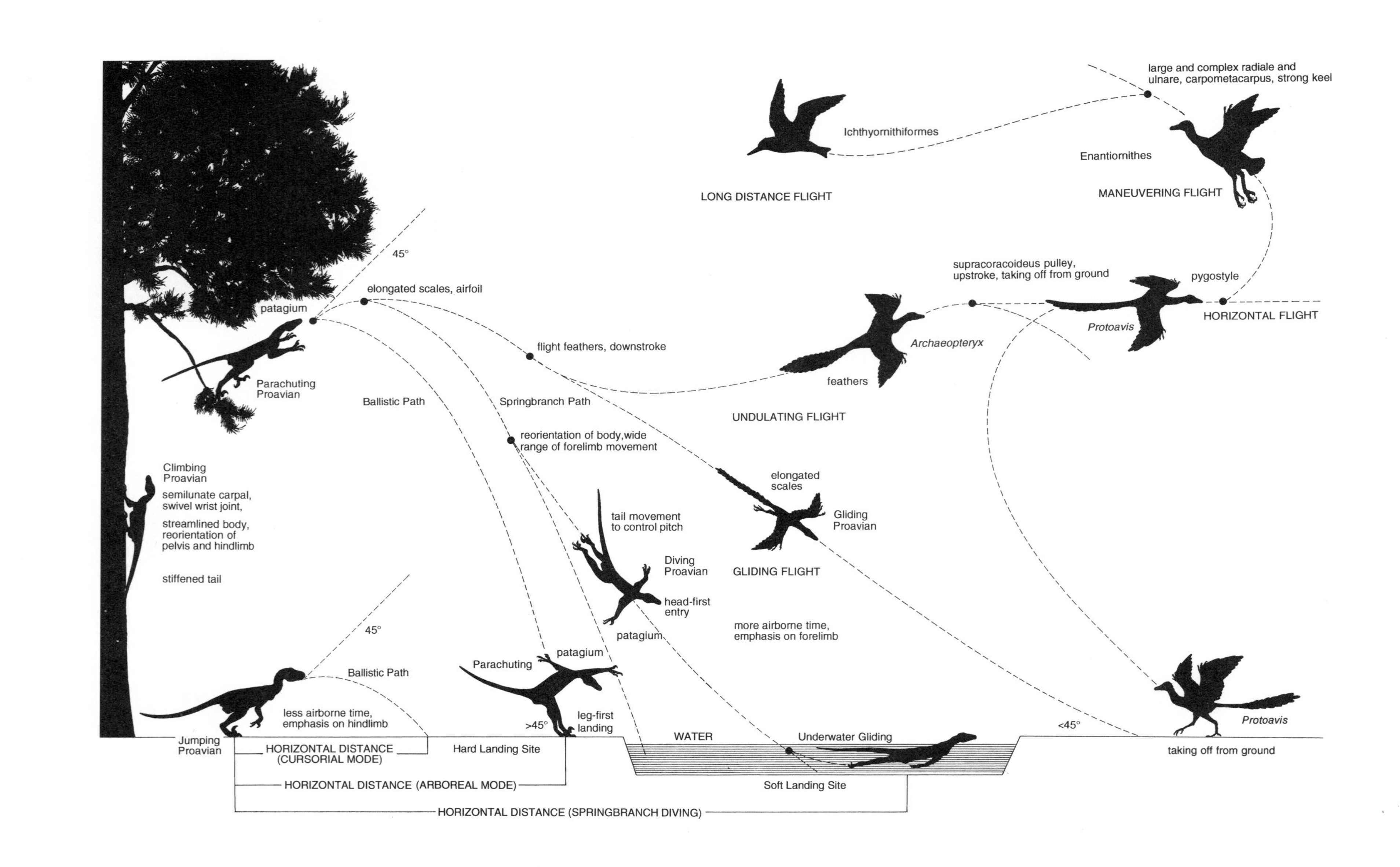

large and complex radiale and
ulnare, carpometacarpus, strong keel
Ichthyornithiformes
Enantiornithes
LONG DISTANCE FLIGHT
MANEUVERING FLIGHT
45°
supracoracoideus pulley,
upstroke, taking off from ground
pygostyle
elongated scales, airfoil
patagium
HORIZONTAL FLIGHT
Protoavis
flight feathers, downstroke
Archaeopteryx
Parachuting
Proavian
feathers
Ballistic Path
Springbranch Path
UNDULATING FLIGHT
reorientation of body,wide
range of forelimb movement
Climbing
Proavian
semilunate carpal,
swivel wrist joint,
streamlined body,
reorientation of
pelvis and hindlimb
stiffened tail
elongated
scales
tail movement
to control pitch
Gliding
Proavian
Diving
Proavian
GLIDING FLIGHT
head-first
entry
45°
patagium
more airborne time,
emphasis on forelimb
patagium
Parachuting
Ballistic Path
less airborne time,
emphasis on hindlimb
>45°
leg-first
landing
<45°
Protoavis
Jumping
Proavian
HORIZONTAL DISTANCE
(CURSORIAL MODE)
Hard Landing Site
WATER
Underwater Gliding
taking off from ground
HORIZONTAL DISTANCE (ARBOREAL MODE)
Soft Landing Site
HORIZONTAL DISTANCE (SPRINGBRANCH DIVING)

the fall. They would launch themselves with a jump. When they were airborne, they would outstretch the forelimbs to maximize the drag forces and would land on their hind feet. The initial takeoff would begin with a symmetrical or two-footed jump from a tree branch, with folded forelimbs to reduce drag. The proavians would launch themselves into the air by using thrust from their legs against a perch at the optimal angle of 45° to enhance the velocity and would follow a ballistic path. As the animals jumped, they accelerated their body to the takeoff velocity

FIGURE 9.14

Hypothetical stages in the evolution of major forms of avian flight in the arboreal mode. The advantage of taking off from a height as opposed to a ground takeoff is shown at the *bottom*. Taking off from a tree branch would increase the airborne time and the horizontal distance. The development of maneuverability of body, airfoil, and feathers is more likely in an arboreal mode that allowed several transitional stages. Stage 1—parachuting: The proavians would climb trees and jump from branches for stealthy attack. The patagium, which had developed proximally in front of the elbow during scansorial adaptation, would serve as an airfoil during landing to slow the fall. Stage 2—springbranch diving: Landing on water may be a safe strategy when jumping from flexible tree branches; as diving was perfected, body reorientation was achieved and the angle of descent was decreased by tail movement. Stage 3—gliding flight: As the duration of airborne time was increased, elongated scales created gliding surfaces on the outstretched wings and tail, thus further extending the gliding path. The elongated scales at the distal end of the wings would increase the wingspan and reduce the wing tip vortex. The proto-wing at this stage would be narrow and long, with a high aspect ratio, so that proavians could glide at much shallower angles and travel longer distances. Stage 4—undulating flight, exhibited by *Archaeopteryx*. Since supracoracoideus muscle could not function as a wing elevator at this stage, the animal must glide to achieve the upstroke position by vertical air resistance. The height lost during the gliding is regained in the next downstroke. Stage 5—horizontal flight, exhibited by *Protoavis*. In this stage the supracoracoideus muscle could act as a wing elevator, so that taking off from the ground became possible. The animal could fly short horizontal distances at a high speed, analogous to chicken flight. Stage 6—maneuvering flight, exhibited by enantiornithine birds with the loss of the long bony tail and hence stability. The trade-off in losing stability is the achievement of maneuverability. Complex maneuvers such as foraging during flight and landing on a perch probably evolved at this stage. Stage 7—long-distance flight, exhibited by Ichthyornithiformes. With the development of powerful wings, the carpometacarpus, a complex radiale and ulnare, and a large keeled sternum, long-distance migration became possible.

and rose to a maximum height, depending upon the strength of the thrust. As soon as they reached the highest point, they curved over in a ballistic trajectory and began to outstretch the wings in dihedral configuration to retard the fall and decrease the angle of descent. Parachuting is a common form of locomotion among many arboreal vertebrates because it is a cheaper way of locomotion than running. It is also an efficient way to escape from another arboreal predator.

STAGE 2: SPRINGBRANCH DIVING A good understanding of jumping technique can be derived from Newton's Third Law. The law states that for every action there is an equal and opposite reaction. When a jumper pushes down the springing branch using its feet, it is being pushed up by the branch equally and in the opposite direction, resulting in a lifting movement. Jumping from a springing branch is analogous to jumping on a trampoline or diving from a springboard, which gives more vertical lift and more airborne time for maneuverability. The recoil of the depressed branch throws the jumper into the air. The height achieved by jumping from a springing branch is far greater than that achieved by jumping from the hard ground (fig. 9.14). Moreover, springbranch jumping would greatly increase the horizontal distance and the duration of the airborne time.

Bipedal landing from a height on hard ground is always hazardous. To protect themselves from the impact of landing, early proavians might have selected a soft landing site such as water, as is done by many modern birds (Rüppell 1977). Water is uniform and yielding to form a great cushioning medium. The abundance of food in water may be another motivation for jumping into water. Young hoatzins, when threatened with danger, jump to water to avoid injury and then climb back to their nest with their manual claws. The ecology of hoatzins provides a glimpse of the habitat of early proavians. Like hoatzins, proavians would probably live in trees overhanging streams and lakes.

Once the safety of landing was assured, a proavian could control the orientation of its body when airborne. Springbranch diving is analogous to the swan dive practiced by springboard divers. As the ballistic path reached a very steep angle for free fall, the proavian would extend its forelimbs to the sides with its head pointing downward, using the pull of gravity as thrust. In this position the legs would be lifted and the tail would play an important role in changing the angle of descent. The proavian would orient its lifting surface, the patagium, at the angle of attack that would give the best lift/drag ratio and alter its course tangentially to a gliding path. To do so, it would change the pitch axis of the body by flexing the tail upward and swinging the wing forward to enter a gliding path.

Pennycuick (1986) pointed out that such a wing design, with a patagium, would have a higher aspect ratio and a smaller area than that of modern gliding animals, so that proavians would glide faster with a relatively flatter gliding path. The height of the bifurcation point between the parachuting and the diving path was crucial in the evolution of flight; it increased with the perfection of gliding flight (fig. 9.14). As the animals approached water for landing, they would fold arms and legs into a parasagittal plane, like a springboard diver, for entry into water. Since the body was buoyed by water, the early proavians might have learned the dynamics of gliding underwater soon after the entry. Once the gliding posture was developed in a buoyant aquatic medium, this movement could be perfected in air to retard the fall.

STAGE 3: GLIDING FLIGHT The longer the duration of airborne time for a proavian, the greater the opportunity for controlling and maneuvering its body axis, extending its gliding path, and developing flight surfaces. For gliding proavians, the outstretched wings and long tail would create drag and vortex wakes during gliding. The vortex would be especially strong at the wing tip, and it could be minimized if the wings were narrowed and lengthened so that the tips on either side of the body were widely separated (Kaufmann 1970). In pterosaurs and bats the wing span was increased internally by lengthening of the

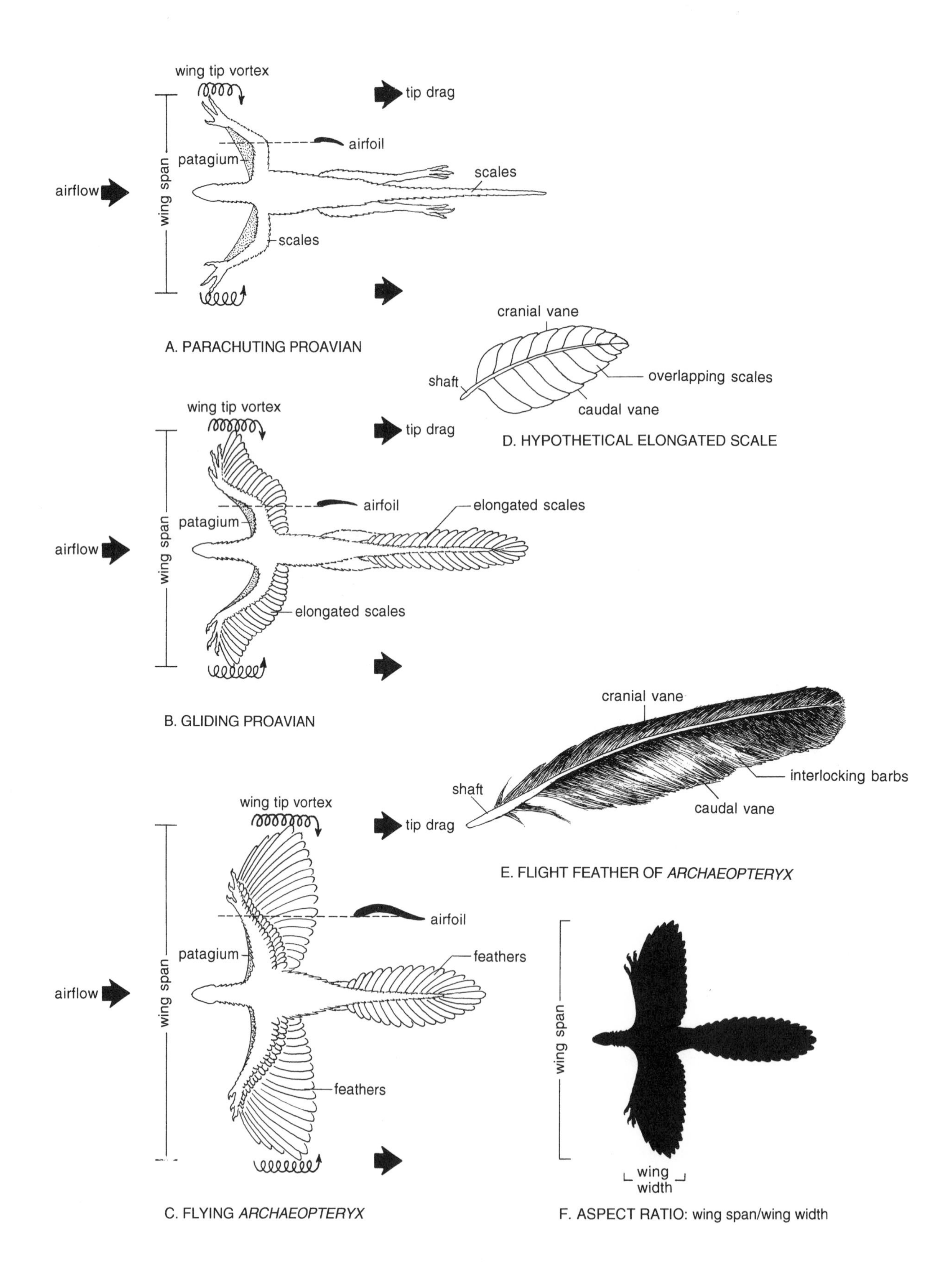
wing tip vortex
tip drag
airfoil
patagium
scales
airflow
wing span
scales
A. PARACHUTING PROAVIAN
cranial vane
shaft
overlapping scales
caudal vane
D. HYPOTHETICAL ELONGATED SCALE
wing tip vortex
tip drag
airfoil
elongated scales
patagium
airflow
wing span
elongated scales
B. GLIDING PROAVIAN
cranial vane
interlocking barbs
shaft
caudal vane
wing tip vortex
tip drag
E. FLIGHT FEATHER OF ARCHAEOPTERYX
airfoil
patagium
feathers
airflow
wing span
feathers
wing span
wing width
C. FLYING ARCHAEOPTERYX
F. ASPECT RATIO: wing span/wing width

manual digits, whereas this problem was solved in proavians by lengthening of the external integuments (fig. 9.15). Enlarging wings would reduce the cost of transport considerably (see fig. 9.9).

As diving was perfected and the angle of descent decreased, the proavians would develop an airfoil surface to lengthen the downward leap and retard the fall. The airfoil would consist of a series of overlapping, elongated, flexible scales along the trailing edge of the forelimbs to minimize drag and vortex and to enhance lift. These elongated scales would be proportionately large at the hand section to increase the wingspan and wing area and decrease the wing loading. The proto-wing at this stage would be narrow and long, with a high aspect ratio so that proavians could glide at much shallower angles and travel longer distances. The long tail would support additional lifting and flight surface, and the elongated, overlapping scales would be arranged horizontally on the lateral and posterior sides of the tail to reduce tail drag. Once these proto-wings with elongated scales evolved in gliding proavians, reduction of weight and enlargement of these integuments would allow the animal to remain airborne for longer periods. Natural selection would lead to the development of light, long feathers to produce a streamlined body and more flexible wings with an increased airfoil, low wing loading, and a low aspect ratio.

FIGURE 9.15

Dorsal views of the hypothetical parachuting proavian (*A*) and gliding proavian (*B*) and of the flying *Archaeopteryx* (*C*) to show how elongated scales and flight feathers in early birds might have increased the wingspan and airfoil surface to decrease the interference of the wing tip vortex. *D,* hypothetical elongated scales developed during gliding flight to increase the airfoil surface and were precursors to feathers. *E,* isolated flight feather of *Archaeopteryx* showing the asymmetrical vanes. *F,* the aspect ratio is the ratio between wing span and wing width and is an important feature of the flight characteristics of birds. A gliding proavian (*B*) with long, narrow wings has a higher aspect ratio than does *Archaeopteryx* (*C*), with broad wings.

STAGE 4: UNDULATING FLIGHT The fourth stage is represented by *Archaeopteryx*, with the development of aerodynamic feathers (Olson and Feduccia 1979). In this stage, avian flight would be an alternating sequence of flapping and gliding, called *undulating flight* (Burton 1990), as feathers began to evolve and enlarge. Each minor increment of protofeather length at this stage would be clearly adaptive for aerodynamic efficiency. Increased wing area would reduce wing loading and gliding speed, allowing safer landings. A slight flapping in a gliding animal could produce a new horizontal thrust force and the lift necessary to balance the body, resulting in a longer glide path (Norberg 1985, 1990). As the wing lengthens, flapping reduces the total cost of energy by as much as 80 percent (Rayner 1991). *Archaeopteryx* was capable of executing a downstroke because of the presence of a robust furcula to house a large pectoral muscle (Olson and Feduccia 1979). Since *Archaeopteryx* lacked a triosseal canal, the supracoracoideus muscle could not function as a wing elevator at this stage. The animal must have glided from a tree trunk or a cliff and sunk to achieve the upstroke position by upwardly directed air resistance. Once the wings were moved to the upstroke position, *Archaeopteryx* was able to execute the downstroke. The height lost during the glide was regained by the next downstroke. By alternating this flapping and gliding sequence, *Archaeopteryx* would fly up and down in an undulating path (see fig. 9.14). This kind of flight path is employed by many living birds (starlings, bee-eaters, swifts, swallows, fulmars, crows, etc.). These birds use undulating flight to save 10 to 20 percent of their muscular energy (Burton 1990).

Archaeopteryx could possibly land on soft ground by setting the wings at a large angle of attack and spreading the long rectrices of the tail to stall. However, takeoff from level ground would be difficult at this stage because of the absence of the supracoracoideus pulley. It is likely that *Archaeopteryx* had to climb trees or cliffs to gain the potential energy to glide.

STAGE 5: HORIZONTAL FLIGHT The fifth stage of avian flight would be a short burst of horizontal flight, analogous to that of chickens, as documented by *Protoavis* (Chatterjee 1995, n.d.). This mode of flight was possible with the evolution of the supracoracoideus pulley as the wing elevator (see figs. 9.13F and 9.14). This new arrangement permitted active dorsoventral flapping of the wing so that the animal could maintain a horizontal path and could take off from level ground. The sternum had developed a keel for the attachment of flight muscles, whereas the furcula, with a large hypocleidium, had acquired a springlike structure (figs. 4.4A, 4.4B, and 4.4M). The wrist bones had developed a primitive interlocking device to secure the manus in place during the downstroke (Vasquez 1992). This implies that *Protoavis* was capable of powered flight, possibly at a cruising speed; slow flight might have been difficult at this stage (Rayner 1989). As in *Archaeopteryx*, the long tail of *Protoavis* provided stability and enabled it to control its motion with respect to pitch and roll. The animal probably kept a straight and level course without much turning, climbing, or diving. The wing loading was high and the flight was short and fast, like that of Galliformes.

STAGE 6: MANEUVERING FLIGHT The sixth stage in avian flight would be the evolution of decreased stability in favor of greater maneuverability. Stability was decreased by transforming the long, bony tail into a pygostyle, as is seen in several enantiornithine birds, such as *Concornis*, *Sinornis*, *Iberomesornis*, and *Enantiornis*. As a result, the center of gravity would shift behind the main lifting surface to increase instability. Loss of the tail would lead to low wing loading, a low stalling speed, and more maneuverability at this stage. The presence of alula feathers in *Eoalulavis* indicates that these birds were able to prevent stalling at low speed during takeoff and landing (Sanz et al. 1996). Complex maneuvers such as foraging during flight and landing on a perch probably evolved at this stage. The wings were short and rounded with a low aspect ratio and high wing loading, which may indicate foraging in trees and bushes. Flight in forests was more expensive than that in the open spaces. The decrease of stability was compensated for by finer coordination between senses and muscles so that rapid and precise movements could be regulated to alter the line of flight. The decrease of stability was an advantage for a flying animal as it allowed greater maneuverability in any direction, especially among trees (see fig. 9.14).

STAGE 7: LONG-DISTANCE FLIGHT This final stage in the evolution of flight is represented by *Ichthyornis, Ambiortus*, and other carinate birds with the development of large, keeled sternums and powerful wings. The wrist was modified in such a fashion that the manus could be locked in place during the downstroke (Vasquez 1992). As a result, various subtle and complex flight-related movements of the wings could be performed at this stage. The wings became narrow with a high aspect ratio and low wing loading, which allowed foraging in open water, sustained flight, and long-distance migration at a cheaper cost. With further enlargement of the brain, the neurosensory control system would correct constantly and almost instantaneously for minor displacements of the wings (Maynard-Smith 1952; Brown 1963). Ichthyornithiformes were the long-distance pioneers, traveling across the oceans and continents. Landing on hard substrate was also perfected at this stage, with the development of the synsacrum, the fused tibiotarsus, and the fused tarsometatarsus.

THE ORIGIN OF FEATHERS

The evolution of feathers and that of flight are closely interlinked and most central in the origin of birds. Unfortunately, fossil feathers and their precursors are sparse in the paleontological record. One of the striking features of *Archaeopteryx* is that its feather appears in every respect to be like that of modern birds, even in the details of microstructure (see fig. 9.15E). Its rachis supports the asymmetrical vane on either side; each vane is formed of interlocking barbules that end in hooks. It has the backward curvature of a flight feather with airfoil function. Because of this perfection, the *Archaeopteryx* feather does not provide any clue to the precursor of feathers.

Feathers grow from pits or follicles of the skin. It is generally believed that feathers originated from reptilian scales, but no intermediate stage between the two has ever been found in the fossil record. Both feathers and reptilian scales contain β-keratin; however, the ϕ-keratin in bird feathers is a unique evolutionary novelty (Gill 1990). Modern birds possess both feathers and scales. The lower legs and toes are covered by horny scales similar to those of reptiles. Rawles (1963) has shown that transplants of embryonic epidermis may produce either scales or feathers depending on the nature of the underlying dermis. Scales and feathers have a common developmental origin; both develop from similar germ buds. The conversion of scales into feathers is known in some living birds. For example, young pigeons have legs covered with scales, but the scales are replaced by feathers in adult pigeons. In some hawks, scales on the legs support some rudimentary feathers. In the bald eagle, the bare tarsus is covered by scales; in the golden eagle, however, the tarsus is covered by feathers. Recent experiments by Zou and Niswander (1966) indicate that blockage of bone morphogenetic protein (BMP) caused the scales on the foot of the chick to develop into feathers.

Two major theories for the evolution of feathers have been presented: (1) in connection with flight and (2) as insulation to retain body heat. Because feathers are uniquely designed for flight, it is not surprising that much speculation about the origin of feathers is related to the origin of flight. Heilmann (1926) suggested that feathers evolved in an aerodynamic context from elongated proavian scales that lay on the trailing edges of the tail and body. These elongated scales would provide an airfoil surface to slow the fall of the proavian during parachuting and gliding.

Parkes (1966) also argued that flight feathers evolved early on wings and tail by providing an increased airfoil. Once flight was achieved, simple, degenerate feathers spread all over the body for further insulation, resulting in the appearance of down feathers. Alan Feduccia (1980, 1985, 1995a, 1996) provided some recent examples in support of this flight model. First, in leaping lemurs such as Malagasy sifakas (*Propithecus*), which possess a gliding membrane, a thick fringe of elongated hairs develops along the caudal forelimbs. Using this analogy Feduccia argued that primitive contour feathers along the caudal edge of the forelimbs and the lateral margins of the tail could provide the initial stage in the evolution of an airfoil. Second, in recent flightless birds, such as the ostrich, rhea, cassowary, and kiwi, feathers are primarily for insulation; these feathers have become degenerate, lost their pennaceous structure and aerodynamic design, and become fluffy and hairlike in structure. Feduccia thus questioned why such complicated structures as feathers would be evolved in the first place if they were originally adapted for temperature control. He maintained that feathers first evolved in the context of flight and later were coopted for insulation.

Opponents argue that contour feathers were primitive, whereas flight feathers appeared late in evolution. They propose that selection for the initial development of feathers in early proavians was for the function of insulation and later was an exaptation for flight (Bakker 1975; Regal 1975; Ostrom 1979; Bock 1986; Bock and Bühler 1995). Since trees are considerably cooler than the ground, an insulating cover of the body was advantageous for arboreal proavians. As the proavians began parachuting and diving into water, body plumage would confer a streamlined shape; it might serve as waterproof insulation and enhance bouyancy (Thulborn and Hamley 1985). Eventually, during gliding, unspecialized contour feathers along the caudal border of the forelimbs and the lateral margins of the tail evolved into asymmetrical remiges and rectices for aerodynamic function.

The recent discovery of a downy compsognathid "*Sinosaur-opteryx*" from China supports the insulation theory (Browne 1996). The specimen was found in the same lacustrine deposits (Jurassic-Cretaceous boundary) in Liaoning province that has yielded *Confuciusornis* and other spectacular fossils. The one-meter-long theropod is exquisitely preserved in the same attitude as the Solnhofen *Compsognathus.* The specimen is covered with short, hairlike, downy feathers from the top of the head to the tip of the tail. Since these feathers lack any aerodynamic design, they probably functioned entirely for thermoregulation. The Chinese specimen hints at a theropod-bird link and offers a glimpse of how early proavians might have had a furry covering. The specimen "*Sinosauropteryx*" is yet to be described, and its evolutionary significance has not been assessed. The coeval *Confuciusornis* indicates the presence of both flight feathers and down feathers (Hou et al. 1995).

The possible sequence of the origin of flight feathers was refined by Saville (1962), Parkes (1966), and Feduccia (1980, 1995a, 1996). Regal speculated that a hypothetical intermediate stage might bridge the gap between the reptilian scale and the avian feather; this protofeather consisted of a long, tapered central shaft that supported enlarged, overlapping scales on its two sides (Regal 1975). The vane was asymmetrical, as in a flight feather, while the scales were flexible and might have been precursors to barbs. These elongated scales would be placed in an overlapping fashion and could be folded over one another when not in use (see fig. 9.15D).

Fossil evidence tends to support this scenario. Sharov (1970) described an interesting archosaur, *Longisquama* from the Early Triassic lake beds of Turkestan, in which overlapping, elongated scales are found along the dorsal appendages and clavicles. Sharov concluded that these elongated scales were attached to the back in life and may have functioned as a kind of parachute, braking the animal's fall as it jumped from branch to branch or from the trees to the ground. He suggested that these scales constitute a structural stage in the evolution of feathers. Maderson (1972) showed elegantly how these elongated scales of *Longisquama* could give rise to the protofeathers of early birds through a series of transformations. Eventually, these protofeathers would increase the wingspan and reduce the wing tip vortex during the gliding stage, as discussed earlier (see fig. 9.15).

The Genesis of Birds

CHAPTER 10

From him are born gods of diverse descent.
From him are born angels, men, beasts, birds.
—Upanishads, ca. 800 B.C.

Birds are unique and distinctive among vertebrates. Natural selection has given them an edge by creating a lightweight skeleton, an aerodynamic coat of feathers, and highly efficient metabolic and respiratory systems allowing them to defy gravity and fly. Yet their basic body plan bears the unmistakable heritage of bipedal archosaurs. The transition from archosaurs to birds is documented by such a limited number of fossils that its early history remains obscure. Several questions arise regarding the origin and early evolution of birds. Which archosaurs are the closest relatives to birds? How do we distinguish an early bird from its immediate forerunner? What was the sequence of acquisition of avian attributes during the Mesozoic period? Having surveyed the anatomical details of such early birds as *Protoavis* and *Archaeopteryx*, we now examine these taxa in a broader context—to explore their roots and relatives and to establish their relationships with other Mesozoic birds.

The Origin of Birds

Archosaurs are a major group of vertebrates that dominated the earth during the Mesozoic and that survive today as birds and crocodilians. Modern crocodiles are the closest relatives of living birds and share a common ancestry. It has long been accepted that birds are descended from archosaurs; they share three derived characters in the skull: the antorbital fenestra, an ossified laterosphenoid, and an external mandibular fenestra. Many early avian features seem to have originated in parallel in various lineages of archosaurs. Shortly after the discovery of *Archaeopteryx* in the 1860s, there was controversy over which group of archosaurs—the crocodiles, thecodonts, or theropods—is most closely related to the avian lineage. Witmer (1991) provided an excellent historical account of this debate.

THE ANKLE JOINT AND THE PHYLOGENY OF ARCHOSAURS

To understand how birds are related to a specific group of archosaurs, we must examine the phylogeny of archosaurs. Recent cladistic analyses recognized two distinct lineages of archosaurs (fig. 10.1A), one leading to crocodilians (Crurotarsi) and the other to birds (Ornithodira) (Gauthier 1986; Sereno 1991). The key character for the basal dichotomy of archosaurs seems to be the different style of ankle joint. The Crurotarsi includes phytosaurs, ornithosuchids, aetosaurs, poposaurids, and crocodylomorphs, all of which have a crurotarsal ankle joint. In this group the astragalus and calcaneum articulate with each other by means of a peg-and-socket joint, allowing rotational movement between them. Functionally, the astragalus is part of the crus and the calcaneum is part of the pes (fig. 10.1B). The calcaneum is very large and bears a prominent heel or tuber at the back for the attachment of the gastrocnemius muscle. The heel acts as a lever for raising the metatarsals during locomotion. In this group, the foot is primitively designed and pentadactyl, with plantigrade pose (fig. 10.1C).

The Ornithodira includes pterosaurs, lagosuchids, ornithischians, sauropodomorphs, and theropods. In this assemblage, the mesotarsal ankle joint is developed between the proximal and distal rows of the tarsals. The calcaneum is highly reduced, lacks the heel, and is immovably attached or fused to the astragalus. Functionally, astragalus and calcaneum are part of the crus, while the distal tarsals are part of the pes. The astragalus is a mediolaterally elongated hemicylinder with an ascending process for locking of the tibia, a feature first shown by Huxley (1870) to be unique to theropods and birds (fig. 10.1D). The mesotarsal ankle joint is associated with improved posture and a digitigrade pes, with the heel raised off the ground, and the middle three metatarsals form a tridactyl pes (fig. 10.1E). The shortened contact with the ground and the permanent elevation of the sole led to cursorial locomotion.

THE PROBLEMS OF PHYLOGENETIC RECONSTRUCTION

The current debate on the origin of birds centers on a radical shift in the phyletic branching point of archosaurs: birds branched off from either the crurotarsal lineage or the ornithodiran lineage. The controversy partly reflects different methods and philosophies in reconstructing evolutionary history. The traditional approach is to trace evolutionary (ancestor-descendant) relationships. The search for the ancestor in a clade is an important task in evolutionary systematics, and from the ancestor various descendants can be derived. The time dimension is emphasized in this approach, and fossil taxa are arranged in a temporal sequence showing progressive evolution. The current trend, however, is to reconstruct the branching pattern of evolutionary history on the basis of synapomorphies. Unlike an evolutionary tree, a cladogram does not incorporate stratigraphic time. In this method, identification of the ancestor is difficult; however, one can recognize two sister taxa that shared the most recent common ancestor. The cladogram closely reflects the evolutionary sequence of the acquisition of novel characters. It tells us not only what happened in evolutionary history, but how and why. Cladistics is now widely accepted for phylogenetic analysis because it provides an explicit and testable hypothesis of biotic relationships. The cladis-

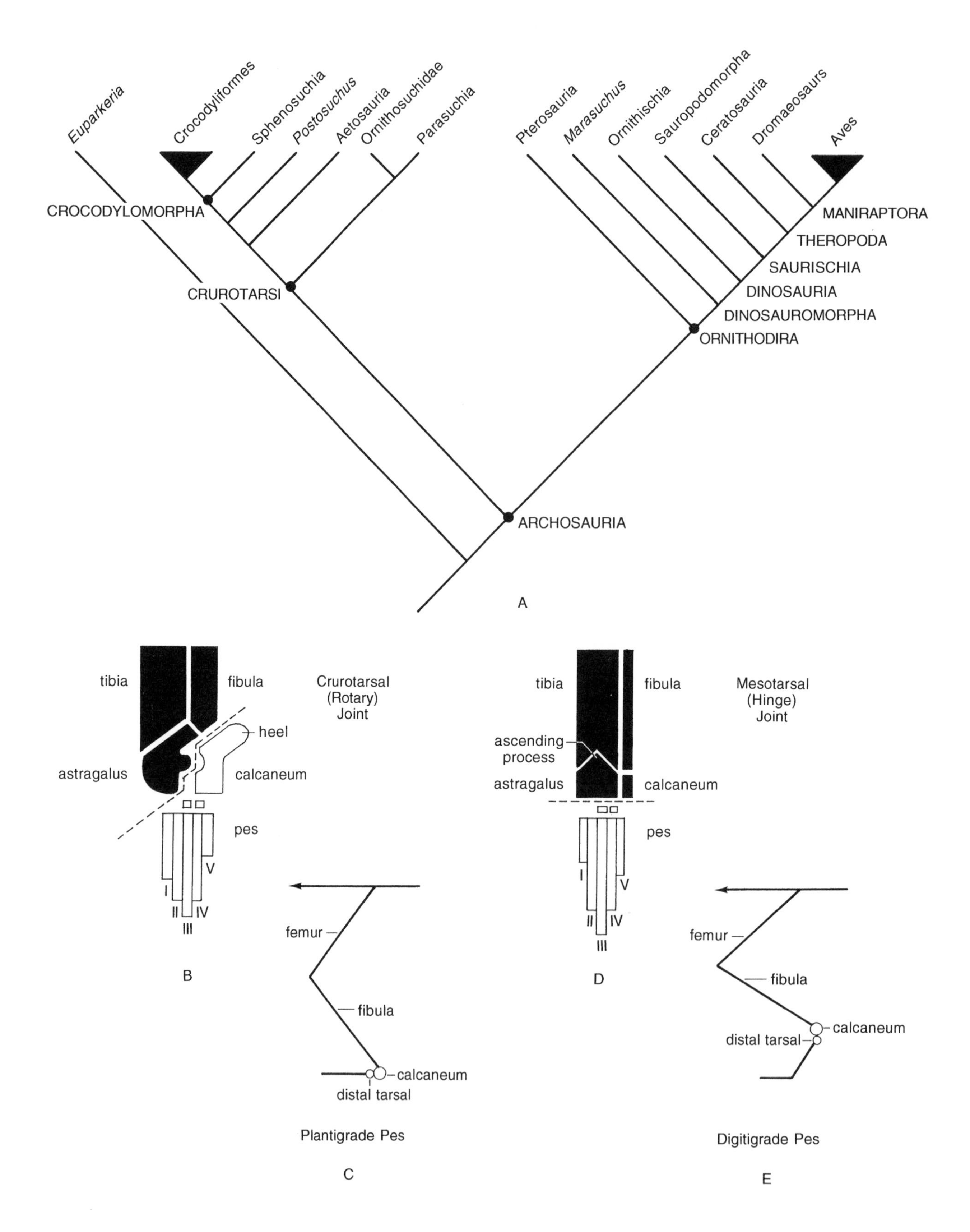
Euparkeria
Crocodyliformes
Sphenosuchia
Postosuchus
Aetosauria
Ornithosuchidae
Parasuchia
Pterosauria
Marasuchus
Ornithischia
Sauropodomorpha
Ceratosauria
Dromaeosaurs
Aves
CROCODYLOMORPHA
CRUROTARSI
MANIRAPTORA
THEROPODA
SAURISCHIA
DINOSAURIA
DINOSAUROMORPHA
ORNITHODIRA
ARCHOSAURIA
A
tibia
fibula
Crurotarsal
(Rotary)
Joint
heel
astragalus
calcaneum
pes
I
II
III
IV
V
B
femur
fibula
calcaneum
distal tarsal
Plantigrade Pes
C
tibia
fibula
Mesotarsal
(Hinge)
Joint
ascending
process
astragalus
calcaneum
pes
I
II
III
IV
V
D
femur
fibula
calcaneum
distal tarsal
Digitigrade Pes
E

tic method is followed here to trace the origin and early evolution of birds. Our primary task is to reconstruct past evolutionary history that happened millions of years ago. Since there is only one history, there is only one true phylogenetic tree, which we must infer from the few clues left in the fossil record. With this tantalizing evidence we try to build hypotheses or models in a logical fashion, reconstructing the genealogy that can be tested. However, our models are tentative at best and will change as new evidence becomes available.

ARCHAEOPTERYX AND THE ANCESTRY OF BIRDS

Comparative anatomy shows that, although *Protoavis* is the oldest bird, *Archaeopteryx* is the most primitive taxon, linked more closely to its archosaurian heritage. *Archaeopteryx* is, of course, more dromaeosaur-like than any living bird. The origin of birds is often associated with the origin of *Archaeopteryx* (Ostrom 1976a). In the Jurassic world, *Archaeopteryx* was an archaic relic that existed contemporaneously with more derived birds. There are three current hypotheses of the origins of birds (fig. 10.2).

THE THECODONT-BIRD HYPOTHESIS The Danish naturalist Gerhard Heilmann (1926) proposed that both birds and

FIGURE 10.1

The phylogeny of archosaurs. *A,* cladogram showing the postulated relationships of the basal dichotomy of archosaurs into Crurotarsi (leading to crocodilians) and Ornithodira (leading to birds) (source: modified from Sereno 1991). *B,* Crurotarsi is characterized by the crurotarsal ankle joint, in which the astragalus forms part of the crus, but the calcaneum, distal tarsals, and combined metatarsals move as a unit on the astragalus and fibula. The peg-and-socket joint between the astragalus and the calcaneum allows rotary motion between them. *C,* schematic lateral view of the hindlimb of the Crurotarsi showing the primitive, plantigrade pose. *D,* Ornithodira is characterized by the development of a hinge joint between the proximal and distal rows of the tarsi. The astragalus and calcaneum are fused together and form functionally part of the crus. The astragalus has a typical ascending process for locking the tibia; the calcaneum is reduced. *E,* schematic lateral view of the hindlimb of the Ornithodira showing improved posture and digitigrade pes.

dinosaurs derived from an ancestral thecodont such as *Euparkeria* (fig. 10.2A). A difficulty with the thecodontian hypothesis is that "Thecodontia" is now regarded as a paraphyletic, artificial grouping of distantly related forms, with some that seem to be closer to crocodylomorphs (Crurotarsi) and others that might be closer to birds (Ornithodira). *Euparkeria,* in particular, is currently considered as an outgroup of archosaurs, a lower branch in the phylogenetic tree that is far removed from the avian lineage. Thus, the thecodontian hypothesis has become less attractive with the development of cladistic phylogeny, which better expresses the interrelationships of different archosaur groups. A new version of thecodontian relationships was suggested by Tarsitano and Hecht (1980). They believe that *Archaeopteryx* evolved from a *Marasuchus*-like animal (fig. 10.2B). *Marasuchus* (*Lagosuchus*), a small, bipedal animal from the Middle Jurassic of Argentina, shows some birdlike proportions in the hindlimbs and has a mesotarsal ankle joint but does not share other synapomorphies with birds. Currently, *Marasuchus* is considered as an outgroup of dinosaurs and a distant relative of birds (Sereno and Arcucci 1993).

THE THEROPOD-BIRD HYPOTHESIS Thomas Huxley (1868b) first proposed the theropod-bird link on the basis of thirty-five shared characters. Huxley argued that Jurassic theropods such as *Compsognathus* show clear general skeletal affinities with *Archaeopteryx,* especially in the hindlimbs and girdles. In recent times, Huxley's theory of the relationship between theropods and birds was revived by John Ostrom (1976a, 1985a, 1991), who believed that birds evolved from a small, unknown coelurosaur (fig. 10.2C). Ostrom documented an impressive list of postcranial similarities between dromaeosaurs and *Archaeopteryx* that are not found in either "thecodonts" or crocodylomorphs. Such a striking resemblance between dromaeosaurs and *Archaeopteryx,* Ostrom argued, must reflect common descent, not convergence. However, Ostrom did not accept birds as living dinosaurs. Subsequent cladistic analyses provided the empirical evidence that birds are not only descended from theropod dinosaurs, but also nested within Theropoda (fig. 10.2E) (Gauthier and Padian 1985; Gauthier 1986; Holtz 1994).

The evolution of birds from dromaeosaur-like ancestors is now widely accepted among cladistic paleontologists. Tantalizing gaps in the fossil record complicate our understanding about the relationship between dromaeosaurs and birds. Both groups show skeletal differences that must have developed after they diverted from a common stock during the Middle Triassic. The oldest dromaeosaurs are known from the Cretaceous sediments, when birds were already diversified. Known dromaeo-

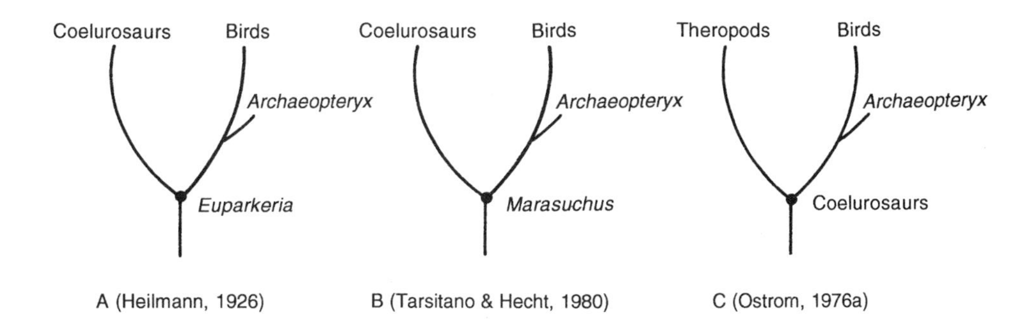

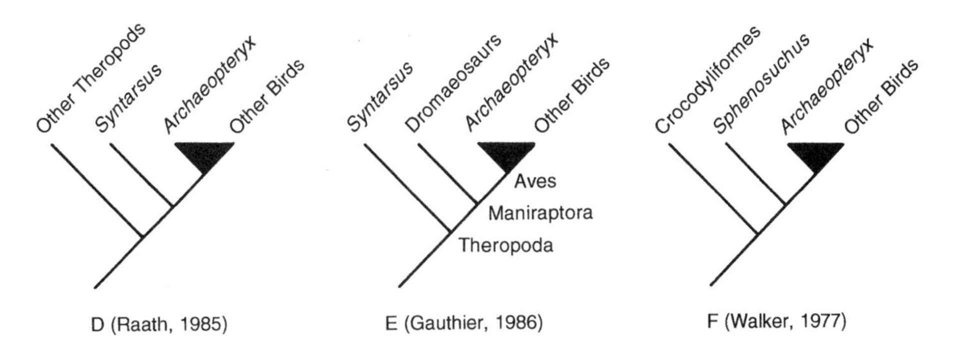

FIGURE 10.2

Various hypotheses regarding the origin of birds. *A–B,* thecodontian hypothesis; *C–E,* theropod hypothesis; *F,* crocodilian hypothesis.

saurs appear too high in the stratigraphic record to be directly ancestral to birds. In structure, however, they are the ideal avian precursors. Every paleontologist is familiar with this paradox, in which an ancestral species clearly primitive in morphology appears stratigraphically above its descendants. This anomaly shows how paleontological grandfathers may coexist with their descendants. This discordance between phylogenetic position and stratigraphic occurrence reflects the paucity of the fossil record. The chances for the preservation of small, climbing proavians in the fossil record are exceedingly slim. We have yet to discover in the Triassic sediments a small protodromaeosaur that diverged before the origin of birds. The recent finds of an early ostrich dinosaur (*Shuvosaurus*) and a little, new theropod from the Post quarry are encouraging, indicating the coexistence of coelurosaurs with early birds.

To overcome this stratigraphic discordance, Raath (1985) proposed an alternative version of the theropod hypothesis. He suggested that *Archaeopteryx* is more closely related to basal theropods such as *Syntarsus* than to Cretaceous dromaeosaurs (fig. 10.2D). Since *Syntarsus* is known from the Triassic-Jurassic boundary, its temporal gap from early birds is minimal. However, the few synapomorphies cited by Raath between *Syntarsus* and birds are not unique; they are also reported in a variety of archosaurs (Chatterjee 1991). If we accept Raath's view, birds must have split away long before the dromaeosaurs achieved their birdlike form. To endorse this view, one would have to accept a very large amount of convergent evolution between dromaeosaurs and early birds. This proposal seems to be unlikely at present.

THE CROCODILE-BIRD HYPOTHESIS Alick Walker (1972, 1974, 1977) proposed a modified version of the thecodontian hypothesis. He suggested that birds are more closely related to crocodiles than to other groups of archosaurs (fig. 10.2F). He derived his hypothesis from *Sphenosuchus*, a primitive crocodylomorph from the Early Jurassic of South Africa that possessed many avian features in the skull, especially in the otic region. Some of these shared derived characters include a tympanic recess, a fenestra pseudorotunda, a bony eustachian tube, elongated cochlear recesses, the loss of the descending process of the squamosal, and quadrate-prootic contact. Walker concluded that both crocodiles and birds were derived from a common ancestor among thecodontians. This view was endorsed by Martin (1985, 1991), with additional evidence from dental morphology. However, Walker (1985) later rejected his own hypothesis and described these similarities between sphenosuchians and birds as superficial or convergent. The small brain size, primitive configuration of the brain, monimostylic quadrate, and akinetic nature of the skull of sphenosuchians provide the most serious argument against the crocodilian hypothesis. Many of the similarities shared by the crocodylomorphs and birds, especially in the ear region, are analogous, indicating similar functional requirements, perhaps associated with vocalization (Chatterjee 1991).

THE PHYLOGENETIC DEFINITION OF A BIRD

Before assessing the systematic positions of primitive birds, such as *Archaeopteryx* and *Protoavis,* we must consider exactly what we mean by the term *bird.* To most people a bird means a feathered, flying vertebrate. Modern birds possess feathers, wings, a distinctive bill, endothermic physiology, elaborate respiratory and circulatory systems, extensive air sacs throughout the body, a vocal organ or syringe, a four-chambered heart, and the lack of a urinary bladder and teeth. Birds produce large external eggs and have developed parental care, a mating system, nesting behavior, territoriality, and vocal ability for communication. Such a wide array of characters is of little value to a paleontologist, who identifies fossils on the basis of skeletons. Accordingly, an osteological diagnosis of birds applicable to fossils is needed. This diagnosis is not easy, however, because the boundary between the early birds and their theropod ancestors is diffuse, making the distinction less than clear-cut.

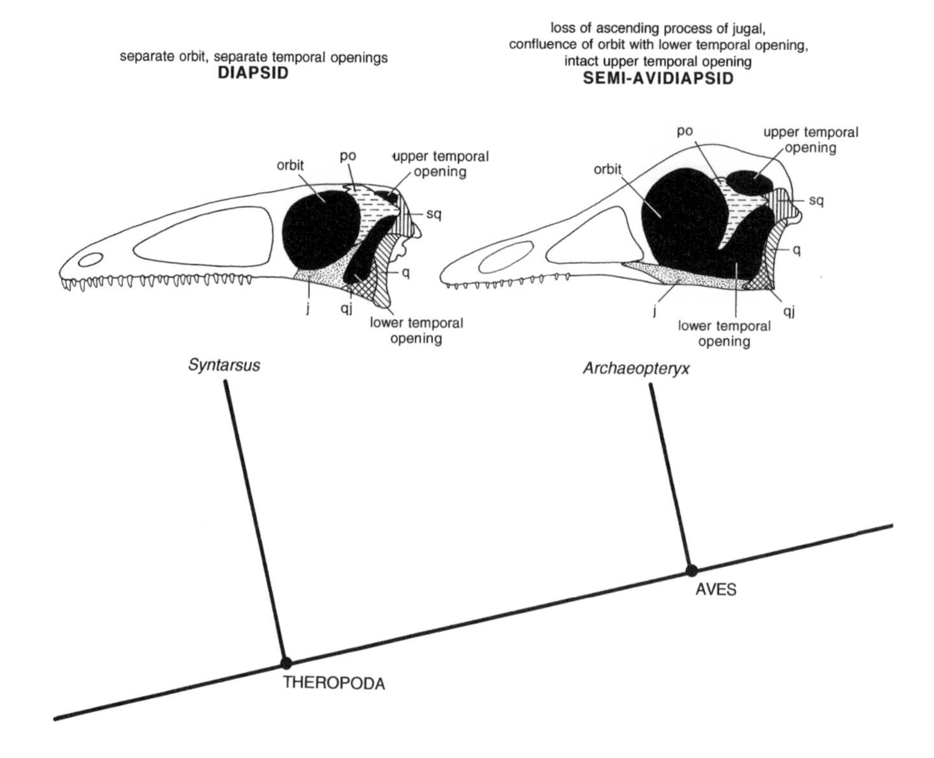

FIGURE 10.3

Lateral views of the theropod skull to show the evolution of the avian temporal configuration. In basal theropods, such as *Syntarsus,* the temporal arches are intact, as in other diapsids; however, the squamosal has lost contact with the quadratojugal in front of the quadrate. In *Archaeopteryx,* a partial modification of the temporal configuration can be seen, with loss of the ascending process of the jugal so that the orbit becomes confluent with the lower temporal opening. This confluence of the orbit with the lower temporal opening is regarded as an important character for defining birds; this type of temporal configuration is termed *semi-avidiapsid*. In *Protoavis* and other birds, the squamosal-quadratojugal bar is further eliminated to free the quadrate. The postorbital bone is lost so that orbit becomes confluent with both upper and lower temporal openings, as in modern birds. This type of temporal configuration is termed *avidiapsid* and is found in all modern birds. Two important avian landmarks are visible with the loss of the postorbital bone—the postorbital process and the zygomatic process; the quadrate develops the orbital process to become streptostylic; in ornithurine birds, the quadrate head is enlarged transversely to contact the prootic medially. The modification of the temporal region in birds is critically linked to the development of streptostyly and cranial kinesis.

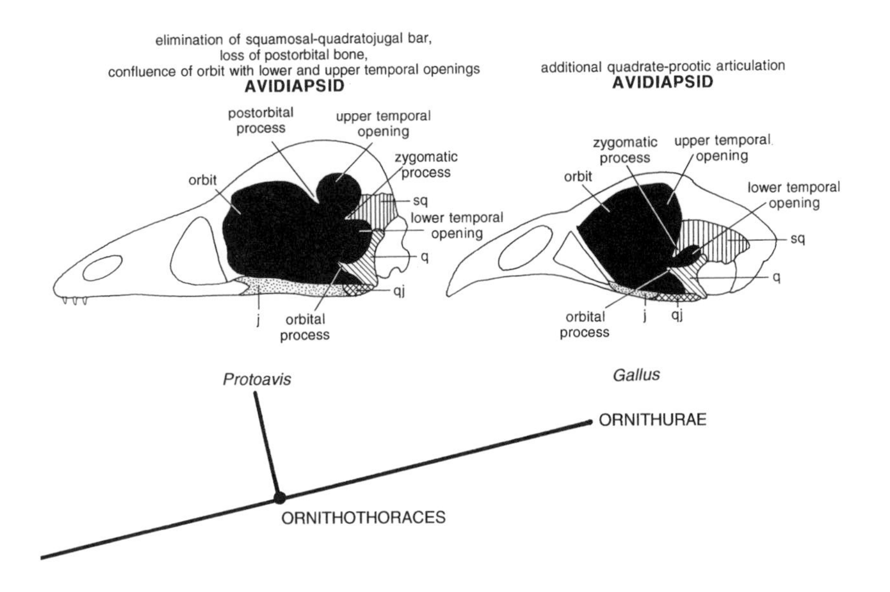

Traditionally, reptiles are classified by the position and number of temporal openings that distinguish the four major groups: anapsid, synapsid, parapsid, and diapsid. Birds evolved from the diapsid archosaurs with the breakdown of two vertical struts in the temporal region: the postorbital-jugal bar and the squamosal-quadratojugal bar (Chatterjee 1991). The elimination of these struts is essential to the mobility of the quadrate and the jugal bar. It seems appropriate that modification of the diapsid temporal opening can further be extended to define birds. In nonavian theropods, such as *Syntarsus*, the diapsid arches are intact. However, the descending process of the squamosal has lost contact with the ascending process of the quadratojugal (fig. 10.3). The squamosal-quadratojugal bar forms a blocking device to prevent forward movement of the quadrate. This type of an intact temporal diapsid pattern is retained in successive theropod clades: tetanurans, ornithomimosaurs, and dromaeosaurs. In basal birds, such as *Archaeopteryx*, we see the beginnings of the modification of the temporal configuration toward the avian lineage. Here, the orbit becomes confluent with the lower temporal opening, with reduction of the ascending process of the jugal. This transitional stage of temporal configuration is termed *semi-avidiapsid* and can be used as an important character to define a bird. In *Protoavis* and ornithurine birds, the squamosal-quadratojugal bar is totally eliminated to free the quadrate. The postorbital bone is lost so that the orbit becomes confluent with both the upper and the lower temporal openings. This condition is termed *avidiapsid.* In this stage, the quadrate develops an orbital process and becomes streptostylic. In ornithurine birds, the head of the quadrate is further modified with additional medial contact with the prootic. The evolution of the temporal configuration in birds is critically linked with the development of streptostyly and cranial kinesis. All modern birds exhibit this avidiapsid condition.

The definition of birds solely on the basis of a modified temporal configuration, although very practical and appealing, has one drawback. Intact skulls of Mesozoic birds are rare. However, the appearance of an orbital process in the quadrate is another important avian feature that can be easily observed. If this criterion is used to define a bird, then *Archaeopteryx* and some basal taxa would be excluded from birds, but *Protoavis* would be included. We need additional postcranial features to define birds when cranial material is missing. Although modern birds possess highly derived postcranial skeletons and can easily be diagnosed, the recognition of basal birds on the basis of postcranial features is not easy. Many of the avian attributes evolved gradually in a mosaic fashion in the skeletons of Mesozoic birds as flight was refined and perfected.

The traditional definition of birds has been based upon the presence of feathers and flight apparatus. If feathers are used to define a bird, *Archaeopteryx* is certainly a bird. Since feathers are rarely preserved in the fossils, their utility in diagnosis is diminished. Moreover, many birds became secondarily flightless, and many flight adaptations in the skeleton were reversed. The definition of a bird is as complex as the definition of a mammal. Where do we draw a boundary between true birds and birdlike forms? Birds can be defined in several different ways, either emphasizing the possession of certain characters (character-based definition) or considering common ancestry (stem-, node-, and apomorphy-based definition). Character-based definitions are problematic. More and more basal birds, with new combinations of characters, have been discovered recently. It is difficult to determine whether these are all birds by following a strictly character-based definition. A more reasonable solution is to use a phylogenetic definition—that is, one based on common ancestry.

It is currently believed that birds form a coherent, monophyletic group, or clade, which evolved from a common theropod ancestor. To justify this conclusion, we must provide a phylogenetic definition of birds on the basis of a set of derived characters. At present, two conflicting definitions of Aves are in use. Traditionally, the clade consists of *Archaeopteryx* and all later evolved birds. In this definition of Aves, we shall include those characters that are determinable in *Archaeopteryx*. However, Gauthier (1986) proposed that Aves should include only living groups of birds and all of the descendants of their common ancestors (the crown group concept of Hennig 1966). This new composition of Aves has met with some criticism because it excludes virtually all of the Mesozoic birds but includes the Cenozoic taxa (Cracraft 1986). To accept Gauthier's proposal means that we have to delete the first two-thirds of avian history. The exclusion of widely accepted avian taxa, such as Enantiornithes, Ichthyornithiformes, and Hesperornithiformes, is unwarranted. There are other problems, too. How do we recognize the most recent common ancestor of Holocene birds? Thus, Gauthier's definition of Aves does not offer a practical solution to the problem. The traditional concept of Aves, including *Archaeopteryx* as a basal member, is more inclusive, pragmatic, and informative and is currently widely used (Chiappe 1995a).

Archaeopteryx achieved a level of structural organization well advanced beyond that of any dromaeosaurs and shows the following avian attributes: feathers; the absence of a prefrontal bone; the semi-avidiapsid condition, in which the orbit is confluent with the lower temporal fenestra; small olfactory lobes; a cerebellar fossa extending to the supraoccipital; the presence of caudal maxillary sinus; the absence of the descending process of the squamosal; fewer than twenty-five caudal vertebrae; rotation of the scapulocoracoid so that the glenoid faces laterally; a flexible scapulocoracoid joint with an acute angle of articulation; ischia separated distally on the midline; a pectineal process on the pubis; a femoral/tibiotarsal ratio of less than 0.80; the attachment of metatarsal I on the distal quarter of metatarsal II; and an anisodactyl pes.

THE ACQUISITION OF AVIAN CHARACTERS

The body plan of birds has been greatly influenced by their ancestry, method of feeding and reproduction, and dual mode of locomotion. Many of the features that we think are birdlike actually appeared in the nonavian theropod ancestors of birds. Since birds are members of the theropod group, we will use the theropod cladogram of Gauthier (1986) to trace the sequence of acquisition of avian characters in a phylogenetic context leading to *Archaeopteryx* (fig. 10.4). Five terminal taxa of theropods were chosen for phylogenetic analysis. These taxa (Ceratosauria, Allosauria, Ornithomimosauria, Dromaeosauria, and *Archaeopteryx*) are defined and discussed below. Character data sets are compiled from Gauthier 1986 and Sereno et al. 1993.

THE CERATOSAUR LEVEL (THEROPODA NODE) The special posture and gait of birds—the obligate bipedality—has already developed in this stage, the Theropoda node, as seen in *Coelophysis, Syntarsus,* and later ceratosaurs (fig. 10.5A). The long, lightweight skeleton is cantilevered over the powerful hindlimbs. The ilium has a low, convex dorsal border and a long preacetabular process. The ischium and pubis are elongate and rodlike. The acetabulum is fully perforated, deep, and cylindrical, with a strong roof to resist the thrust from the head of the femur (see fig. 10.10). The femur and tibia are strong bones; the fibula is relatively slender and is closely attached to the tibia at the cnemial crest. The limbs and centra are hollow. The astragalus and calcaneum are fused to form a simple hinge with the mesotarsal joint. The astragalus has an ascending process that fits into the lower end of the tibia. The foot is exceedingly bird-like and digitigrade, with the middle three metatarsals complete and forming a tridactyl pes. The fifth metatarsal is reduced to a splint and lacks any digit. The first metacarpal is shortened, attached about halfway down metatarsal II; the pedal phalangeal formula is 2-3-4-5-0 (fig. 10.11).

The ceratosaur skull is long, pointed, and lightly built, with sharp teeth lining the jaws. The orbits are large and face sideways (see fig. 10.3). The lacrimal has a broad exposure on the skull roof. The quadrate head has moved considerably forward and fits into a separate squamosal socket without any paroccipital contact. The quadrate has developed opisthostylic mobility

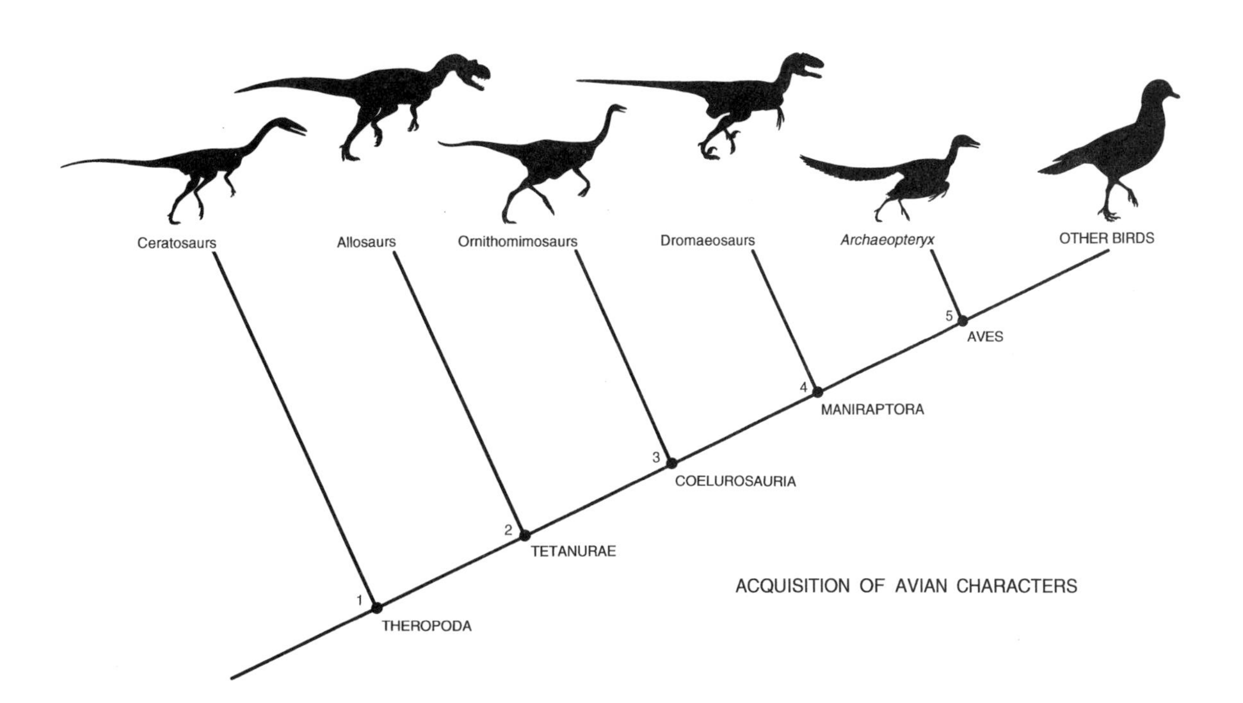

FIGURE 10.4

Cladogram of the major groups of Theropoda (source: simplified from Gauthier 1986); the skeletal structures of each major group are shown separately in figure 10.5. These simplified cladistic relationships of theropods will be used in later figures to show the sequence of acquisition of avian characters.

FIGURE 10.5

The skeletal features of selected taxa of theropods from the previous cladogram. *A,* ceratosaur (*Coelophysis*); *B,* allosaur (*Allosaurus*); *C,* ornithomimosaur (*Ornithomimus*); *D,* dromaeosaur (*Velociraptor*); *E, Archaeopteryx; F,* modern bird (*Columba*). Bipedalism and an erect posture are characteristics of theropods. There is a progressive change in the forelimb and shoulder girdle to form the flight apparatus. In nonavian theropods, the tail is fairly long and the caudal vertebral count ranges from forty to fifty. The long tail indicates that the caudofemoral muscle, the main femoral retractor, was powerful in these groups. Rotation of the femur at the acetabulum was the primary mechanism of hindlimb locomotion. In dromaeosaurs, the tail was further strengthened with the development of ossified tendons to form a stiff, dynamic stabilizer (Ostrom 1979). In *Archaeopteryx,* the tail is truncated considerably and shows twenty to twenty-three vertebrae; truncation of the tail may be linked to reduction of the caudofemoral musculature (Gatesy 1990) so that the primary mechanism of hindlimb locomotion shifted from hip to knee joint. In later birds (*Columba*), the tail is further reduced to fewer than fifteen vertebrae with the development of a pygostyle; the tail is completely decoupled from hindlimb locomotion.

so that the lower jaw can slide back and forth during capturing and eating prey. There is an intramandibular joint between the dentary and postdentary bones. The quadratojugal is L-shaped. The palate is vaulted; the choana has been shifted backward so that the palatal processes of the maxilla form a false palate rostral to it. The two vomers are fused to meet the premaxillae. The otic capsule of the braincase is primitively built; the metotic foramen occurs just behind the fenestra ovalis (fig. 10.7). The brain is small and arranged in the reptilian fashion, in which the optic lobes lie behind the cerebrum in the dorsal aspect (fig. 10.6).

The ceratosaur neck is long and slender, with a distinctive S-shaped curvature. The trunk is short, the hip is strengthened by five coalesced sacral vertebrae, and the tail is fairly long and slender to counterbalance the neck and head (fig. 10.5). The scapula is a narrow, vertical bone with a flaring at its dorsal edge, and the coracoid is small and fused to the scapula (fig. 10.8). *Segisaurus* had separate but paired clavicles. The forelimbs are very short and played no role in locomotion. The humerus has a prominent deltopectoral crest. In the wrist, there are two or three proximal carpals; distal carpals 1 and 2 are fused into a compound bone (fig. 10.9), permitting a wide range of hand motion. Hand length equals radius length. There are four digits in the hand, but the fourth digit is vestigial, consisting of the metacarpal and a rudimentary proximal phalanx. The fifth digit is lost, so that the manual phalangeal formula becomes 2-3-4-1-x.

THE ALLOSAUR LEVEL (TETANURAE NODE) The Tetanurae node includes allosaurs and other theropods. The teeth are lost at the back of the jaw so that the tooth row ends rostral to the orbit (fig. 10.5B). The maxillary fenestra is developed within the antorbital fenestra. In the otic region, the beginning of the development of the metotic process can be seen. As a result, the vagus foramen has been diverted backward, tunneling through the metotic process and emerging at the occiput near

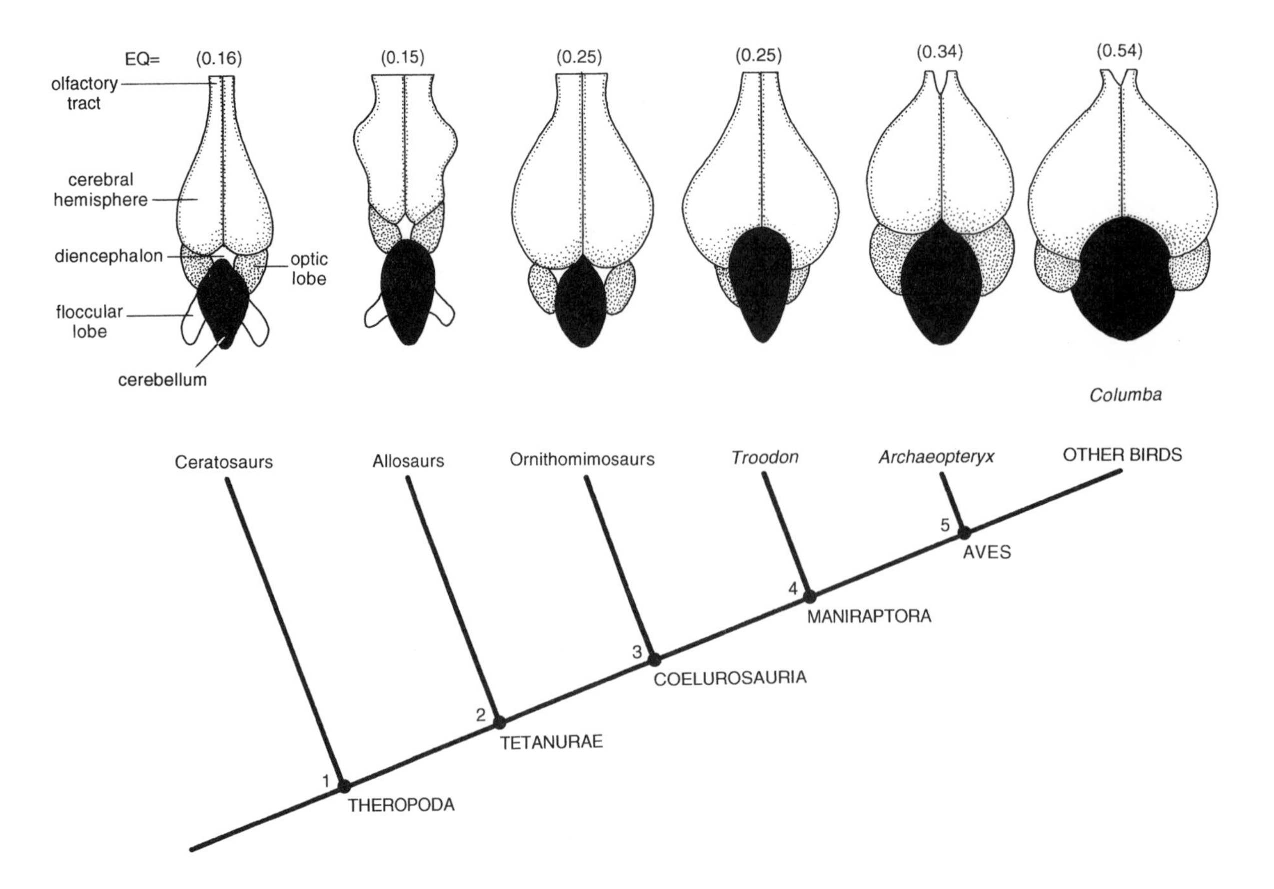

FIGURE 10.6

Diagrammatic dorsal views of the brains of major groups of theropods to show the evolution of the avian condition in a cladistic relationship. There is a progressive increase of the encephalization quotient (EQ) from 0.16 to 0.54 as the brain becomes more and more enlarged. The EQ is generally high in predatory birds and attains 1.6 in crows (*Corvus*) (Chatterjee 1991). In ceratosaurs (*Syntarsus*) and allosaurs (*Allosaurus*), the brain was small and primitively designed, with optic lobes immediately behind the cerebral hemispheres. In ornithomimosaurs (*Dromiceiomimus*), the brain was enlarged in the avian fashion, and the optic lobes were displaced laterally and somewhat ventrally. This trend toward progressive brain enlargement continued in maniraptorans (*Troodon*), *Archaeopteryx*, and later birds (*Columba*).

Both ceratosaurs and allosaurs were fully terrestrial animals, like modern dogs, and lived in a two-dimensional world. The sudden enlargement of the brain in coelurosaurs may indicate their capability in an arboreal habitat, like modern cats, living in a three-dimensional world. In *Archaeopteryx* and later birds (*Columba*), the cerebellum becomes progressively enlarged for balance and coordination, the optic lobes are enlarged for visual acuity, and the olfactory bulbs are highly reduced, indicating less dependence on smell. The brains of various nonavian theropods were made from endocasts. (Source: *Syntarsus* after Raath 1977; *Archaeopteryx* and *Columba* after Bühler 1985.)

the hypoglossal foramen (fig. 10.7). At the occiput, there is a sinus canal that separates the supraoccipital from the epiotic (fig. 10.7, Tetanurae). A tubular cochlear recess houses the lagena. The caudal tympanic recess is present at the front of the paroccipital process in some taxa. The quadrate has developed parastylic motion for mandibular spreading. The scapula is straplike, whereas the posteroventral margin of the coracoid tapers. The hand is large and would be half the length of the radius and humerus. It has three fingers; the fourth and fifth have been lost. Metacarpal I is very short; the manual phalangeal formula is 2-3-4-x-x. The ischium bears an obturator process; the pubis shows an expanded foot. The tibia has a winglike anterior trochanter, whereas the astragalus shows a tall, broad ascending process in front of the tibia. The pes is four-toed, with a reduced metatarsal I; the phalangeal formula is 2-3-4-5-x.

THE ORNITHOMIMOSAUR LEVEL (COELUROSAURIA NODE) The Coelurosauria node is represented by ornithomimids and other large-brained theropods (fig. 10.5C). The long, slender, toothless jaw and large eyes give this skull a very birdlike appearance. The braincase is highly inflated, with an encephalization quotient of 0.25, which is in the lower range of living birds (fig. 10.6). The C-shaped bony tube around the floccular recess is well developed for the rostral vertical semicircular canal. The rostral tympanic recess is formed between the prootic and basisphenoid. The occipital condyle is smaller than the foramen magnum. The anterior cervical centra are broader than they are deep cranially, with kidney-shaped anterior surfaces. The cervical ribs are fused to the centra. The coracoid is ventrally elongated. Both the clavicles and sternal plates are fused. The forelimb is elongated and exceeds half the length of the hindlimb. The manus is long with a reduced metatarsal I; its phalangeal formula is 2-3-4-x-x. The hindlimbs are extremely long, powerful, and gracile, and the tibia is longer than the femur. The fourth trochanter is subdued, but the posterior trochanter is developed. The slender metatarsals are clamped

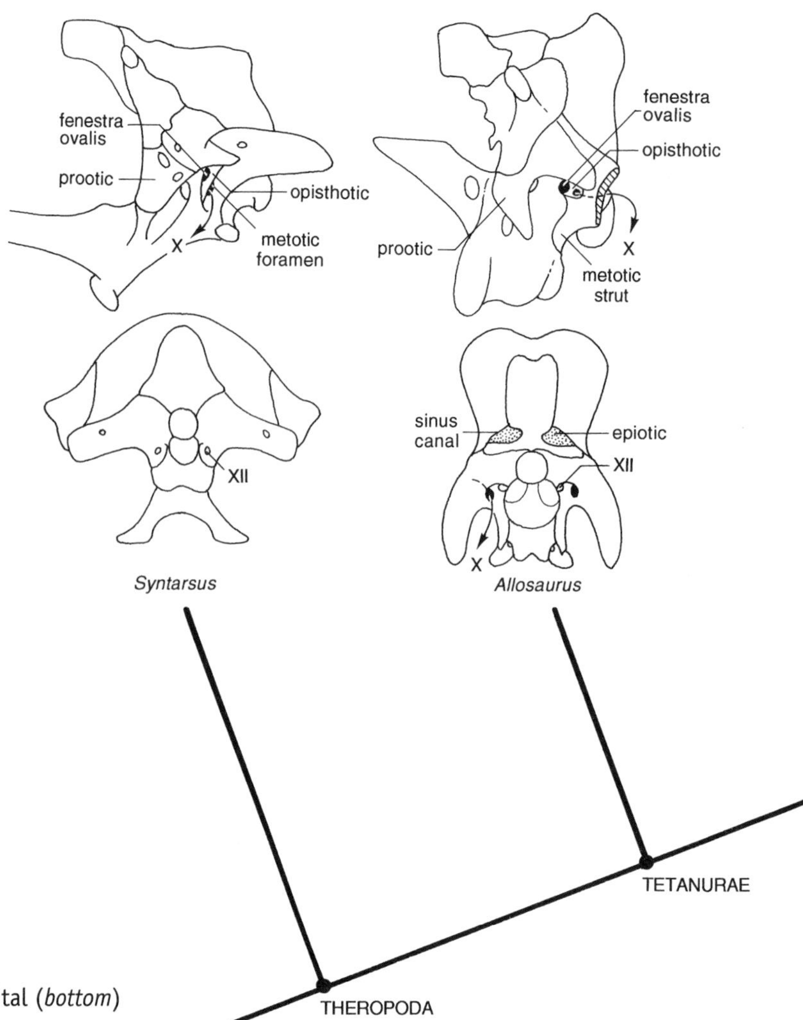

FIGURE 10.7

Diagrammatic lateral (*top*) and occipital (*bottom*) views of the braincase of major groups of theropods to show the evolution of the avian middle ear. In ceratosaurs (*Syntarsus*), the otic capsule is primitively built; it shows two large foramina on the lateral wall, separated by a stout bar of opisthotic. The rostral one is the fenestra ovalis, which receives the footplate of the stapes. The caudal one is the metotic foramen, which provides an exit for the IX–XI cranial nerves and possibly the internal jugular vein. In allosaurs (*Allosaurus*) and ornithomimosaurs (*Struthiomimus*), a subscapular cartilage, the metotic strut, is added to the exoccipital, thus enclosing the rostral part of the metotic foramen. As a result, the vagus (X) foramen in these groups has been diverted backward from the metotic foramen behind the metotic strut and emerges at the occiput, lateral to the hypoglossal (XII) foramen. In Maniraptora (*Dromaeosaurus*), with the elongation of the cochlea, the perilymphatic duct is shifted to a new aperture, the fenestra pseudorotunda, at the position of the metotic foramen. The vagus foramen has severed its lateral connection through the metotic strut and takes a shorter route; it is now directed medially to the endocranial cavity. The opisthotic is further reduced in *Archaeopteryx* and later birds (*Columba*) and becomes largely internal to form a slender bar.

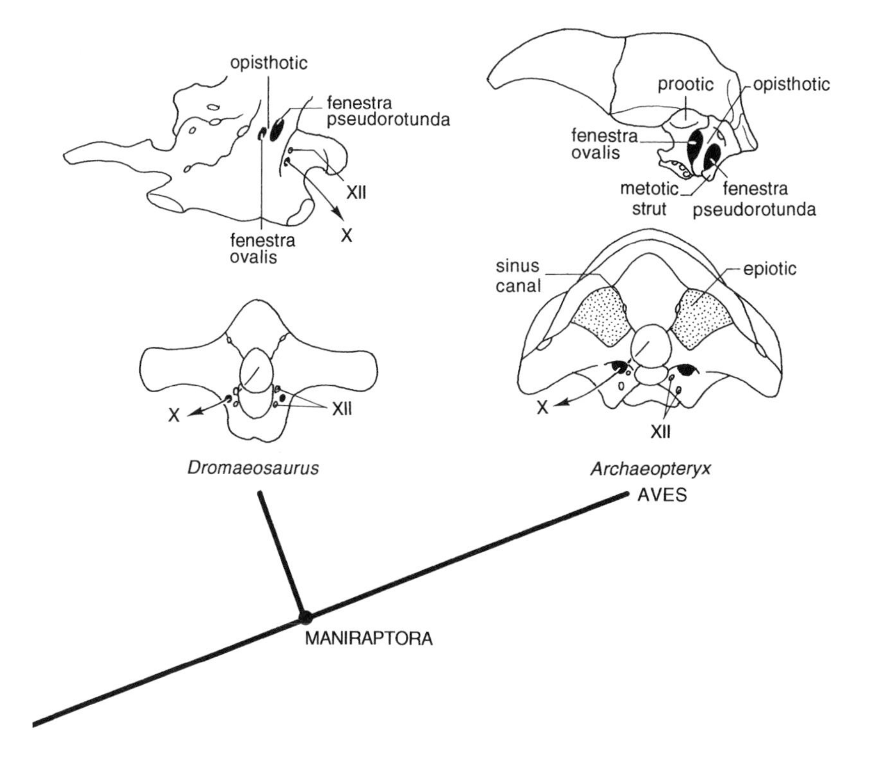

together, and the central metatarsal is pinched proximally. Both pedal and manual phalangeal formulas correspond well with those of *Archaeopteryx.*

THE DROMAEOSAUR LEVEL (MANIRAPTORA NODE)
This stage, the Maniraptora node, is exhibited by dromaeosaurs and other lightly built theropods with a stiff, dynamic tail, strengthened by ossified tendons (fig. 10.5D). In the skull, the prefrontal is reduced. The otic region is considerably modified in the avian fashion; a separate pseudorotunda can be seen behind the fenestra ovalis. The vagus foramen has severed its connection with the metotic foramen and takes a shorter route; it is now directed medially to the endocranial cavity (fig. 10.7, Maniraptora). The bony eustachian tube is developed. The dorsal tympanic recess is formed at the lateral surface of the prootic. The coracoid is ventrally elongated, with a subtriangular profile. It has developed the biceps tubercle on the lateral surface of the coracoid (fig. 10.8). The semilunate carpal has provided the swivel wrist joint (fig. 10.9) so that the hand can be folded tightly against the body in a Z fashion when not in

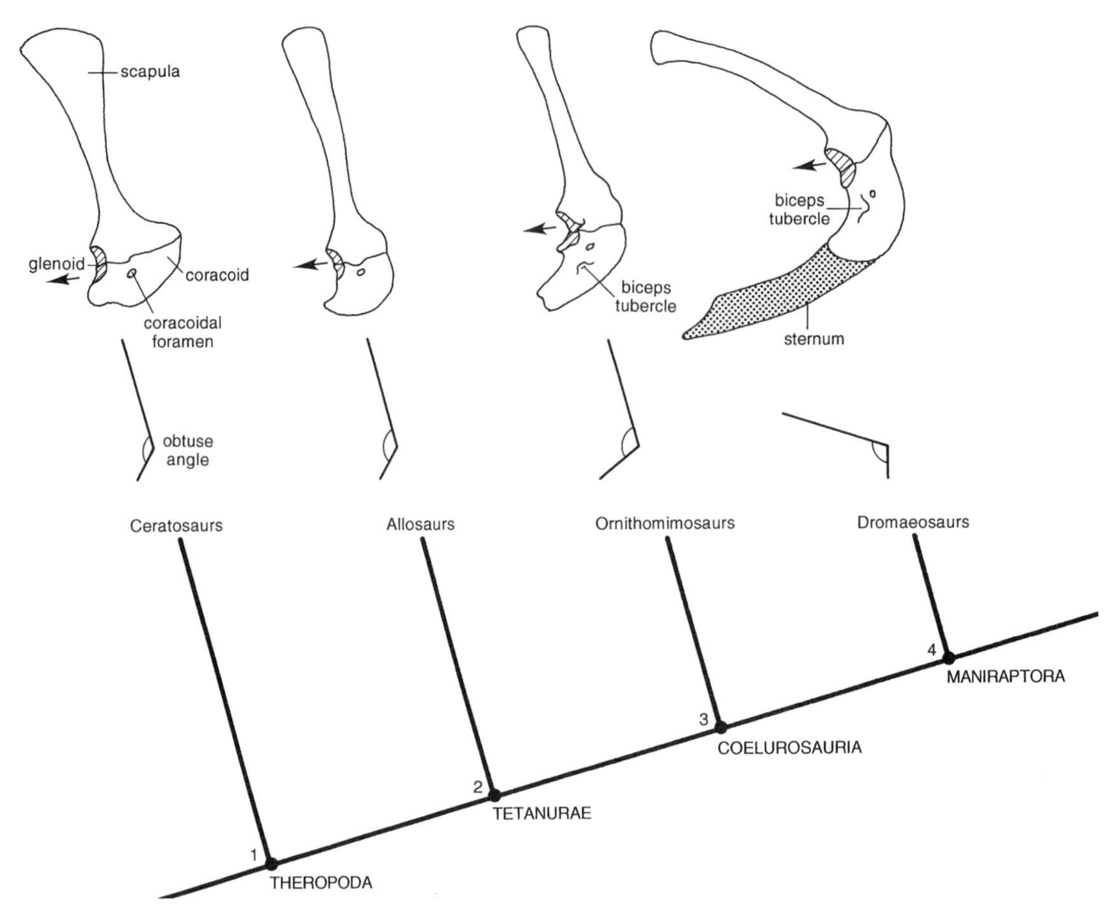

FIGURE 10.8

Right lateral diagrammatic view of the shoulder girdle of theropods to show evolution of the avian condition. The cladogram exhibits a progressive change in the angle of articulation between the scapula and the coracoid from ceratosaurs (*Syntarsus*), through allosaurs (*Allosaurus*), ornithomimosaurs (*Struthiomimus*), dromaeosaurs (*Deinonychus*), and *Archaeopteryx*, to modern birds (*Columba*). An obtuse angle of scapulocoracoid articulation is found in nonflying outgroups, but the acute angle is characteristic of flying birds. In this progression, the coracoid becomes more and more deep and the biceps tubercle migrates rostrally and dorsally to form the avian acrocoracoid process (Ostrom 1976b). The sternum is known in dromaeosaurs and might have evolved for climbing. Although the sternum was unossified in most specimens of *Archaeopteryx,* there is a small sternum, as was reported recently in *Archaeopteryx bavarica* (Wellnhofer 1993). A large sternal keel is characteristic of flying birds, with a furcula between the two shoulder girdles. The keel provides an area for the attachment of flight muscles. The distribution of the furcula among nonavian theropods is uncertain. It is certainly present in ceratosaurs such as *Segisaurus* (Camp 1936) and oviraptorosaurs (Barsbold 1983).

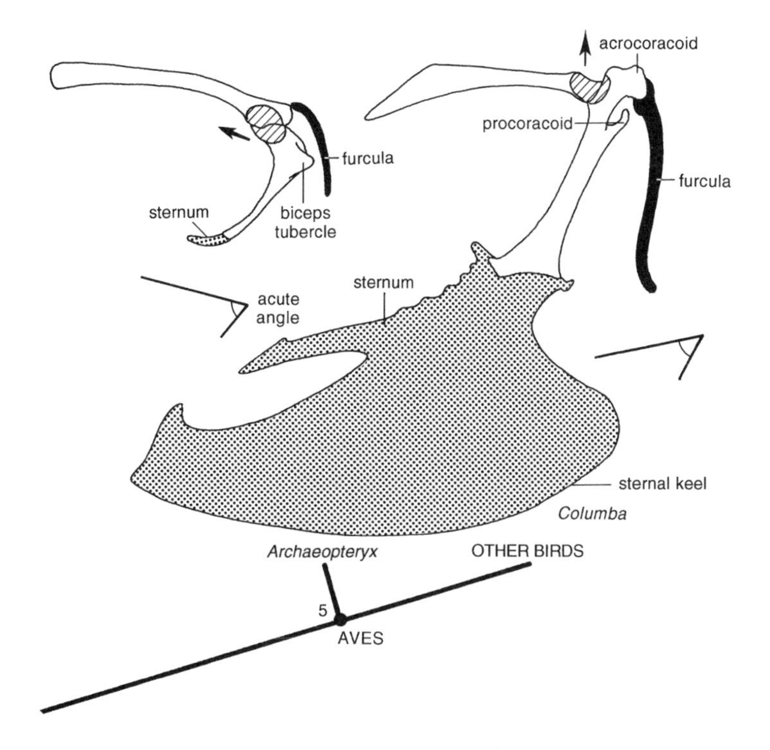

use. The anterior process of the the ilium is considerably enlarged and the pubis has rotated backward (fig. 10.10).

THE *ARCHAEOPTERYX* LEVEL (AVES NODE) The Aves node is represented by *Archaeopteryx* and other primitive birds with the development of wing and tail feathers. The overall size is considerably reduced for flight. Miniaturization seems to be an important evolutionary novelty in birds, leading to the loss of many bones. In the skull the prefrontal is lost. The ascending process of the jugal is reduced so that the orbit is confluent with the lower temporal fenestra. The squamosal is diminished, with the loss of the descending process (fig. 10.3). The brain is considerably enlarged where the cerebellum extends to the supraoccipital (fig. 10.6). The olfactory lobes are atrophied. The palate shows a caudal maxillary sinus. The tail is somewhat truncated, and the caudal vertebrae number fewer than twenty-five (fig. 10.5). The shoulder girdle is rotated so that the glenoid now faces laterally. The scapulocoracoid joint becomes separate and flexible. The coracoid is further elongated and sharply reflexed backward to make an acute angle with the scapula. The

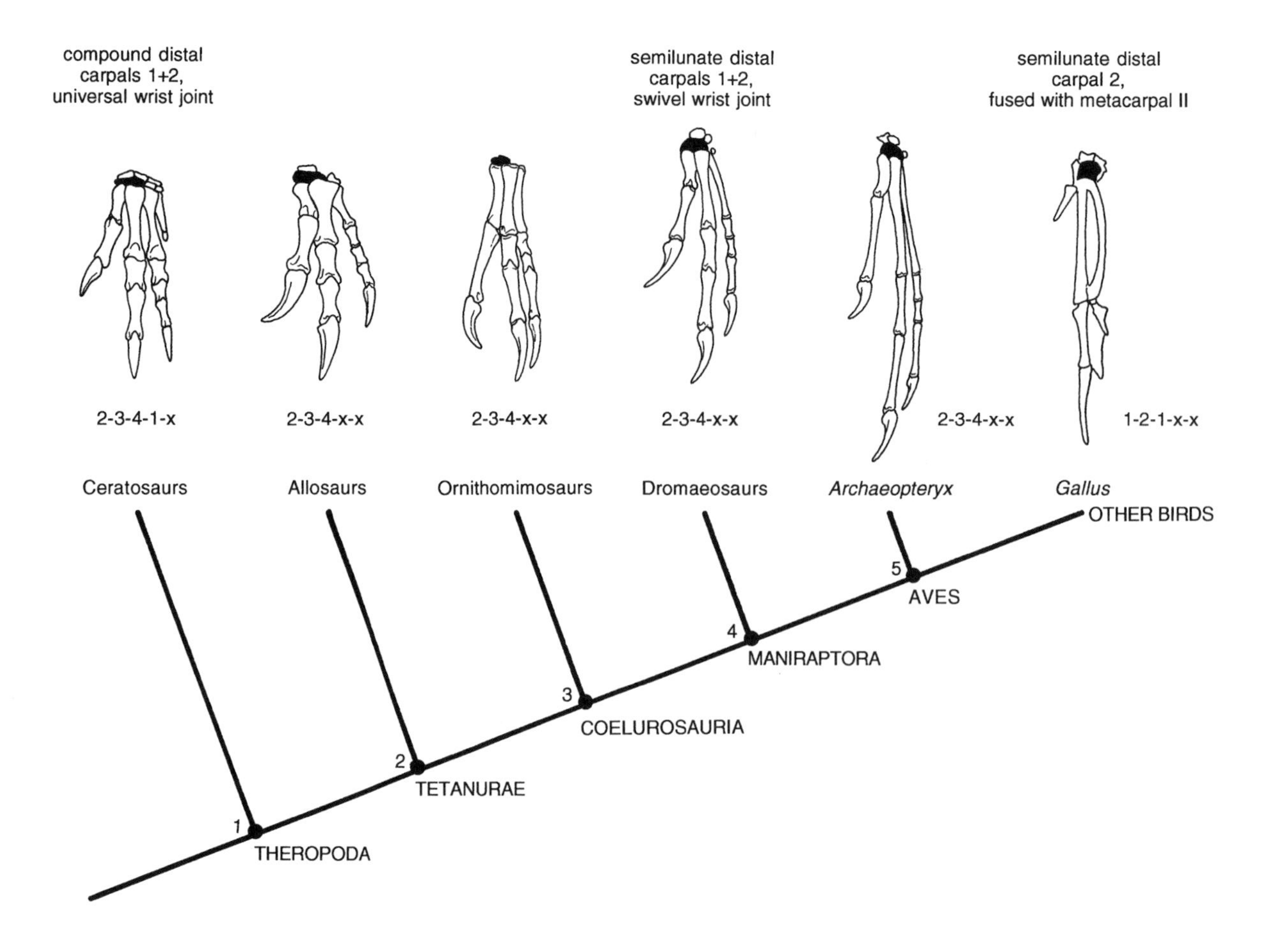

FIGURE 10.9

Diagrams of the left carpus and manus of theropods to show the evolution of the avian hand. The modification of distal carpals 1 and 2 (shown in *black*) is the key to the structure of the avian wing. In nonmaniraptoran theropods (such as ceratosaurs, allosaurs, and ornithomimosaurs), the carpal bones are small and numerous, permitting a wide range of motion (universal joint). In these groups, distal carpals 1 and 2 are fused into a compound bone that articulates distally with the first two metacarpals. In maniraptorans such as dromaeosaurs and *Archaeopteryx*, this compound bone develops a curved, rolling articular surface (the semilunate carpal of Ostrom 1979) for the proximal carpals, thus permitting a restricted movement of the hand in the plane of the forearm. This swivel wrist joint evolved initially for climbing trees and later was co-opted for flight. It allows folding of the manus along the side of the body when not in use during terrestrial locomotion. In later birds, the semilunate carpal consists of a single bone, distal carpal 2, which becomes fused with metacarpal II to form part of the carpometacarpus. The cladogram also indicates a progressive reduction in the number of digits and phalanges toward the avian condition. The phalangeal formula code is after Padian (1992); 0 indicates metapodials supporting no phalanges, and x indicates digits that are completely lost.

biceps tubercle is directed forward toward the glenoid (fig. 10.8). The forelimb is considerably longer than the hindlimb. The pubis has developed a pectineal process at its contact with the ilium. The ischium lacks the ventral symphysis, and this lack forms the large pelvic outlet. The femur is 80 percent of the length of tibia. Metatarsal I is attached more distal to metatarsal II; the pes becomes anisodactyl (fig. 10.11).

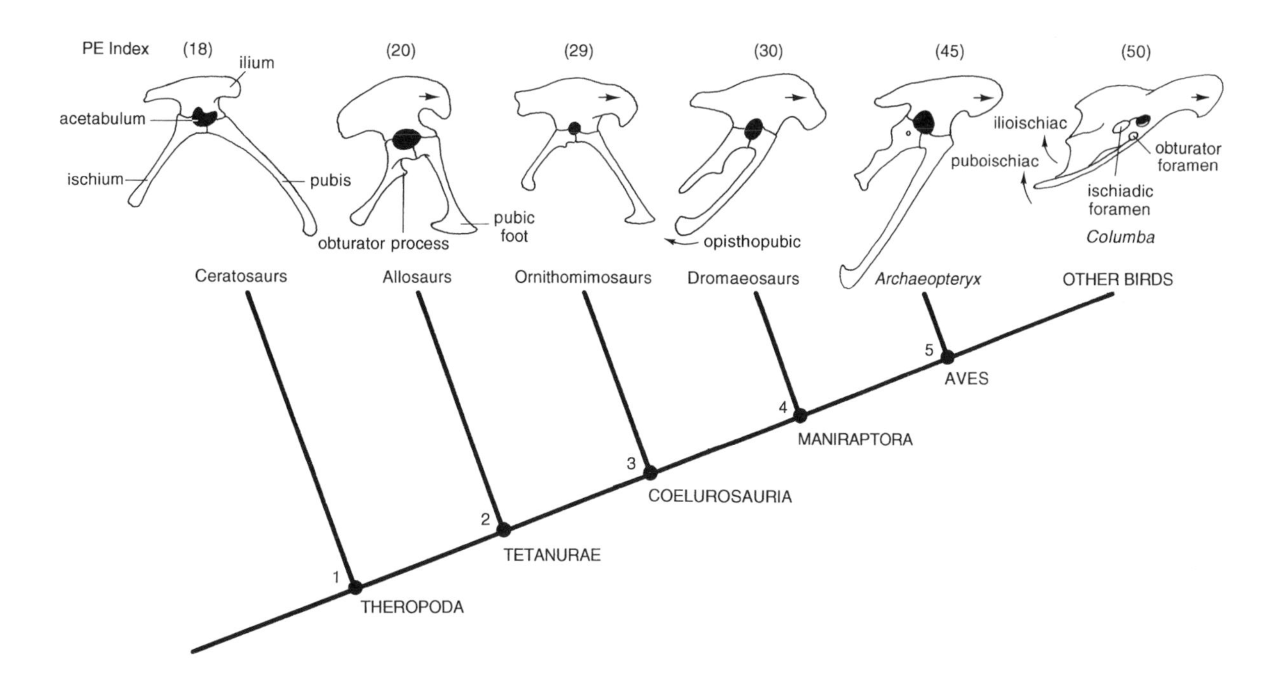

FIGURE 10.10

Right lateral views of the pelvic girdles of theropods to show the evolution of the avian hip. The cladogram shows the progressive enlargement of the preacetabular process, which is quantified in the preacetabular elongation (PE) index ([preacetabular length of ilium rostral to pubic peduncle × 100]/total length of ilium). The cladogram also depicts the development of the pubic foot in the allosaur stage and gradual rotation of the pubis in a backward direction (the opisthopubic condition) in the dromaeosaur (*Velociraptor*) stage. The backward rotation of the pubis may be linked to the climbing adaptation. In more derived birds, the ischium is also rotated backward to fuse with the ilium and enclose the ilioischiadic foramen.

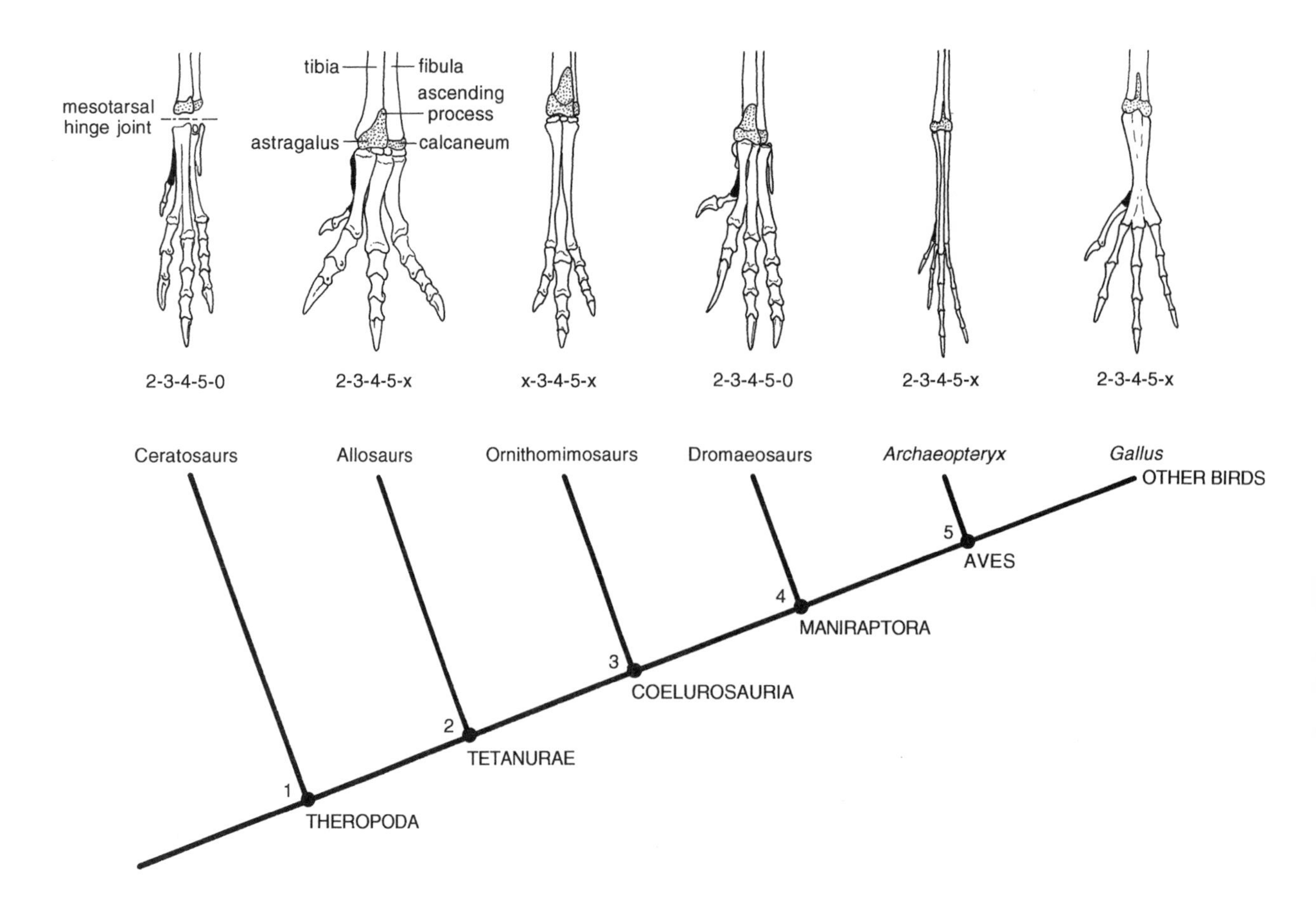

FIGURE 10.11

Diagrams of the left tarsus and pes to show the evolution of the avian foot. In theropods, the ankle joint is mesotarsal; the astragalus develops a characteristic ascending process in front of the tibia, a feature first observed by Huxley (1868). The ascending process provides extra support for locking the tibia with the astragalus. The progressive enlargement of the ascending process in the theropod lineage may be linked to improvement in the cursorial adaptation. In hesperornithiforms and neognathous birds, the ascending process ossifies as a separate element (the pretibial bone) in young individuals but becomes fused with the calcaneum in adults; in nonavian theropods and palaeognaths, on the other hand, the ascending process is not separate but is always fused with the astragalus. There is a progressive distal shift of metatarsal I from basal theropods to birds. In birds, metatarsal I (in *black*) is attached farther distal to metatarsal II, and the first digit becomes anisodactyl. The primitive phalangeal formula is retained in the first four digits in birds, but the fifth digit is lost.

The Early Evolution of Birds

Once we establish *Archaeopteryx* as the most primitive of known birds, our next task is to trace the phylogenetic relationships among Mesozoic birds. Recently, Cracraft (1986), Martin (1987, 1995b), Chiappe and Calvo (1994), and Chiappe (1995a) attempted to hypothesize the interrelationships of early birds. These data were further refined with the incorporation of *Protoavis* as an early member of this adaptive radiation (Chatterjee 1991, 1995, n.p.; Kurochkin 1995). Several stages of the early radiation of birds can be recognized (fig. 10.12).

THE STAGES IN THE EARLY RADIATION OF BIRDS

THE AVES NODE The Aves node is the basal stage of avian radiation, represented by *Archaeopteryx*. The acquisition of several avian characters at this stage was discussed in a preceding section (see "The *Archaeopteryx* Level").

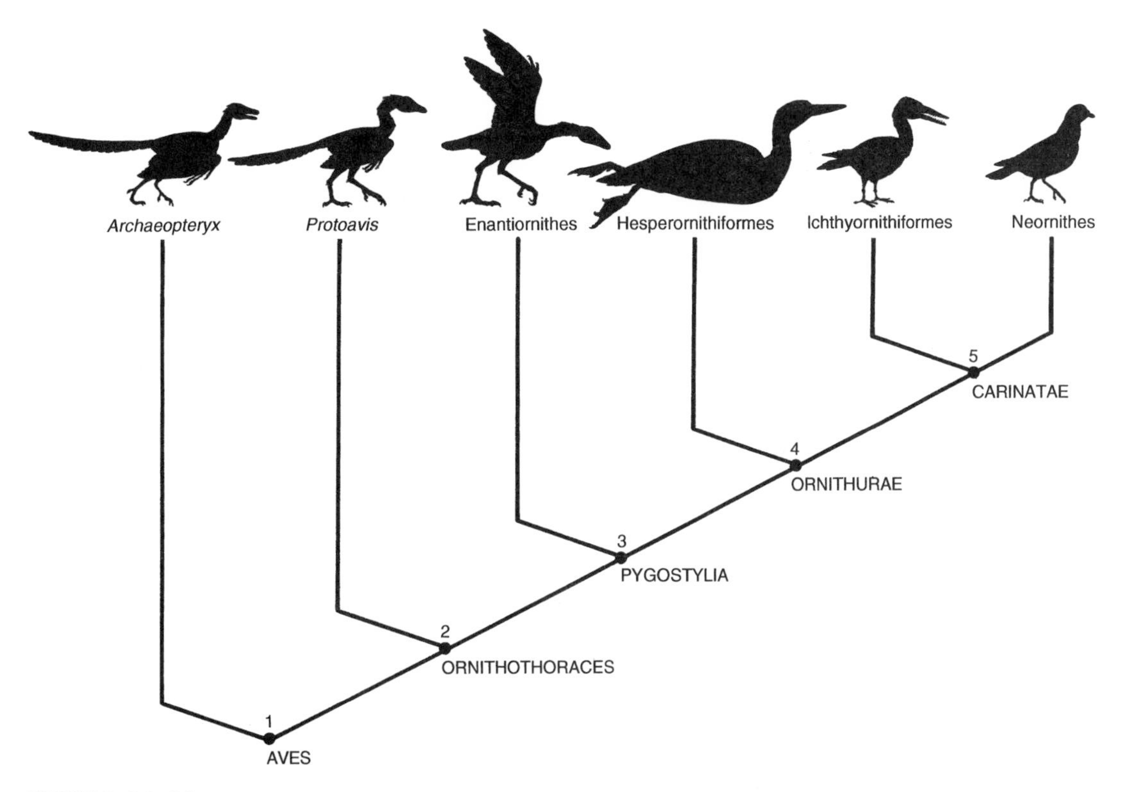

FIGURE 10.12

Cladogram of the major groups of Aves.

THE ORNITHOTHORACES NODE Chiappe and Calvo (1994) recognized the Ornithothoraces node, which demonstrates a flight apparatus designed for powered flight. This level is represented by *Protoavis* and other birds (Chatterjee 1995). At this stage, a large number of avian attributes are evolved (fig. 10.13). There are several modifications from the *Archaeopteryx* stage in the skull. The teeth are considerably reduced in number and are restricted to the tip of the jaws. The temporal configuration is further modified with loss of the postorbital bone and the ascending process of the jugal; the orbit has become confluent

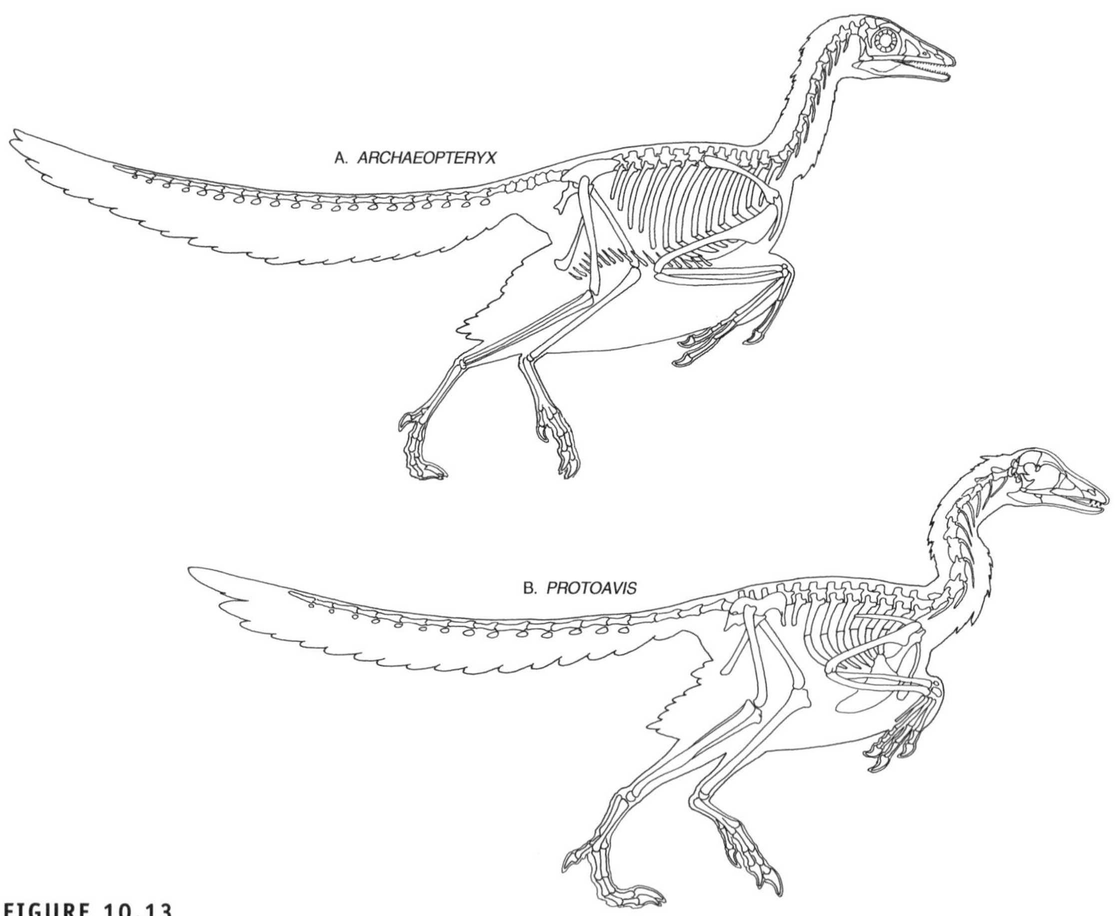

FIGURE 10.13

Anatomical comparisons between *Archaeopteryx* and *Protoavis*. Both animals were small, about the size of a crow, with a long, bony tail; however, *Protoavis* was more strongly built because of its predatory adaptation.

with the two temporal openings. The squamosal-quadratojugal bar in front of the quadrate is eliminated to make the quadrate mobile. The squamosal bears a distinctive zygomatic process. Along with streptostyly, the quadrate has developed several other avian attributes, such as the free orbital process, ventral condylar articulation with the pterygoid, and a lateral cotyle for the quadratojugal (fig. 10.14). The quadrate articulates with the mandible by means of three condyles. The tympanic diverticulum has not only invaded the braincase in the dorsal, rostral, and caudal regions, but also pneumatized the quadrate and articular bones. In the palate, the ectopterygoid is lost so that the pterygoid can move back and forth in response to streptostyly. There are several modifications in the cervical vertebrae: the beginning of a heterocoelous centra and the development of the hypapophysis; the neural spines in this region are considerably reduced.

The scapula shows a pneumatopore near the acromion. The coracoid is strutlike and elongated ventrally, with an expanded sternal end. It has developed the acrocoracoid process to form the triosseal canal for the supracoracoideus pulley in conjunction with the scapula and furcula. There is a distinctive procoracoid process. The furcula is springlike, with a large hypocleidium. The sternum is longitudinally oriented, with a distinct ventral keel for attachment of the flight muscles. The humerus shows a bicipital crest and a strongly defined proximal head. Distally, the condyles for the radius and ulna are asymmetrical and well developed. The metacarpals show quill knobs for attachment of the primary feathers. In the wrist, the radiale is fairly large, and the ulnare has an interlocking articulation with the ventral ridge of metacarpal III.

The ilium is fused with the ischium to enclose the ilioischiadic fenestra and renal fossa. The ischiadic peduncle of the ilium is atrophied. Both ischia and pubes are separated distally at the midline. The tibia shows both cranial and lateral cnemial crests, and metatarsal V is absent in the pes.

The most remarkable thing about *Protoavis* is that, although

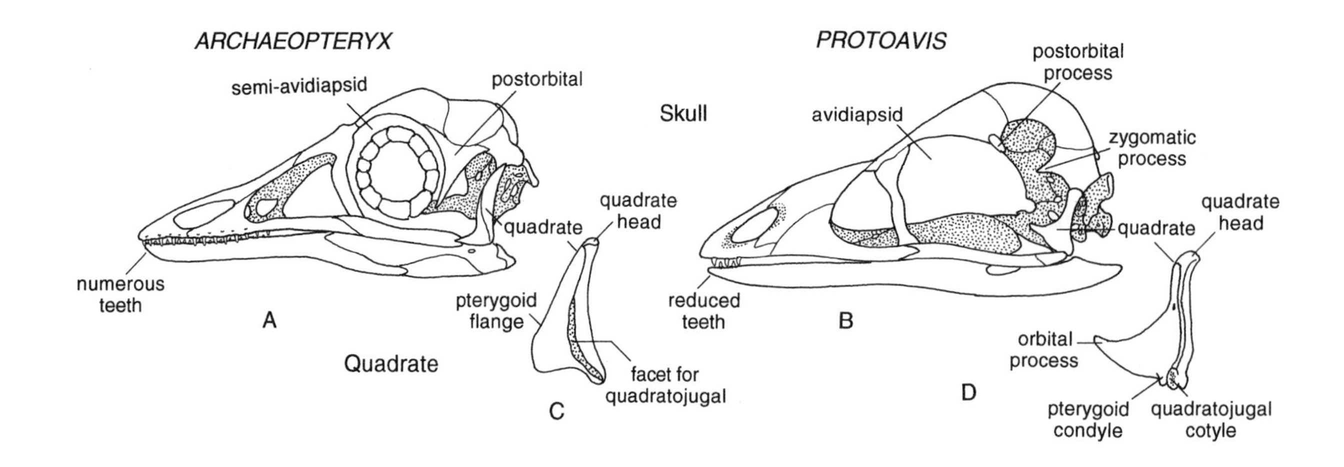
ARCHAEOPTERYX
PROTOAVIS
Skull
semi-avidiapsid
postorbital
quadrate
quadrate head
numerous teeth
A
pterygoid flange
facet for quadratojugal
Quadrate
C
avidiapsid
postorbital process
zygomatic process
quadrate
quadrate head
reduced teeth
B
orbital process
D
pterygoid condyle
quadratojugal cotyle

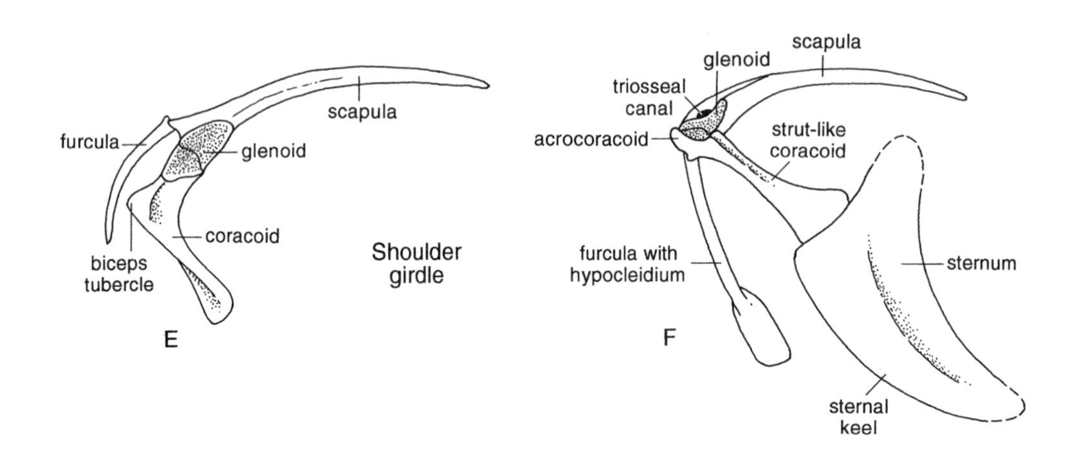
scapula
furcula
glenoid
coracoid
biceps tubercle
E
Shoulder girdle
glenoid
scapula
triosseal canal
acrocoracoid
strut-like coracoid
furcula with hypocleidium
sternum
F
sternal keel

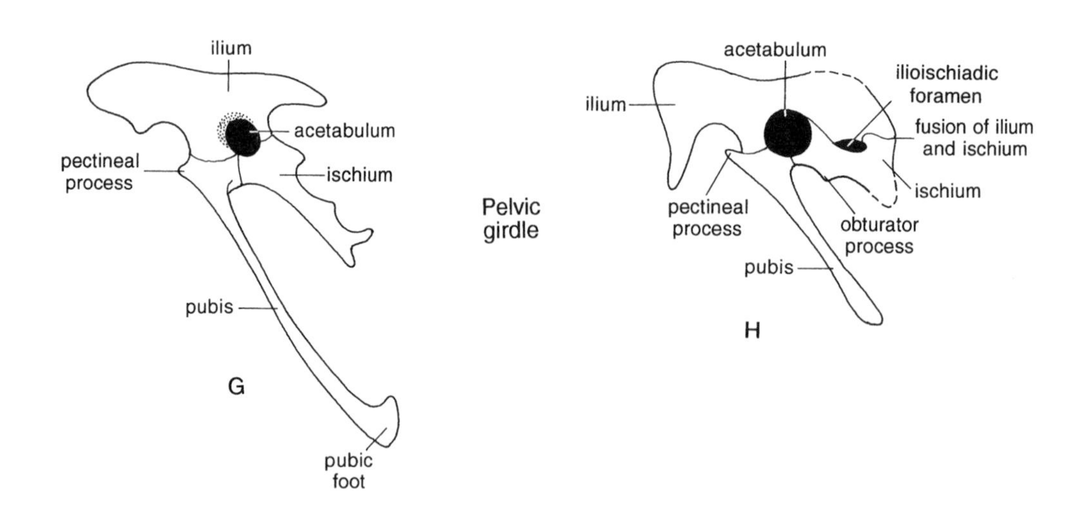
ilium
acetabulum
pectineal process
ischium
pubis
G
pubic foot
Pelvic girdle
acetabulum
ilium
ilioischiadic foramen
fusion of ilium and ischium
ischium
pectineal process
obturator process
pubis
H

it predates *Archaeopteryx* by 75 million years, it is considerably more advanced than *Archaeopteryx*. *Protoavis* evolved specializations beyond the point at which it could be precluded from being ancestral to all later birds. Its antiquity and derived morphology indicate that *Protoavis* is more closely related to modern birds than is *Archaeopteryx*. This new assessment would place *Archaeopteryx* in a side branch of the main evolution of birds.

THE PYGOSTYLIA NODE The newly defined Pygostylia node includes Enantiornithes and other birds that show considerable truncation of the tail and the development of a pygostyle. In the skull, the nasal process of the premaxilla extends posteriorly to contact the frontal, thus displacing the nasals lat-

FIGURE 10.14

Anatomical comparisons between *Archaeopteryx* and *Protoavis*. *A*, lateral view of an *Archaeopteryx* skull. A partial modification of the temporal configuration (semi-avidiapsid) can be seen, with loss of the ascending process of the jugal, so that the orbit becomes confluent with the lower temporal opening. The upper temporal arch remains intact because of the presence of the prootic bone. Teeth are numerous and extend backward to the level of the antorbital fenestra. *B*, lateral view of a *Protoavis* skull. The temporal configuration is further modified (avidiapsid), like that of modern birds, with loss of the postorbital bone so that the orbit is confluent with both the upper and the lower temporal openings. Two avian landmarks are visible on the cheek region, the postorbital process (formed by the frontal and laterosphenoid) and the zygomatic process at the rostral end of the squamosal. The teeth are considerably reduced and are restricted to the tip of the jaws. *C*, left lateral view of the quadrate of *Archaeopteryx*. Its quadrate is similar to that of nonavian theropods. *D*, the quadrate of *Protoavis* shows various avian hallmarks, such as the orbital process, pterygoid condyle, and cotyle for the quadratojugal; these modifications are linked with the acquisition of streptostyly. *E*, left lateral view of the shoulder girdle of *Archaeopteryx*. The sternum is unossified in *A. lithographica*, but a small, flat sternum has been reported in *A. bavarica*. *F*, the shoulder girdle of *Protoavis* shows several features indicating that it was adept at powered flight: strutlike coracoid, triosseal canal, keeled sternum, and springlike furcula. *G*, left lateral view of the pelvis of *Archaeopteryx*. *H*, the pelvis of *Protoavis* shows fusion of the ilium and ischium to enclose the ilioischiadic fenestra; this fusion suggests that the pelvis was strongly built to withstand the impact of landing from heights.

erally. The premaxillae are fused in the adult individual. The external naris has been shifted considerably backward from near the rostral process of the jugal. The medial vertical plates, such as the mesethmoid, and the orbital septum are ossified. The caudal tympanic recess extends backward at the basioccipital. The caudal vertebrae number fewer than fifteen. The distal carpals are fused to metacarpals to form the single-element carpometacarpus. The femur shows a capital ligamental fossa at its head and a deep rotular groove at the distal end. The hypotarsus is developed on the proximal end of the tarsometatarsus.

THE ORNITHURAE NODE This stage, the Ornithurae node, is represented by Hesperornithiformes, *Patagopteryx*, and higher birds. Here, the quadrate head is expanded medially and shows both squamosal and prootic articulations. In the palate the caudal maxillary sinus shows a cup-shaped depression. The synsacrum is enlarged, with the incorporation of more than eight vertebrae. The uncinate process of the rib is ossified. The humerus shows a well-developed ventral tuberosity separated from the head by a deep capital groove. A large extensor process is developed on the carpometacarpus. The pelvic elements are fused, where the ilium, ischium, and pubis are more or less parallel. The femur shows a trochanteric crest. The fibula is greatly reduced in length distally. The tarsometatarsus is completely fused with the distal tarsals. A vascular distal foramen is developed between metatarsals III and IV.

THE CARINATAE NODE The Carinatae node is represented by Ichthyornithiformes, *Ambiortus*, and other highly volant birds with the development of a large carina on the sternum. The humerus shows a brachial depression at the distal end. The external condyle of the ulna is developed into a semilunate ridge.

THE NEORNITHES NODE This stage, the Neornithes node, is represented by *Gobipipus*, the Antarctic loon, and modern birds. The individual bones in the skull are fused. The lacrimal/

jugal contact is breached. The ascending process of the maxilla is reduced. The quadrate head is bifurcated by penetration of the dorsal tympanic recess. The basipterygoid processes are atrophied.

THE MACROEVOLUTIONARY PATTERN

To evaluate the detailed phylogenetic relationships of well-known taxa of Mesozoic birds, I conducted a cladistic analysis (Chatterjee n.d.). Eighty-four morphological characters of Mesozoic birds were examined. These characters were coded as primitive (0) or derived (1), with character polarity determined from outgroup analysis. We examined the data matrix using the PAUP software package. The analysis found a single, most parsimonious tree. The tree was 140 steps long, with a consistency index of 0.6. The tree is simplified and slightly modified in figure 10.15.

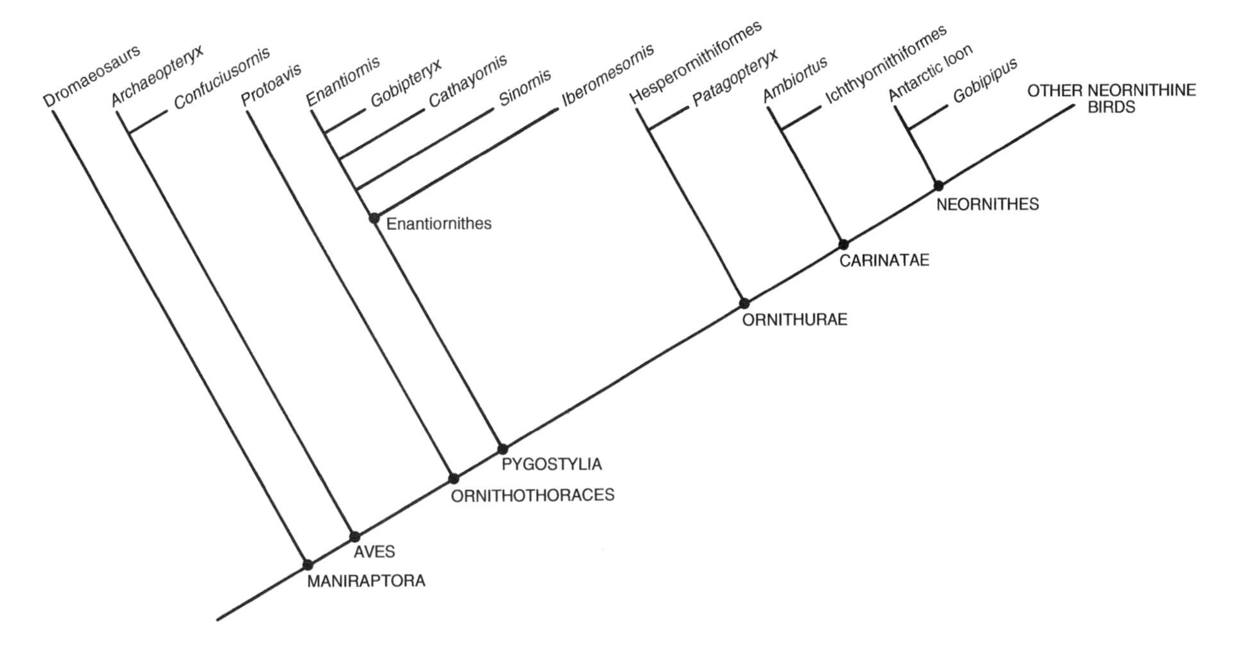

FIGURE 10.15

Cladogram of the phylogenetic relationships of the well-known taxa of Mesozoic birds (source: modified from Chatterjee 1995, n.d.).

To understand the evolution and early radiation of birds, I have converted the cladogram into a phylogenetic tree on the basis of stratigraphic data (fig. 10.16). The branching character of evolutionary lineages portrays the tree of avian life. This tree not only depicts the interrelationships of the core taxa, but also indicates the probable historical time of the acquisition of avian characters. However, the topology of the tree is far more complex than is generally believed. It emphasizes the discordance between the phylogenetic position and stratigraphic occurrence among many avian taxa. Although this discordance may reflect the low sampling level of the fossil record, it also manifests heterochrony in avian evolution.

The fossil record of birds is poorly documented. There are numerous gaps in the fossil record as known species do not grade into one another in a stratigraphic context. This stratigraphic discordance indicates that a large number of intermediate species are missing. It is generally assumed that the oldest species in a stratigraphic sequence is more likely to display the primitive condition. This is particularly true when the fossil record is dense. However, when the fossil record is spotty, as is the case with avian evolution, there seems to be a conflict between antiquity and morphological primitiveness. For example, *Archaeopteryx* is clearly more primitive in morphology, but it appeared much later than did *Protoavis.* This apparent contradiction indicates that *Archaeopteryx* and *Protoavis* lie in two separate lineages and is probably linked to the unequal survivorship of sister taxa in the fossil record. *Archaeopteryx* seems to be a late example of the ancestral type, a "living fossil" in the Late Jurassic avian world. It did not give rise to modern birds but instead lay on a distinct sideline that underwent little diversification. *Protoavis,* on the other hand, was a highly specialized bird from its first appearance and was a sister-group of Pygostylia or enantiornithine birds. Phylogenetic calibration of the fossil record indicates that birds must have originated during the Middle Triassic or even earlier (fig. 10.16).

For more than a century, the genealogy of Mesozoic birds

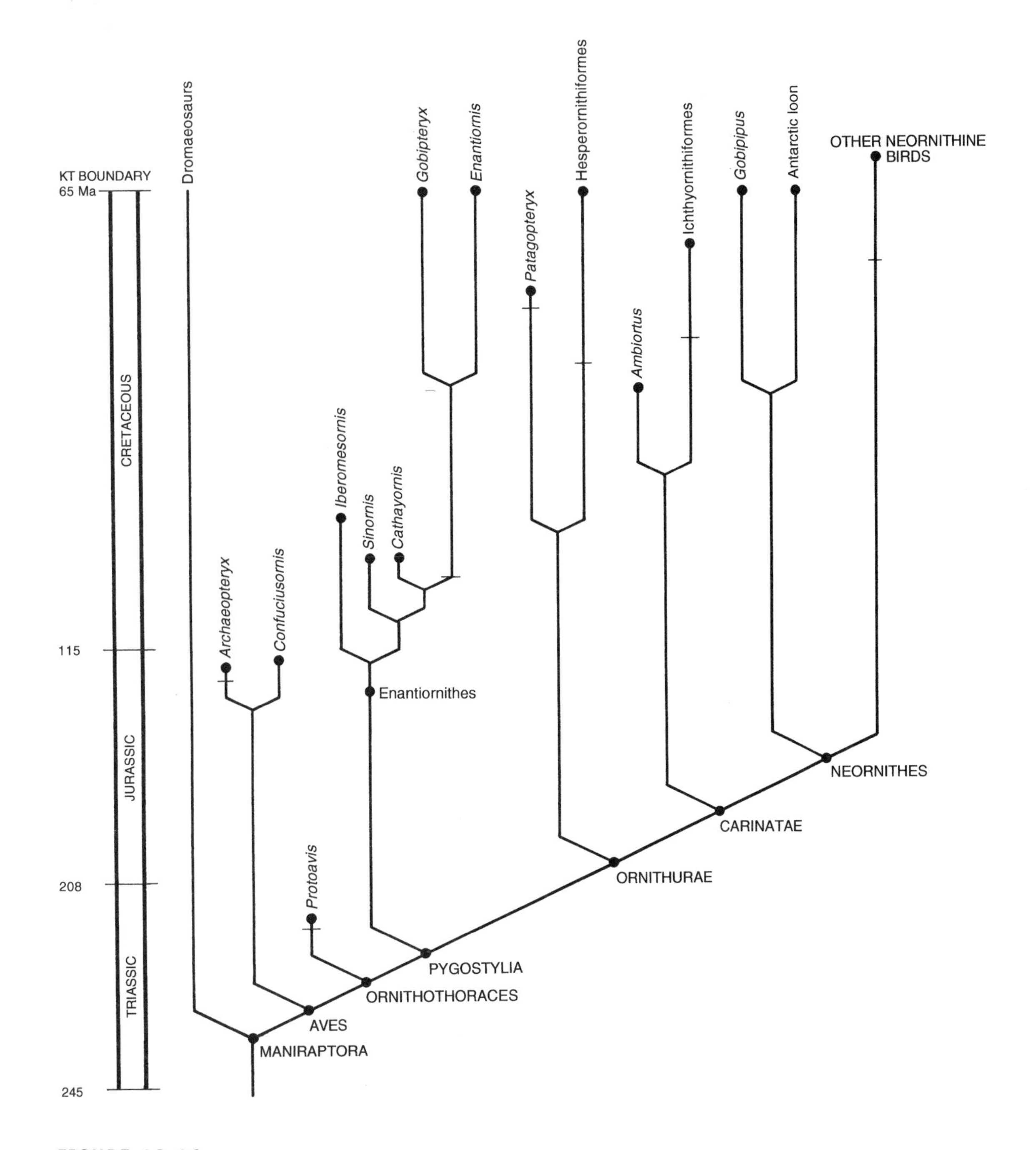

FIGURE 10.16

Phylogenetic tree showing the interrelationships of well-known taxa of Mesozoic birds; derived from the cladogram in figure 10.15.

was thought to be a simple Y-shaped tree, in which two derived taxa—hesperornithiforms and ichthyornithiforms—split from the *Archaeopteryx* stem. A simple ladder-like progression was the focus of the early history of birds. This simple succession of early avian history has been replaced by a much more complex pattern and by much less certainty about ancestor-descendant relationships. The phylogenetic tree of Mesozoic birds, as deduced from recent discovery, indicates a punctuated series of adaptive radiations. These bursts of diversification occurred very rapidly within a population. The radiation style in the early history of birds is a copiously branching bush, not a simple twig (fig. 10.16).

Heterochrony in Avian Evolution

It has long been known that there is a close correspondence between the development of the individual from egg to adult (ontogeny) and the ancestry of species (phylogeny), but the nature of this relationship is far more complex than is usually believed. Soon after the publication of *The Origin of Species,* the German biologist Ernst Heinrich Haeckel (1866), a most ardent supporter of Darwinism, proposed his famous biogenetic law: "ontogeny recapitulates phylogeny." Haeckel believed that, in the course of development, each kind of animal repeats all of the stages of its evolutionary history. Thus, adult characters of ancestors become the juvenile characters of descendants, but they are crowded back into earlier stages of ontogeny. Haeckel also coined the term *heterochrony* to explain unusual change in the developmental timing of a particular organ. Later, Gavin de Beer (1958) clarified some of the misconceptions of recapitulation and showed other relations between embryonic development and evolutionary descent. For example, in many evolutionary lineages, adult descendants retain ancestral juvenile characters. These descendants superficially look like "degenerate" forms that passed through fewer morphological stages during their development than did their ancestors. De Beer termed this evolutionary pattern *paedomorphosis,* meaning "child formation." Recently, there has been a great deal of interest in synthesizing the relationship between ontogeny and phylogeny, with better definitions and formalisms (Gould 1977; Alberch et al. 1979; McKinney and McNamara 1991; Livezey 1995). These studies have revealed how ontogeny, the internal influence, can play a major role in the evolutionary process along with external influences, such as natural selection. During ontogeny, structures can be added or subtracted from those of ancestors, or changes can be accelerated or retarded.

In summary, *heterochrony* implies differences in the developmental timing of features relative to the same events of the ancestor. It may affect a wide range of developmental phenomena. Heterochrony produces two forms of morphological expression: (1) peramorphosis, or derived overdevelopment, in which the characters of new descendants are produced by addition to the ancestral ontogeny (recapitulation), and (2) paedomorphosis, or derived underdevelopment, in which the retention of ancestral juvenile characters is maintained in later ontogenetic stages of descendants.

There is a great deal of correlation between the phylogeny and ontogeny of birds. Differences in ontogenic trajectories between species may reflect evolutionary divergence. Peramorphosis is the dominant process in the morphological novelties during the early evolution of birds from their theropod ancestors. However, paedomorphosis seems to have been significant in the pectoral reduction of flightless birds from their flying ancestors. It provides an alternative explanation of the convergent evolution of flightless birds among different lineages throughout the Mesozoic and Cenozoic periods.

PERAMORPHIC TRENDS

Many features of the juvenile structure of birds show resemblance to the adult structure of theropod ancestors. These features include the retention of sutures between the cranial bones; premaxilla-nasal contact at the midline; an amphicoelous centra; an obtuse angle of articulation between the scapula and the coracoid; separate clavicles; an unossified uncinate process; four metacarpals; separate carpometacarpals; manual claws; a propubic pelvis; an unfused ilium, ischium, and pubis; a short preacetabular process; the absence of an obturator foramen; a fibula as long as the tibia; a separate tibiotarsus; a separate tarsometatarsus; and a separate ascending process on the astragalus.

PAEDOMORPHIC TRENDS

Thulborn (1985) identified in modern birds the following paedomorphic features from their theropod ancestors: very large orbits, an inflated braincase, and retarded dental development. To this list may be added miniaturization of size.

Paedomorphic trends are more frequent in the pectoral reduction of flightless birds. Physiologically, flight is an expensive endeavor. Flightlessness is a recurrent theme in avian evolution, probably to save energy. A variety of birds, both palaeognaths and neognaths, have become secondarily flightless. Among palaeognaths, the ratites are probably the most celebrated flightless birds. Among neognaths, both gruiforms (cranes, rails, and allies) and penguins include a large number of flightless species. Flightlessness has evolved independently among three genera of grebes, nine different lineages of waterfowls, and several genera of auks. James and Olson (1983) documented several species of ibises, geese, rails, and pigeons in the Pacific islands that reverted to flightless forms during prehistoric times because of geographic isolation and the relative absence of predators. It is generally believed that modern ratites, such as ostriches, rheas, and emus, evolved from their flying ancestors as a result of paedomorphism (de Beer 1956).

All birds are flightless when they are small chicks, and the hatchlings of flying birds show features similar to those that characterize adult flightless birds. Paedomorphosis has been a major component of the evolution of flightless birds, especially in the reduction of the flight apparatus, leading to juvenilized morphology. These features include retention of the cranial sutures; a flat sternum without a ventral keel; small wings; disproportionately short distal wing elements; an obtuse angle of articulation between the scapula and the coracoid; a broad unossified region between the ilium and the ischium; and downy, juvenile-like feathers.

Thus, many paedomorphic characters in flightless birds are, in fact, derived and represent apomorphic reversals in a phylogenetic context, resembling primitive or ancestral stages. Such reversals may simply reflect neotenic solutions to particular adaptive problems. Increased body size and great sexual dimorphism are frequently associated with flightlessness and may enhance foraging and the partitioning of feeding niches. McKinney and McNamara (1991) pointed out that paedomorphism in flightless birds is not the whole story of their heterochrony. Another component of evolution has been generally ignored. Along with paedomorphism, there is a trend to peramorphism in flightless birds, where the skull, trunk, and legs become proportionately large and overdeveloped; the legs become more specialized for fully terrestrial or fully aquatic locomotion. Because of the loss of flight, selection favors alternative and efficient means of locomotion involving the hindlimbs. McKinney and McNamara argued that the locomotor adaptations of flightless birds show a combined effect of dissociated heterochrony—pectoral paedomorphosis and pelvic peramorphosis. It is ironic that flight, which was the prime stimulus for the origination of the avian clade, has been discarded in favor of flightlessness time and again in both aquatic and terrestrial groups exploiting new adaptive niches.

The Early Radiation of Birds

Once established as a new group of aerial vertebrates, Mesozoic birds proliferated in both form and function, underwent a major adaptive radiation, and occupied different niches. The great variety of birds on different continents might be linked to plate movements during the later Mesozoic, allowing different groups to evolve in isolation from one another, encouraging speciation and diversification. Mesozoic birds developed a great variety of body forms, including those specialized for foot-propelled diving and flightless forms. Many of these adaptive lineages became extinct at the KT (Cretaceous/Tertiary) boundary, but at least two modern orders, Gaviiformes and Charadriiformes, can be traced back to the Late Cretaceous. We can broadly define four ecological types among the known Mesozoic birds: basal land birds, shore birds, foot-propelled divers, and flightless terrestrial birds.

THE BASAL LAND BIRDS

The first birds, such as *Protoavis*, inhabited riverine forest environments in the tropics, as were supported by the Dockum paleoecology. They were mostly tree dwellers, with the versatile features that this habit required. The forest habitat must have played a crucial role in the evolution of early birds from their arboreal ancestors. *Archaeopteryx* was also a climbing or perching bird living on trees. If we look at the phylogenetic tree (fig. 10.16), the explosive radiation of terrestrial birds in the Mesozoic becomes apparent. Many enantiornithine taxa, such as *Sinornis, Iberomesornis, Cathayornis, Concornis, Noguerornis,* and *Enantiornis*, were partly ground-dwelling, presumably for feeding, and partly tree-dwelling, for hiding, sleeping, and nesting. Their highly recurved pedal claws and anisodactyly indicate their perching and grasping ability. Because of their rounded wings and woodland habitat, they would probably fly short distances from tree to tree. Some of the early enantiornithine birds had toothed jaws and would spend their days on the open water of rivers and lakes in search of fish but fly up to the trees during the night. The sediments that entombed these birds were largely lacustrine. Embryonic evidence from *Gobipteryx* indicates that some of these birds were precocial and laid eggs on the ground.

THE SHORE BIRDS

A large number of Cretaceous birds adapted to marine coastal habitats where food and vegetation were usually abundant. Marsh (1880) speculated that the mode of life of *Ichthyornis* and *Apatornis* was similar to that of modern gulls and terns.

Their powerful wings, long jaws and recurved teeth for capturing fish, and small feet suggest that these birds were well adapted for a piscivorous volant life in the Cretaceous seaways. In more open areas, selection favored efficient flight adaptation. It is likely that *Ambiortus* from the Early Cretaceous of Mongolia enjoyed a similar lifestyle.

Olson and Parris (1987) reviewed and described several genera of Charadriiformes—*Graculavus, Telmatornis, Anatalavis, Laornis,* and *Palaeotringa* from the Late Cretaceous (or Early Paleocene) of New Jersey—on the basis of fragmentary material. These diverse shore birds are essentially modern-looking and ranged in size from the smallest of the modern Burhinidae to that of a large crane. They lived along the shores and marshes of the Cretaceous sea of eastern North America. It is probable that *Gansus* from the Early Cretaceous of China (Hou and Liu 1984) and *Lonchodytes* (Olson and Feduccia 1980) from the Late Cretaceous of Wyoming are other examples of shore birds.

THE FOOT-PROPELLED DIVING BIRDS

Some land birds probably went to the sea to exploit that abundant food resource. Many became highly specialized foot-propelled diving birds, which evolved independently several times in avian history. In the Cretaceous sea, we see evidence of two lineages—Hesperornithiformes and Gaviidae. The most striking adaptations of these birds are for propulsion by the hindlimb: the development of a long and narrow pelvis, a short femur, a long tibiotarsus with an extensive cnemial crest, and a laterally compressed tarsometatarsus. The legs were positioned far back on the body, which made moving on land difficult. Bones became secondarily apneumatic to overcome buoyancy. The feet were lobed and specialized for underwater diving.

There are thirteen known species of hesperornithiforms, ranging in size from that of a small chicken to that of a large penguin (Martin 1987). All were flightless, with unkeeled sterna, unfused clavicles, and degenerate wing bones. The oldest known hesperornithiform is *Enaliornis* from the Early Cretaceous of

England. By the Late Cretaceous, they had become widely distributed in the Northern Hemisphere. They retained teeth on the maxilla and dentary, but the premaxillae became toothless. Their jaws were adapted for capturing fish. Their dense, heavy, nonpneumatic bones enabled them to swim slightly below the water surface. The Hesperornithiformes probably represent an evolutionary dead end without any descendants.

Loons, on the other hand, seem to have originated in the Southern Hemisphere. The oldest known loon, *Polarornis,* is from the Late Cretaceous Lopez de Bertodano Formation of Antarctica (Chatterjee 1989). It has a spear-shaped bill, a stocky neck, and an extended cnemial crest on the tibiotarsus. This early loon was not as highly specialized for diving as are modern genera because of the retention of a relatively long femur. On the other hand, the skull is so similar to that of the living loon that it is referable to the modern family Gaviidae. *Neogaeornis* from the Late Cretaceous of Chile has been reinterpreted as a loon (Olson 1992). Except for these Late Cretaceous finds, both fossil and living loons are restricted to the Northern Hemisphere.

THE FLIGHTLESS TERRESTRIAL BIRDS

Patagopteryx from the Late Cretaceous of Patagonia is a ground-dwelling bird about the size of a rooster (Chiappe 1995b). It has atrophied wings but well-developed hindlimbs. It had obviously lost the ability to fly, as have living ratites. Other enigmatic Cretaceous genera from Mongolia, such as *Mononykus* and *Avimimus,* are also interpreted as flightless forms. Recently, a large flightless bird from the Upper Cretaceous of France has been reported from its synsacrum, which is comparable in size and proportions to that of the giant Eocene *Diatryma* (Buffetaut et al. 1995).

The Cretaceous Crisis

Thou was not born for death, immortal Bird!
No hungry generations tread thee down;
The voice I hear this passing night was heard
In ancient days by emperor and clown.
—John Keats, "Ode to a Nightingale," 1819

CHAPTER 11

Extinction is forever. It is a process that has occurred since life appeared on earth. It is estimated that 99 percent of the plant and animal species that have ever lived on earth are now extinct. There are several kinds of extinctions, but the specific cause of an extinction is still elusive. The average life span of a species in the fossil record is about four million years. Many major groups of plants and animals that were once important parts of the global biota became extinct and left no descendants. Other species became extinct during the speciation event, when the ancestral species died out but the descendant species continued. The ancestral species is then said to have undergone *pseudoextinction*. Extinction with no descendants and pseudoextinction are both called *background extinctions*, which have occurred throughout geological history. In rare cases, 50 percent or more of the unrelated living species died out fairly rapidly at a certain time of earth's history. Such episodes, which are called *mass extinctions,* have affected terrestrial as well as marine organisms. The biodiversity of earth has been repeatedly punctuated by mass extinctions. All mass extinctions, however, have been followed by at least a partial evolutionary recovery, during which the number of species on earth has increased again. Fossils provide direct evidence that millions of species have disappeared from Earth.

Five major episodes of mass extinction have occurred during the past 600 million years: Late Ordovician (440 Ma), Late Devonian (305 Ma), Late Permian (245 Ma), Late Triassic (210 Ma), and Late Cretaceous (65 Ma). (Ma denotes mega-annum, millions of years ago.) Of these mass extinctions, the most famous is the KT (Cretaceous/Tertiary) extinction, when the dinosaurs and two-thirds of all marine animal species were wiped out. The sudden extinction of nonavian dinosaurs has puzzled both scientists and the public for more than a century. Having survived for 160 million years, nonavian dinosaurs seemed in-

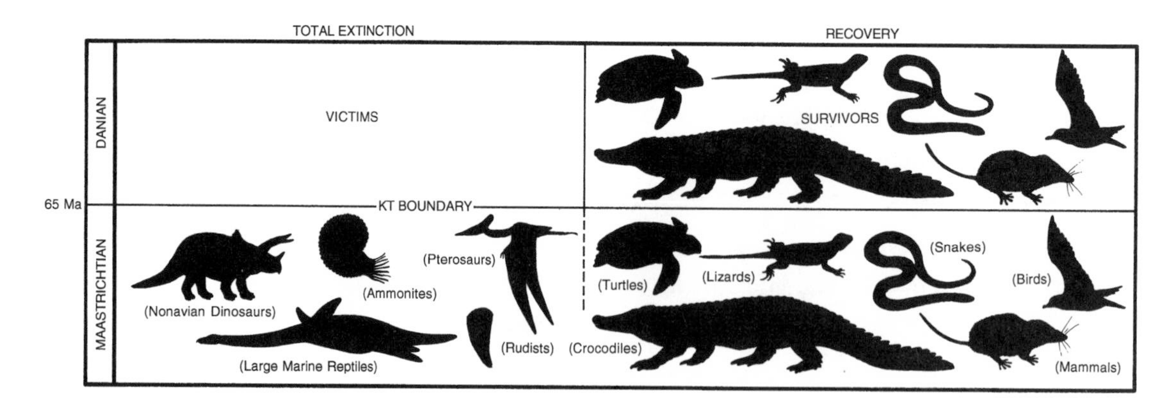

FIGURE 11.1

Victims and survivors after the KT extinction. The primary victims were nonavian dinosaurs; pterosaurs; large marine reptiles, such as plesiosaurs and mosasaurs; and various invertebrates, such as ammonites and rudists. Lizards, snakes, turtles, crocodiles, birds, and mammals endured this catastrophe and radiated. After this crisis both birds and mammals underwent explosive evolution in the Tertiary.

destructible. During that relatively brief period, all land animals weighing more than 25 kg disappeared from the planet. All pterosaurs, plesiosaurs, and mosasaurs, as well as several families of birds and marsupial mammals and hundreds of plants, were also suddenly wiped out. The small calcareous planktons that float at the ocean surface and ammonites and rudists from the depths also vanished. The earth was devastated. Life was ravaged by one of the worst catastrophes imaginable.

There were survivors, of course. Neornithine birds, placental mammals, crocodiles, turtles, lizards, and snakes all survived as groups—despite the extinction of some species (fig. 11.1). The KT extinction had opened the door for the age of mammals and the rise of birds. What was this catastrophe, which led to such an unprecedented ecological crisis? Over the years, many theories, some bizarre and some plausible, have been offered to explain the mystery behind the extinction of dinosaurs. Any explanation of biotic crises must focus on agents of destruction that affected environments, organisms, and climates.

By the Late Cretaceous, there were harsh environmental changes due to plate movements, mountain buildings, volcanic emissions, and sea regressions. Exactly what caused the biotic crisis remains a mystery. Currently, two competing models have been proposed to explain this apocalyptic disaster at the KT boundary: the meteorite impact hypothesis and the volcanic hypothesis. The impact theory postulates that environments were lethally altered or destroyed at the end of the Cretaceous by a collision with a large meteorite, leading to biotic crisis. The volcanic theory argues that pollution in the atmosphere and oceans due to the massive outpourings of Deccan flood basalt in India had a devastating effect on the ecology.

The Impact Model

In 1980, the Alvarez group advanced a startling theory to explain the sudden demise of dinosaurs—the most successful land animals ever to arise on Earth. They discovered an abnormally high concentration of iridium at the KT boundary level of Gubbio, Italy. Soon comparable iridium anomalies were found globally at different KT boundary sections. Since iridium is a very rare element in the earth's crust but is fairly abundant in chondritic meteorites, the Alvarez team proposed that the iridium spike at the KT boundary is cosmic in origin, implying the strike of a large meteorite. They hypothesized that a giant meteorite, about 10 km in diameter, had crashed into Earth with a velocity of 90,000 km/hour, causing the worldwide catastrophic event. This impact lofted so much debris into the atmosphere that it created a nuclear winter, which caused much of life on Earth to perish. A blackout of the sun would kill plants and destroy the food chain. The iridium layer was caused by the impact and vaporization of the bolide.

The impact theory was reinforced by additional evidence, such as shocked quartz, tektites (Bohor, Modreski, and Foord 1987), and carbon soot particles (Wolbach et al. 1988) at the KT boundary in different parts of the world. Shocked quartz is a distinctive signature of an impact event, as it can form only at a force of more than 10 gigapascal that travels through quartz-bearing grains of the target rock to produce microscopic shock lamellae. Tektites have also been detected at this boundary. They represent droplets of molten rocks thrown into the atmosphere during the impact event. Sediments at the KT boundary often include a layer of soot particles, which may be the residue of vegetation burned during widespread wildfire caused by the impact. The wildfire would have consumed oxygen and poisoned the atmosphere with carbon monoxide. This massive impact would generate a 100 million megaton blast—1,000 times more powerful than the explosion of the world's entire nuclear arsenal. The impact force would have melted not only the meteorite but also the target rock. This melt ejecta would have spread outward in large waves, until it had formed a crater 180 km across. Oblique impacts would produce even larger craters (Silver and Schultz 1982; Sharpton and Ward 1990).

The strongest evidence in favor of the KT impact is the craters that it left on the solid surfaces of the earth. For over a decade, impact sites have been linked to the mass extinction at

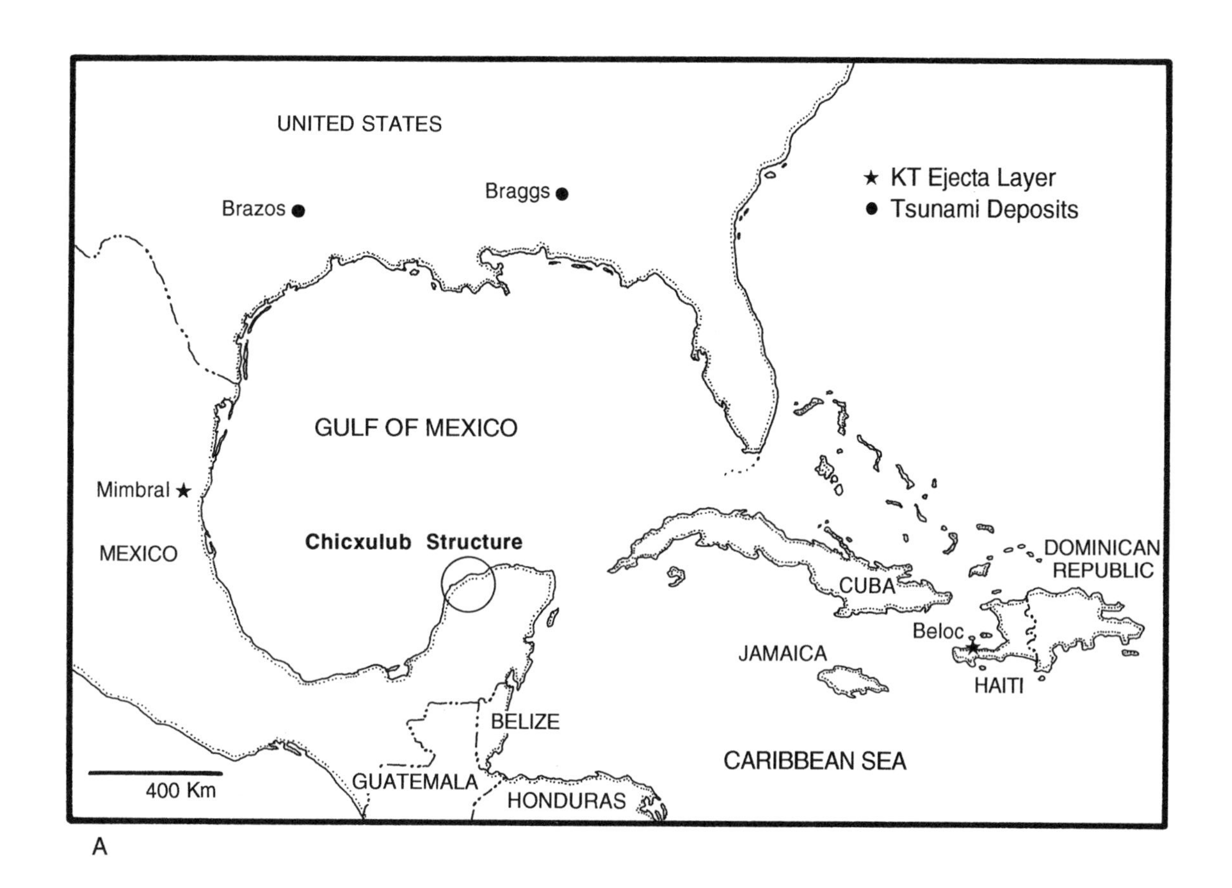

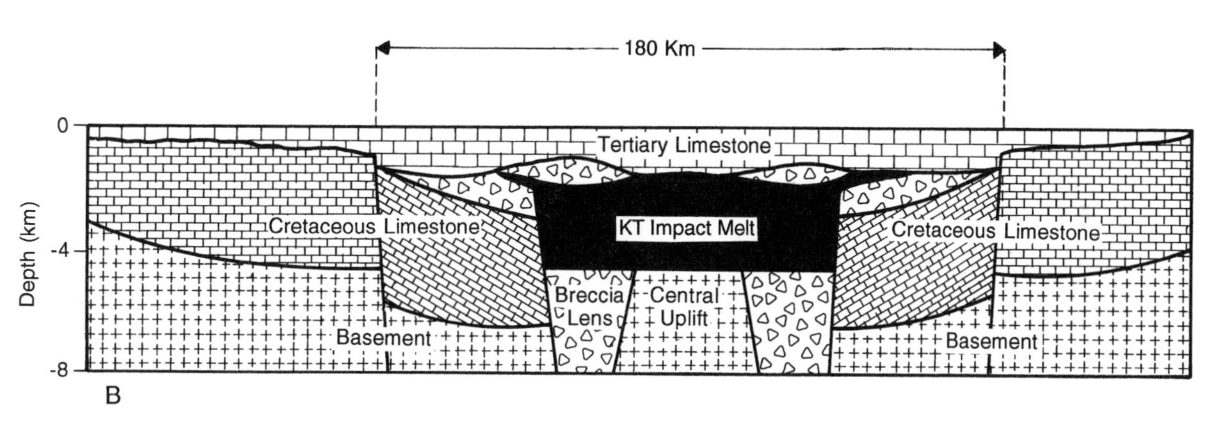

FIGURE 11.2

A, location of the Chicxulub Crater on the northern edge of the Yucatán Peninsula, Mexico, showing the distribution of proximal impact deposits; *B,* cross section of the Chicxulub crater (source: simplified from Hildebrand et al. 1991).

the KT boundary. Recently, two buried impact structures related to the KT boundary have been discovered: the Chicxulub Crater (an approximately 180-km-diameter circular structure) at the Yucatán Peninsula of southern Mexico (Penfield and Camagro 1982; Hildebrand et al. 1991) and the Shiva Crater (an oval structure with the longest diameter being approximately 600 km) at the Bombay coast of India (Chatterjee 1992b, 1996; Chatterjee and Rudra 1996). The Chicxulub structure is buried under 1,100-meter-thick carbonate sediments and is defined by magnetic and gravitational anomalies. The presence of shocked quartz, impact melt, brecciation, and iridium anomaly within the crater itself is compatible with an impact origin for the Chicxulub structure (Hildebrand et al. 1991). If Chicxulub is

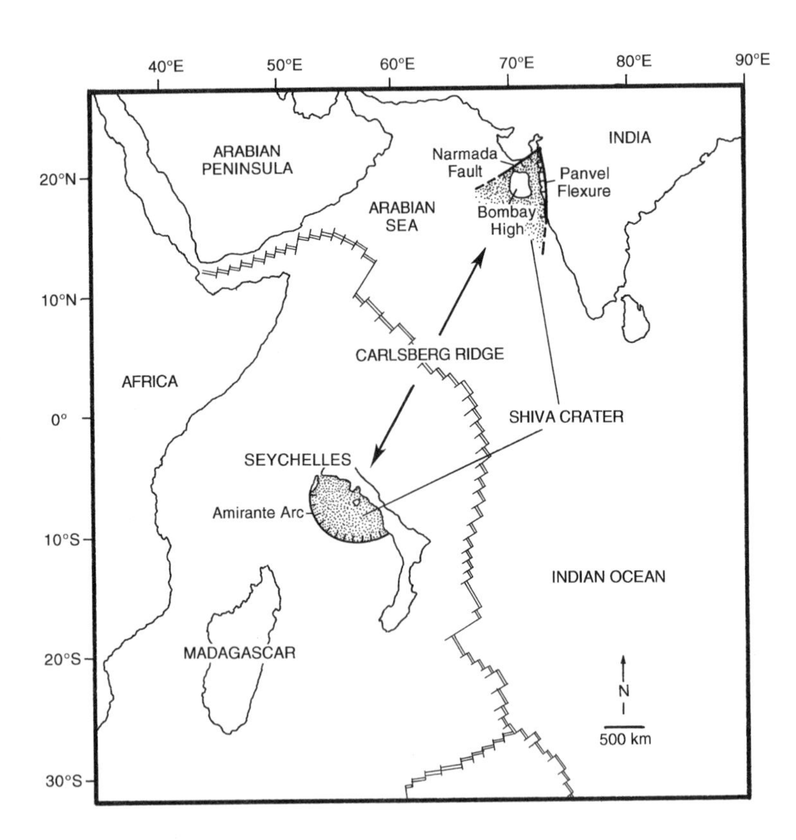

FIGURE 11.3

Sketch map showing the present-day location of the split Shiva Crater in reference to India and the Seychelles on either side of the Carlsberg Ridge. The crater was joined 65 million years ago, when the Seychelles were part of India, before the spreading of the Carlsberg Ridge.

indeed a very large KT impact scar, it should be surrounded by extensive deposits of impact ejecta and tsunami deposits. This is indeed the case. The distribution of proximal ejecta components and tsunami deposits at the KT sections in Haiti, Mexico, Texas, Alabama, the Caribbean, and adjacent areas strengthens the argument for a point of collision at Chicxulub (fig. 11.2).

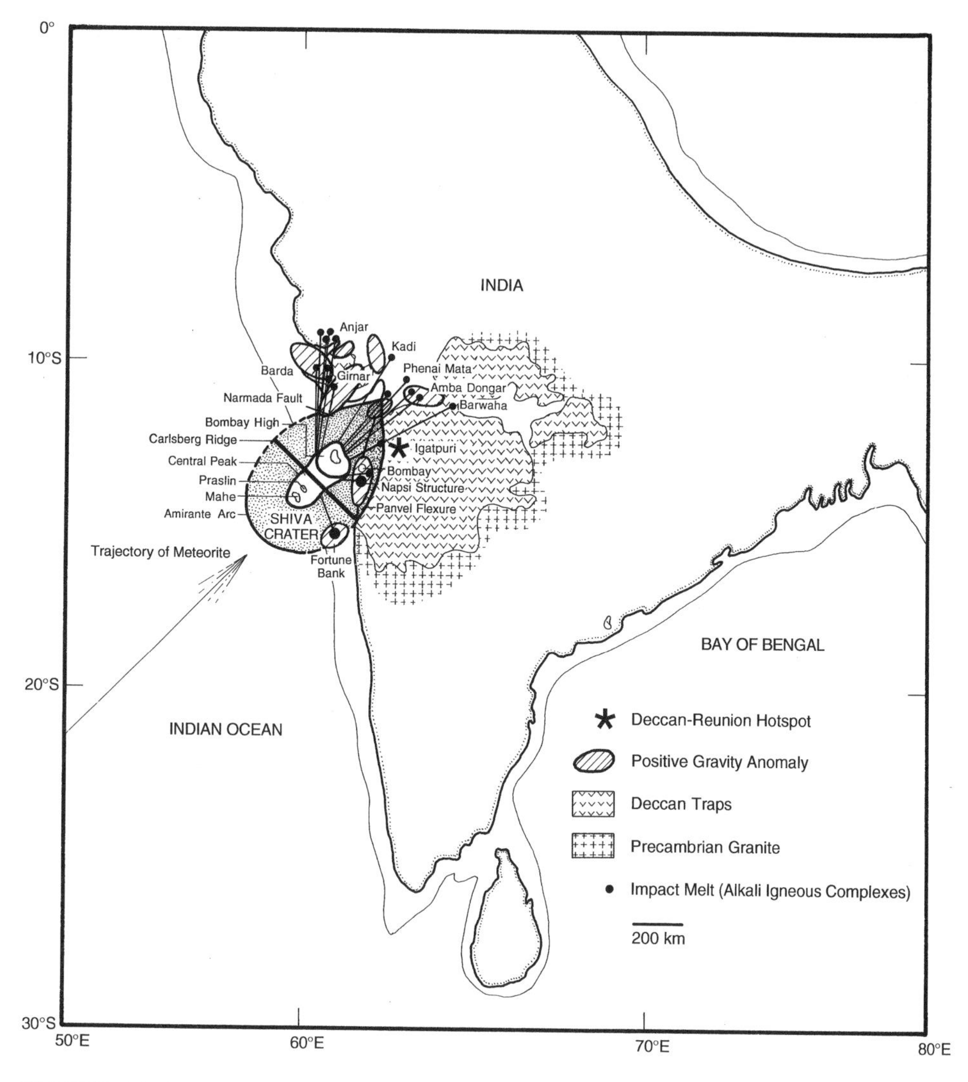

FIGURE 11.4

Location of the Shiva crater at the India-Seychelles rift margin during the KT boundary; *arrow* indicates the trajectory of the meteorite (source: after Chatterjee and Rudra 1996).

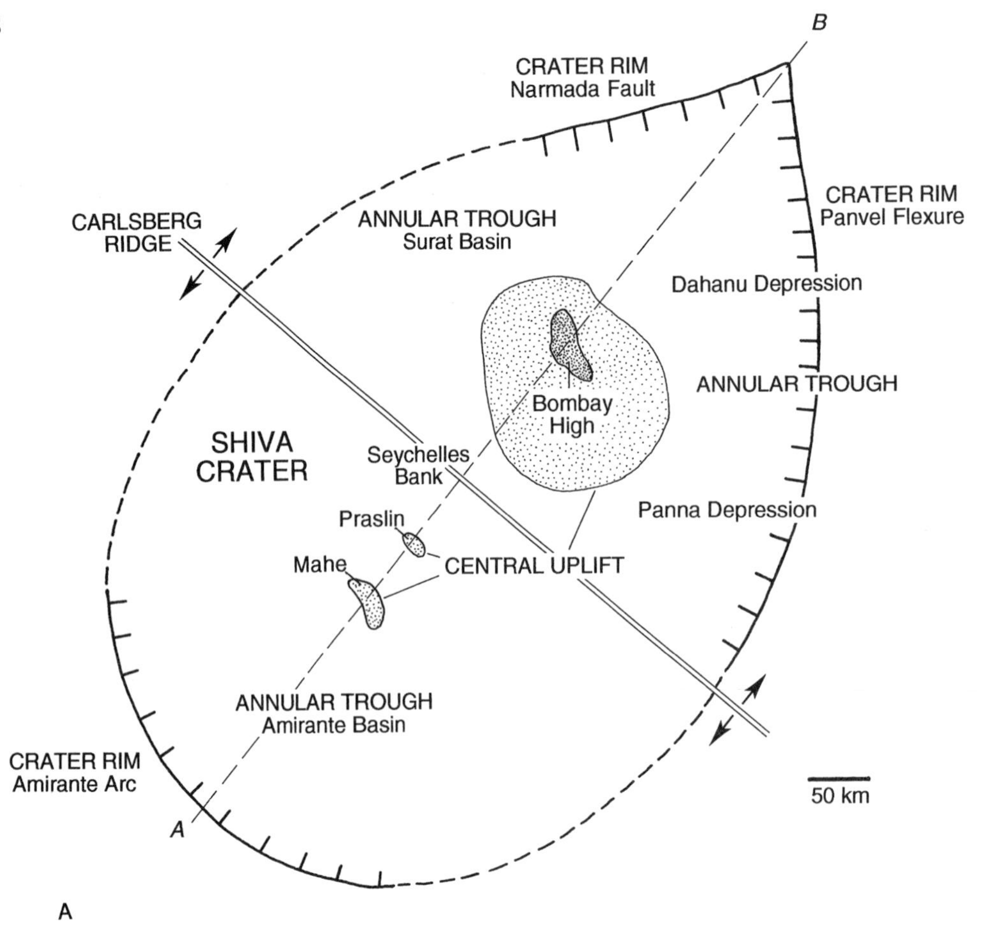
B
CRATER RIM
Narmada Fault
CRATER RIM
Panvel Flexure
CARLSBERG
RIDGE
ANNULAR TROUGH
Surat Basin
Dahanu Depression
ANNULAR TROUGH
Bombay
High
SHIVA
CRATER
Seychelles
Bank
Panna Depression
Praslin
Mahe
CENTRAL UPLIFT
ANNULAR TROUGH
Amirante Basin
CRATER RIM
Amirante Arc
50 km
A

A

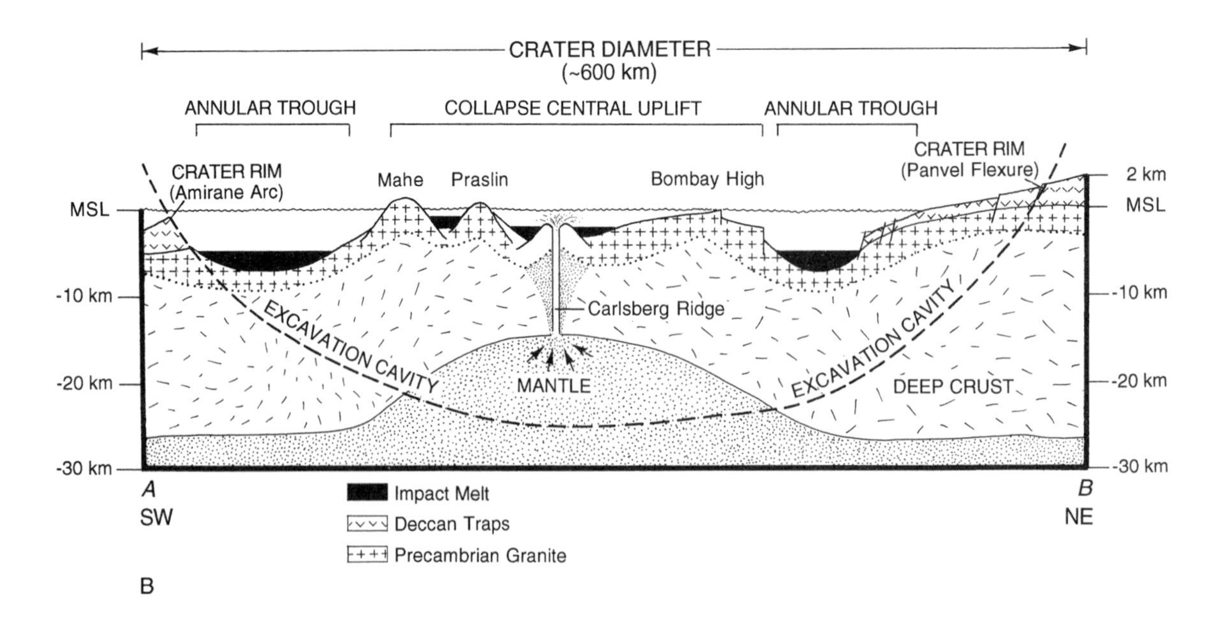
CRATER DIAMETER
(~600 km)
ANNULAR TROUGH
COLLAPSE CENTRAL UPLIFT
ANNULAR TROUGH
CRATER RIM
(Amirane Arc)
Mahe
Praslin
Bombay High
CRATER RIM
(Panvel Flexure)
2 km
MSL
MSL
-10 km
-10 km
Carlsberg Ridge
EXCAVATION CAVITY
EXCAVATION CAVITY
-20 km
-20 km
MANTLE
DEEP CRUST
-30 km
-30 km
A
SW
B
NE
Impact Melt
Deccan Traps
Precambrian Granite

B

Recently, another potential KT impact structure, the Shiva Crater, was recognized from subsurface data at the Indo-Seychelles rift margin, but sea-floor spreading and Deccan volcanism had obliterated much of its morphology. Today, the microcontinent Seychelles is separated from the western coast of India by 2,800 km because of the spreading the Carlsberg Ridge, but at the KT boundary time they were joined together (fig. 11.3). The crucial evidence in support of this impact structure comes from the Bombay High Field, a giant oil basin offshore from western India, which represents the central uplift of the crater. The KT boundary age of the crater is inferred from its Deccan lava floor, the Paleocene age (P1) of the oldest sediment within the basin, isotopic dating of the presumed melt ejecta (approximately 65 Ma), and the Carlsberg rifting event (29R—29 reverse paleomagnetic chron) that split the crater into two halves and separated India from the Seychelles. Today, one part of the crater is attached to the Seychelles, and the other part is attached to the western coast of India. Plate tectonic reconstruction at 65 Ma revealed that the Shiva Crater is oblong, about 600 km long, 450 km wide, and 12 km deep; it represents the largest impact structure of the Phanerozoic age (fig. 11.4).

The crater shows the morphology of a complex impact scar with a distinct central uplift in the form of a multiple peak complex, an annular trough, and a faulted outer rim. The central uplift is represented by the Bombay High and by Praslin and Mahe islands, which together form a linear cluster of peaks;

FIGURE 11.5

Morphology of the Shiva Crater. *A,* plan view of the Shiva Crater showing a central uplift (Bombay High, Praslin and Mahe granitic core); an annular trough (Surat Basin, Dahanu Depression, Panna Depression, and Amirante Basin); and a slumped outer rim (Narmada Fault, Panvel Flexure, and Amirante Arc). The oblong crater is about 600 km long, 450 km wide, and more than 12 km deep; it is bisected by the Carlsberg Ridge. *B,* schematic cross section of the Shiva Crater along line *AB* in the plan view; the postimpact Deccan lava flows are removed to show the morphology of the crater and the possible sites of impact melt sheets (source: after Chatterjee and Rudra 1996).

the annular basin is represented by the Surat Basin, the Panna Depression, and the Amirante Basin. The crater rim is formed by the Panvel Flexure, the Amirante Arc, and the Narmada Fault (fig. 11.5). The oblong, teardrop shape of the Shiva Crater indicates that the projectile, about 40 km in diameter (comparable to the size of the earth-crossing asteroid Ganymed), hit obliquely at the continental shelf of western India in a southwest-to-northeast trajectory. Much of the solid ejecta deposits were covered by the postimpact Deccan volcanic flows. The impact melt rocks were emplaced radially within the Deccan Traps in the form of alkaline volcanic plugs along the northeast downrange direction. Their age matches precisely the KT impact event (Basu et al. 1993). Artificial craters produced by low-angle (approximately 15°) oblique impacts in the laboratory (Schultz and Gault 1990) mimic the shape and ejecta distribution of the Shiva Crater.

If both the Chicxulub and the Shiva craters are real and if both were formed precisely at the KT boundary time, is there any genetic link between them? The synchrony and near-antipodal positions of the Shiva and Chicxulub craters at 65 Ma may indicate multiple impacts (fig. 11.6). The crater pairs could have originated from sequential collisions of a binary asteroid on a rotating globe. The recent crash of twenty-one fragments of comet Shoemaker-Levy 9 on Jupiter provides an interesting analogy. At the time of the comet's breakup, all of the fragments were bound to each other. These fragments did not collide with Jupiter simultaneously, however, but collided sequentially over five days (Shoemaker and Shoemaker 1994). Similarly, if the original meteorite had broken into several fragments and a larger one formed the Shiva Crater on Earth, the second impact, after almost 12 hours, could have created the Chicxulub Crater as the earth rotated counterclockwise around its axis (fig. 11.6). If this were the case, one could predict that additional KT impact scars, if discovered in the future, should lie in this great circle (the Alvarez Impact Belt) joining the Shiva and the Chicxulub craters (Chatterjee and Rudra 1996).

Various authors have predicted a third impact site on the Pacific plate. Frank Kyte (personal communication) of UCLA made the dramatic discovery of a tiny fragment (approximately 3 mm) of the KT bolide in a drill core from Deep Sea Drilling Platform (DSDP) 576 in the western North Pacific. The bolide chip held micrometer-size metallic grains that are up to 87 percent nickel and rich in iridium. From the geochemical signature of the chip, Kyte speculates that the KT projectile is probably an asteroid, not a comet, that slammed into the earth at a shallow angle. This is the first direct evidence regarding the carrier for

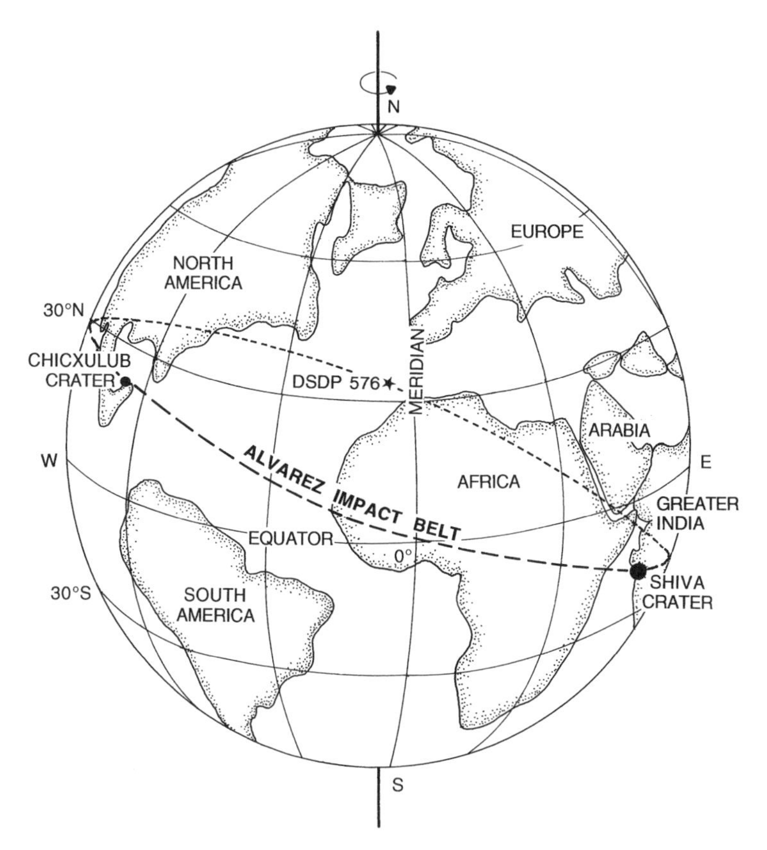

FIGURE 11.6

Possible genetic links between the Shiva and Chicxulub craters. Both craters might have originated when two fragments from a large meteorite crashed on a rotating earth over the course of 12 hours. The possibility that a third impact site may lie on the Pacific plate is evident from a tiny fragment of the bolide chip and meteoric spinel found at DSDP 576.

iridium. The chip may have broken off the asteroid before it crashed on the Pacific plate. At 65 Ma, the Pacific impact site was located midway between the Chicxulub and Shiva craters and on the Alvarez Impact Belt (fig. 11.6). Kyte, Bostwick, and Zhou (1994) also identified an additional five KT boundary sites on the Pacific plate, characterized by iridium anomaly, spherules, and shocked quartz; these sites cluster around DSDP 576.

The point of collision of the KT projectile on the Pacific plate is further supported by other evidence. Robin et al. (1994) concluded from spinel compositions at the KT boundary sites of the Pacific plate that multiple impacts might have occurred from a single disrupted bolide, where the largest objects would have impacted in the Pacific and Indian oceans. Their spinel distribution and proposed impact sites in the Pacific fit nicely with the Alvarez Impact Belt. Finally, all of the KT boundary localities containing iridium anomaly and shocked quartz cluster symmetrically on either side of this belt (Alvarez and Asaro 1990). This model is also supported by the biogeographic selectivity of extinction at low latitudes along the Alvarez Impact Belt.

The Volcanic Model

Not everybody believes that impacts killed the dinosaurs and other organisms at the KT boundary. Critics have advanced a volcanic alternative. The end of the Cretaceous was also a time of massive continental flood basalt volcanism, especially at the Deccan Traps of India. Some scientists argue forcefully that such cataclysmic Deccan volcanism may have been the main contributing factor for the biotic crisis at the KT boundary (Officer et al. 1987; Hallam 1987; Stanley 1987; Courtillot 1990). These gigantic lava floods erupted when India rode over the Reunion hotspot during its northward journey. The Deccan Traps cover 800,000 km^2 (about the size of Texas) of west-central India and extend seaward more than 500 km beyond the modern coastline (fig. 11.7). Recent radiometric dating suggests that the age span of Deccan volcanism ranges from 67 to 64 million years, but that the main pulse of volcanism may have occurred close to the KT boundary 65 million years ago (Duncan and

Pyle 1988; Vandamme and Courtillot 1992). The Deccan volcanoes erupted intermittently for almost 3 million years.

The proponents of the volcanic model argue that the KT extinction was neither global nor instantaneous, but occurred over an extended period, and their evidence is that different organisms disappeared at different levels at or near the KT boundary. Such a stepwise extinction pattern could best be explained by prolonged emissions of volcanic pollutants. Large amounts of iridium have been discovered to be spewing from the Hawaiian and Reunion volcanoes (Olmez, Finnegan, and Zoller 1986), suggesting that the iridium anomaly at the KT boundary could also have had a volcanic origin. There is no doubt that such a massive volcanic outburst over an extended period would have deleterious environmental consequences. Proponents of the volcanic model claim that many of the supposed impact signatures at the KT boundary layer, such as iridium enrichment, shocked quartz, microspherules, clay mineralogy, and soot particles, could have volcanic explanations. The impact proponents disagree. They point out that Deccan volcanism was not explosive and could not account for the global distribution of the iridium anomaly, tektites, and shocked quartz at the KT boundary layer (Alvarez 1986). Moreover, the gradual extinction pattern seen among some organisms may be an artifact of preservation and poor sampling quality. Signor and Lipps (1982) showed theoretically how a sudden catastrophic extinction would appear to have been gradual in the fossil record if the record was not dense. The Signor-Lipps effect weakens the case for a gradual extinction.

Killing Mechanisms

Since India was ground zero for both impact and Deccan volcanism, this would be an ideal place to search for evidence of the crisis on local biota. The vast thickness of the Deccan lava flows were not extruded all at once. Between the lava flows are fluvial or lacustrine deposits of Lameta sediments that contain abundant remains of plants, invertebrates, dinosaurs, and their eggs. Damming of local drainage by lava flows had created new lakes in subaerial environments. Such lake shores were centers for dinosaur communities and became the favorite nesting sites. We could not detect any evidence of biotic crisis in these fossil assemblages during episodic volcanic activity. On the contrary, the fossil evidence indicates that dinosaurs thrived during the recurrent Deccan eruptions (fig. 11.7). Moreover, no bones or eggs from this locality show any pathological abnormality. The extrusion of Deccan volcanism certainly affected the local

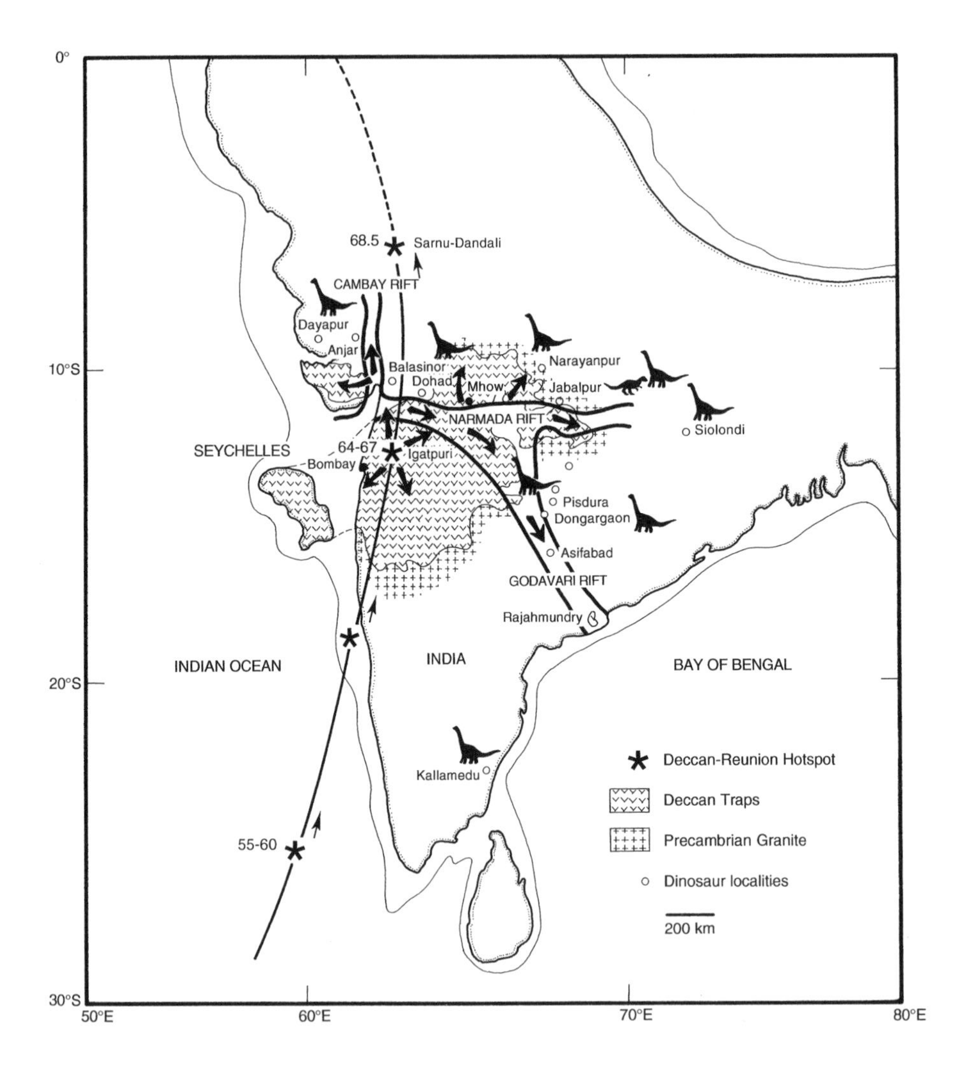

FIGURE 11.7

Sketch map showing the localities of the Maastrichtian dinosaur around the Deccan volcanic province. The position of India is reconstructed as it was during the Late Cretaceous period.

flora and fauna by habitat destruction and pollution, but it had little direct effect on the major decimation of terrestrial organisms. Deccan volcanism consisted largely of nonexplosive tholeiitic eruptions similar to the Kilauea and Reunion emissions, though on a grander scale. Recurrent eruptions do not disrupt much of the rich biota on the islands of Hawaii and Reunion. Similarly, Deccan volcanism cannot be the proximate cause of the mass extinction. Within the intertrappean beds, magnificent fossil accumulations indicate that life was resilient in the hostile environment of Deccan volcanism.

However, Deccan volcanism must have played a significant role in the initial breakdown of the marine food chain. With the emissions of Deccan pollutants, the countdown of disruption of oceanic biota started. Deccan eruption must have pumped millions of tons of volcanic ash, sulfur dioxide, carbon dioxide, and hydrochloric acid into the atmosphere and caused environmental crisis. These pollutants would darken the sky, halt photosynthesis, trigger global climatic perturbations, and lead to acid rain and ozone depletion that might be harmful to marine organisms. One million years before the KT extinction, acidification of sea water started, which would lead to the destruction of both phytoplanktons and calcareous organisms.

The volcanic catastrophe would have been intensified a million times by a collision with asteroids at the KT boundary. The impact would inject a large volume of dust into the stratosphere, darken the skies, halt photosynthesis, ignite global fires, decrease the alkalinity of the ocean surface, and devastate the biosphere. Millions of organisms would die instantly from the direct effect of the impact—shock heating of the atmosphere by the expanding fireball. Most of the world's vegetation would be caught in wildfire. The fireball would fuse megatons of oxygen and nitrogen of the lower atmosphere to combine with steam, forming nitric acid, turning the alkaline sea to acid. This would have destroyed the myriad of calcareous planktons, collapsing the food chain in the sea. Since the impact occurred at the coastal region, huge tsunamis produced by the impact would destroy shallow marine habitats across the globe. Many tsunami deposits linked to the KT impact have been discovered along the Gulf Coast (Hildebrand et al. 1991). The thick deposits of anhydrite on the coastal regions, when impacted, would have created a huge sulfuric acid aerosol cloud, thereby generating greater environmental stress than would have been possible with impact dust alone. The acid aerosol cloud would contribute to a rapid decline of global surface temperature and halt photosynthesis, heralding a "nuclear winter" and killing most of the plants and animals that survived the initial cataclysm (Sharpton and Ward 1990).

Dinosaurs disappeared precisely at the KT boundary iridium anomaly at the Anjar section of the Gujrat of western India. We collected boundary samples from the Anjar section for iridium analysis, where associated skeletons of titanosaurs occur. The iridium analysis was done by Dr. Moses Atrrep, Jr., of Los Alamos National Laboratory. He detected iridium at 348 parts per billion in the samples—indicating significant enrichment consistent with other KT boundary material. This is the first unequivocal evidence indicating that the dinosaur extinction occurred precisely at the time of the impact. We believe that the main culprit for the KT extinction was the titanic lethal effect generated by the impacts—vapor plume, global fire, thermal pulse, evaporation of the photic zone and concomitant sea regression, chondritic metal toxicity, acid rain, and volatilization of target rocks. The impacts probably caused more dramatic changes to KT environments globally and a more traumatic crisis to the ecosystem than could volcanic emissions. Impacts also perturbed the environment so suddenly and catastrophically that most organisms could not adapt to these changes in such a short time and perished instantly. In contrast, prolonged volcanism allowed enough time for some organisms to adapt and for others to disappear gradually. Selectivity and the stepwise extinction pattern can be better explained by the volcanic model. Deccan volcanism could have been an accomplice, but not the main killer, in this massive destruction of life. It seems that impact was the proximate cause to the biotic crisis at the KT boundary, whereas long-term volcanism produced harmful changes that increased the climatic stress and enhanced the extinction process. Both the impacts and Deccan volcanism contributed heavily to the breakdown of stable ecological communities and disrupted the biosphere. The KT extinction was a compound crisis, induced both by impact and volcanism (Silver and Schultz 1982; Sharpton and Ward 1990).

A cause-and-effect connection between impact and Deccan volcanism has been the subject of extensive discussion and speculation. Since the Shiva Crater was proximate to the Reunion hotspot that created the vast outpouring of the Deccan lava at the KT boundary, the near-coincidence of timing and space is striking (see fig. 11.7). Could Deccan volcanism have been triggered by the Shiva impact? Probably not. Deccan volcanism seems to have begun at least 1 Ma before the impact event, making a causal link unlikely. However, the impact might have shaken the earth's mantle violently, enhancing the spectacular Deccan volcanic outburst precisely at the time of the KT boundary.

The Extinction of Birds at the KT Boundary

The KT event is surely the best-known mass extinction because it struck the nonavian dinosaurs, but several groups of birds that lived in the Maastrichtian time also became victims of this crisis. Our knowledge of avian victims and survivors at the end of the Cretaceous is still sketchy. The primary victims included the various archaic lineages such as Hesperornithiformes, Ichthyornithiformes, and Enantiornithes, along with other, less well-known groups whose general appearance was quite different from that of extant birds. At the end of the Cretaceous, we also see the appearance of modern birds. Among these, only two modern orders have been recognized in the late Maastrichtian, the Gaviiformes from Antarctica and Chile and transitional Charadriiformes, such as Graculavidae from New Jersey and Wyoming. Both groups continued to radiate past the KT boundary.

The decline in avian diversity at the KT extinction is caused by both a high extinction rate and a low origination rate. Throughout the Mesozoic both birds and pterosaurs competed for aerial niches but adapted to different habitats. Pterosaurs were a successful group of reptiles that appeared at the dawn of the age of dinosaurs, flourished throughout the Mesozoic, and then disappeared at the end of the Cretaceous. Most of the Maastrichtian pterosaurs became large and relied heavily on sea breezes and thermals for soaring and gliding. They had poor terrestrial ability and occupied relatively narrow ecological niches along cliffs in warm, near-shore marine environments, where their diet relied heavily on fish and other sea creatures. In contrast, contemporary birds were relatively small, adapted to a wide range of habitats in both terrestrial and marine realms, and had ecological superiority over pterosaurs. Their small size and versatile lifestyles may have enhanced survival during the crisis. Unwin (1988) pointed out that, during the Late Cretaceous, as bird diversity rose, that of the pterosaurs steadily declined. The success of birds over pterosaurs at the end of the Cretaceous may be linked to differential survival strategy. Large-bodied pterosaurs had two disadvantages: they generally had smaller populations and lower reproductive rates than smaller-bodied bird species. When struck by catastrophe, they were slower to recover. In contrast, birds, being small, were able to reproduce quickly and expand their numbers. They endured massive extinction at the end of the Cretaceous but rebounded from this crisis and underwent an explosive radiation of modern forms in the Tertiary. Their dinosaur heritage, however, cannot be ignored. The pinnacle of dinosaur evolution may have culminated in the ascendancy of birds. "There is a grandeur in this view" that dinosaurs did not completely vanish from the earth after the Cretaceous catastrophe. The sole surviving lineage of dinosaurs is still around us—we call them *birds* (Bakker 1975).

Recovery during the Tertiary Period

From the ashes a fire shall be woken,
A light from the shadows shall spring;
Renewed shall be blade that was broken:
The crownless again shall be king.
—J. R. R. Tolkien, *The Lord of the Rings,* 1965

CHAPTER 12

The KT extinction left an impoverished fauna on the land, but an evolutionary rebound during the early Tertiary once again brought back biodiversity. The Cretaceous crisis was an important component in the evolution and shaping of Tertiary life. Two modern orders of birds, Charadriiformes and Gaviiformes, transcended the KT extinction and continued into the Cenozoic. From this evolutionary bottleneck, birds underwent an explosive evolution in the beginning of the Tertiary, diversifying and adapting to many different ecological niches.

Avian Ascendancy in the Cenozoic

The KT extinction created opportunities for faunal change by removing nonavian dinosaurs from the land and pterosaurs from the sky and enabling mammals and birds to radiate and multiply in vacant niches. The extinction was like the cosmic dance of the Hindu god Shiva, who holds in one hand the flame of destruction and, in another, a drum, whose sound is the sound of creation. Shiva is the god not only of destruction. He also creates a new world out of the wreckage of the old. In the early Tertiary, this new world began with the rise of mammals and birds. Both groups flourished and blossomed into a myriad of new species in the aftermath of the extinction. The success of birds in the Tertiary was critically linked to the disappearance of pterosaurs in the same way that the success of mammals was linked to the destruction of dinosaurs. Today, birds are certainly highly prolific vertebrates, sharing with the mammals a long evolutionary history, species diversity, endothermic physiology, and a wide range of dietary habits.

The rise of birds in the Cenozoic era is phenomenal (fig. 12.1). By the beginning of Cenozoic times, birds had acquired their present sophisticated body plan. The Cenozoic was a re-

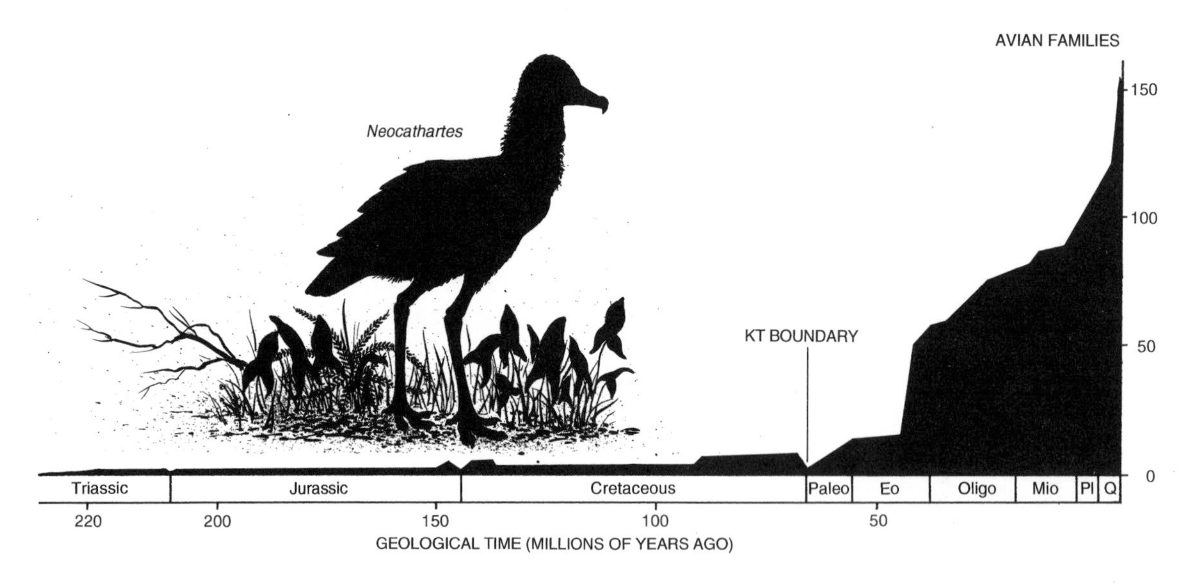

FIGURE 12.1

This graph charts the rise in avian diversity. Note the pattern of explosive evolution of birds after the KT extinction (source: modified from Unwin 1988). Inset, *Neocathartes,* a scavenging bird from the Eocene of Wyoming, was a forerunner of the New World vulture of today. *Paleo,* Paleocene; *Eo,* Eocene; *Oligo,* Oligocene; *Mio,* Miocene; *Pl,* Pliocene; *Q,* Quaternary.

naissance—a time of peak radiation of bird types, when as many as sixteen new orders developed worldwide. The Paleocene-Eocene world, a period of less than 15 million years, was warm, had almost subtropical climates from pole to equator, and supported a great diversity of birds. These birds are generally structural intermediates to modern families and occupied quite different niches than those of their living descendants. Many of these early birds show telltale signs of their ancestry of modern groups.

Our knowledge of the Paleocene birds is limited; most records come from France and Mongolia. The French fauna contains the gigantic flightless bird *Gastrornis,* an early ratite, and a wide variety of owls. The oldest owl fossil, *Ogygoptynx,* is known from the Paleocene of Colorado. The most striking event in early Cenozoic avian history was the rapid radiation of large, flightless, cursorial birds, which invaded the ecological niches left by the extinct nonavian dinosaurs. Such ecological replacement is an example of *evolutionary relay.* During this period mammals were generally small and birds made a brief bid for supremacy. Flightlessness has evolved in many lineages of fossil and modern groups, often associated with geographic isolation and the relative absence of predators. During the early Tertiary, large, flightless, cursorial birds were widely distributed globally in the expansive open plains. They evolved convergently from different stocks and disappeared at different times. Among these flightless forms, we see the radiation of both carnivores and herbivores.

Probably the most impressive birds to inhabit the Paleocene-Eocene landscape of the Northern Hemisphere were the giant diatrymas (*Diatryma, Gastrornis,* and related forms). Their remains have been found in Wyoming, New Mexico, New Jersey, France, and Germany. Diatrymas were large birds, standing over 2 meters tall and weighing perhaps 190 kg. They had a huge head, an enormous beak, vestigial wings, and long, powerful legs (fig. 12.2A). Against the widespread view of diatrymas as cursorial predators, Andors (1995) offered a novel interpreta-

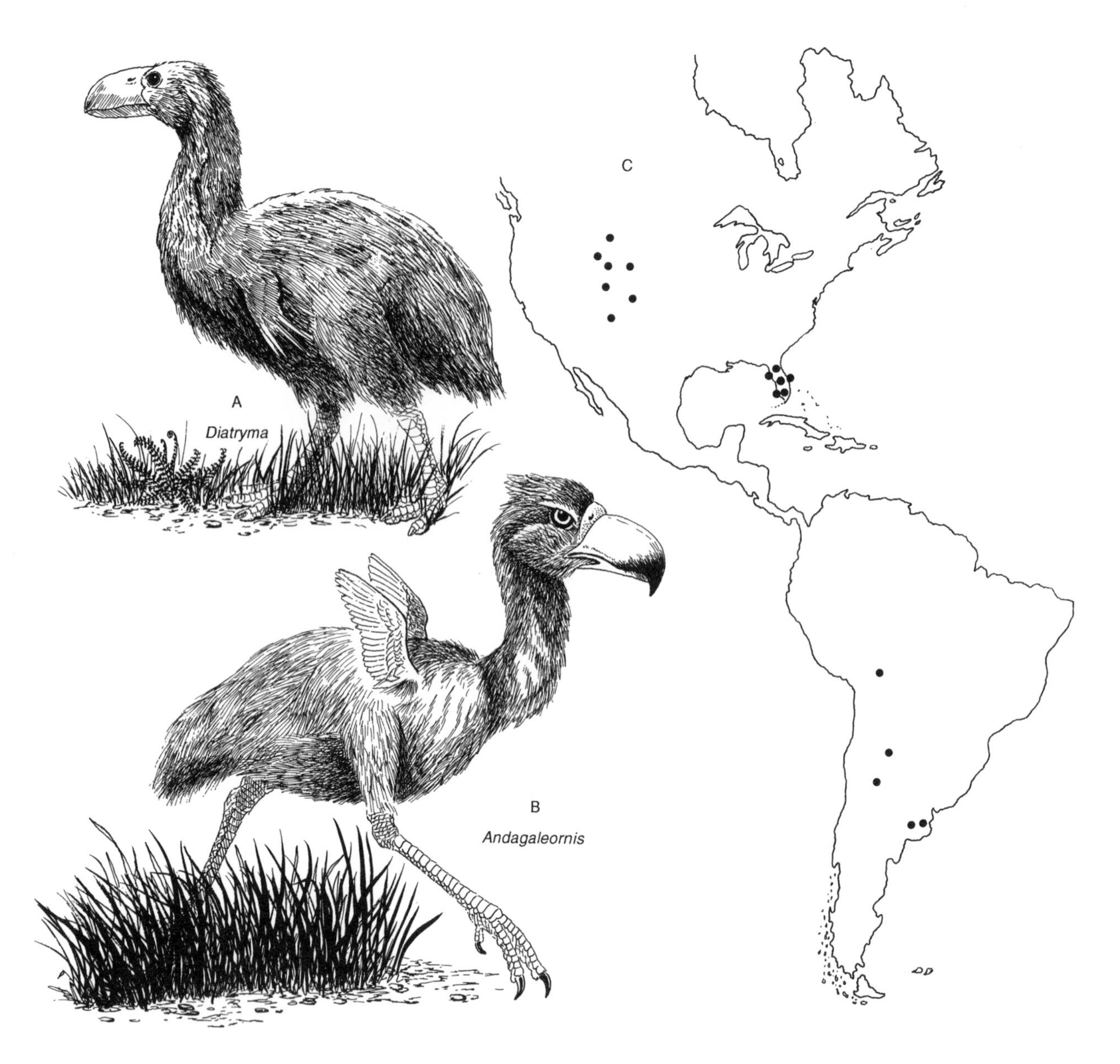

FIGURE 12.2

Restorations of giant flightless birds of the Tertiary period. *A,* the flightless, herbivorous ground bird, *Diatryma,* known from the Eocene of the western United States and Europe. *B,* the terror bird phorusrhacid, *Andagaleornis* of the Pliocene of South America. *C,* distributions of diatrymas in North America and of phorusrhacids in South America.

tion of their mode of life. He suggested that these birds were largely herbivores, similar to the living takahe of New Zealand, and were related to anseriform birds.

In South America, the formidable phorusrhacids were the dominant carnivores throughout the Tertiary and assumed the role of theropods (Marshall 1994). They were first known from the Paleocene of Brazil. During this time South America was geographically isolated from North America and lacked large predatory mammals. Geographic isolation and lack of competition favored the dominance of phorusrhacids. These giant birds were lightly built and about 1.5 meters tall, with a powerful hooked beak highly adapted for tearing flesh and crushing bone (fig. 12.2B). Living on the pampas, these birds must have terrorized the diminutive mammals of the time. The decline of these predatory birds began some 2.5 million years ago, when the placental carnivores invaded their range from North America. Phorusrhacids are also known from the Eocene deposits of Europe and the Antarctic peninsula. The seriemas of South America represent the living relatives of phorusrhacids.

The superiority of large flightless birds in the early Tertiary period was short-lived. With the passage of time, there was active competition between birds and mammals for the role of dominant land vertebrates. Mammals won the contest because of the variety and efficiency of their teeth, with designs for cutting, shearing, piercing, gnawing, grinding, grasping, and processing food. The evolution of precise food-processing mechanisms played a crucial part in mammals' supremacy over ground birds. Many ground birds were unable to protect their chicks, hatched on the ground, from small, fast, predatory mammals. From then on, birds largely became flying vertebrates and avoided direct confrontation with mammals for food and resources. By the Eocene most modern orders of nonpasserine birds had appeared. During this time, shorebirds, flamingo-like birds, and crane- and rail-like forms were extremely diverse. Many of these birds are structural intermediates and provide evolutionary links between modern families.

There are spectacular bird fossils from the early Eocene Green River Formation of Wyoming and London Clay of England, the middle Eocene Oil Shales of Messel, Germany, and the Eocene-Oligocene phosphorite deposits of Quercy, France. The Green River sediments were deposited in a series of intermontane basins of Wyoming and Utah, such as Lake Uinta, Lake Gosiute, and Fossil Lake, during the orogeny of the Rocky Mountains. These sediments have produced a rich array of avian fauna, such as the frigate bird *Limnofregata;* the galliform *Gallinuloides;* the perching birds *Primobucco* and *Neanis; Presbyornis,* a transitional bird between duck and flamingo; several undescribed species of coraciiforms and caprimulgiforms; as well as feathers and footprints (fig. 12.3). These specimens are represented by complete or nearly complete articulated skeletons. Many of these birds were nonaquatic, fell accidentally into alkaline lakes, like those found in the Rift Valley of Africa, and were preserved in exquisite detail (Grande 1980; Feduccia 1980). Several raptors were also recorded during this period; the most famous is *Neocathartes,* a long-legged, terrestrial vulturine bird that scavenged along the shores of large inland lakes (fig. 12.1).

The Messel Oil Shale bed near Frankfurt, Germany, is another famous site for middle Eocene flora and fauna. During the middle Eocene, Europe was an island that lay 10° of latitude farther south than it does today. The Messel setting was a forest environment around a lake, where many species of birds lived in bushes or trees and died accidentally in the lake beds. The Messel shale has yielded excellent articulated skeletons of birds with great evolutionary significance (fig. 12.4). Occasionally, feathers are found preserved. Some species were large, spectacular ground birds, such as the ancestral ratite *Palaeotis,* diatrymas, phorusrhacoids, and seriemas. Other forms include hawks, galliforms, ibises, rails, plovers, owls, swifts, woodpeckers, and rollers. Surprisingly, typical aquatic birds are rare in or absent from the Messel assemblage, except for the stilt-like *Juncitarsus,* which sheds new light on the origin of flamingos (Peters 1992, 1995).

The extensive terrestrial avifauna of the phosphorites of Quercy, France, includes galliforms (*Paraortyx, Palaeocryptonyx, Priotryx,* and *Palaeotyx*), gruiforms (*Idiornis* and *Elaphrocnemus*), the caprimulgiform *Aegialornis,* and the trogon-like *Archaeotrogon* (Mourer-Chauviré 1992). Large seabirds with toothlike projections on the jaws, such as *Odontopteryx,* and pelecaniforms like *Prophaethon* were known from the London clay deposits. Odontopterygiformes were albatross-like birds, essentially fish-eaters, which survived until the Pliocene. The

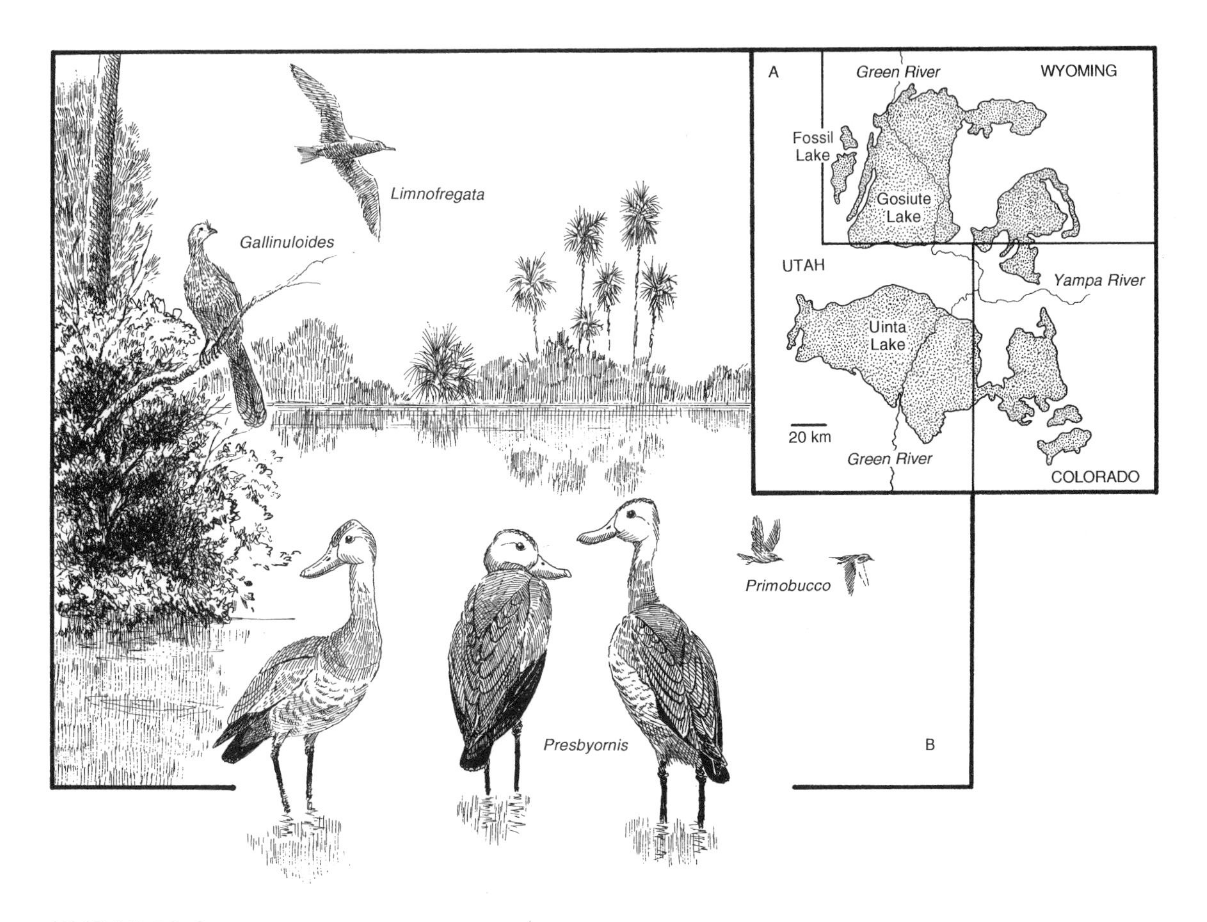

FIGURE 12.3

Green River bird assemblages of the Eocene epoch. *A,* surface exposures of the Eocene Green River Formation (source: simplified from Grande 1980). *B,* reconstruction of the Green River avian community showing *Presbyornis, Gallinuloides, Limnofregata,* and *Primobucco,* which lived in and around large lakes (source: Olson 1977; Feduccia 1980).

penguins are a very ancient order; their earliest fossils have been found from the late Eocene of the Antarctic peninsula.

During this period worldwide cooling began, when global temperatures dropped as much as 10°C. This is the beginning of the explosive radiation of birds. By this time they had acquired their basic forms as flyers, swimmers, and runners. The earliest known passerines (altricial songbirds of perching habits) come from the Oligocene deposits of France. The Oligocene period brought further progress toward recent forms of bird life. Nearly all families of nonpasserines may have existed at this time. There were grebes, flamingos, albatrosses, cranes, limpkins, rails, woodpeckers, hawks, eagles, and vultures. The early Miocene was marked by slight warming, but by the middle Miocene Antarctica was covered by an ice sheet. The global climate began to cool and become much drier, leading to the widespread growth of grassy savannas. During this time, the majority of avian families and many genera of contemporary birds appeared. The Miocene avifauna is extremely rich. Various passerine birds, such as crows, thrushes, wagtails, shrikes, and wood warblers, also evolved. A giant pelecaniform bird, *Osteodontornis*, with jaws containing toothlike bony projections, is also known from the Miocene of California. The largest known flying bird, *Argentavis*, a member of the extinct family Teratornithidae, was also recorded from the late Miocene deposits of Argentina. *Argentavis* had a wingspan of about 6 to 8 meters and stood about 2 meters tall (fig. 12.5).

In the Pliocene, cooling climates dominated the Northern Hemisphere. An extremely rich avian assemblage is known from South Africa. The spread of savannas favored the appearance of many ground birds, such as ostriches, tinamous, and goatsuckers. The flightless auks and many families of passerine birds are known from this epoch. By the mid-Pliocene, the Arctic icecap had formed, heralding an ice age. By the end of the Pliocene, essentially all modern bird genera were established.

The Pleistocene record is extremely rich; all recent types had

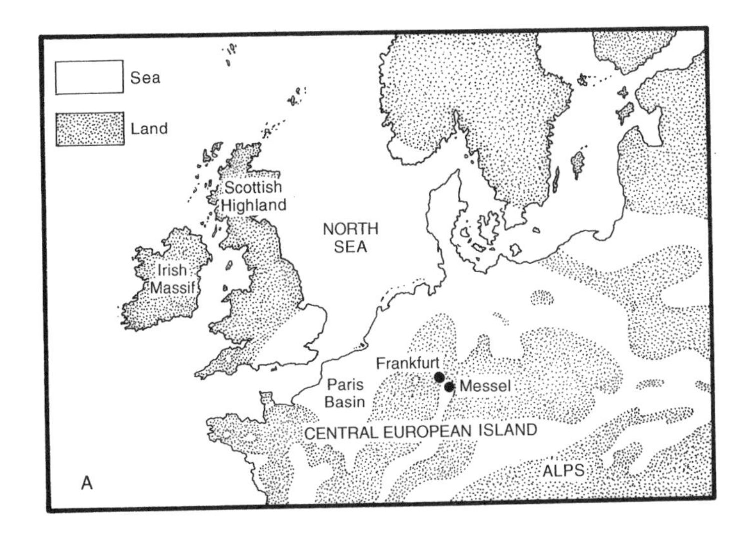

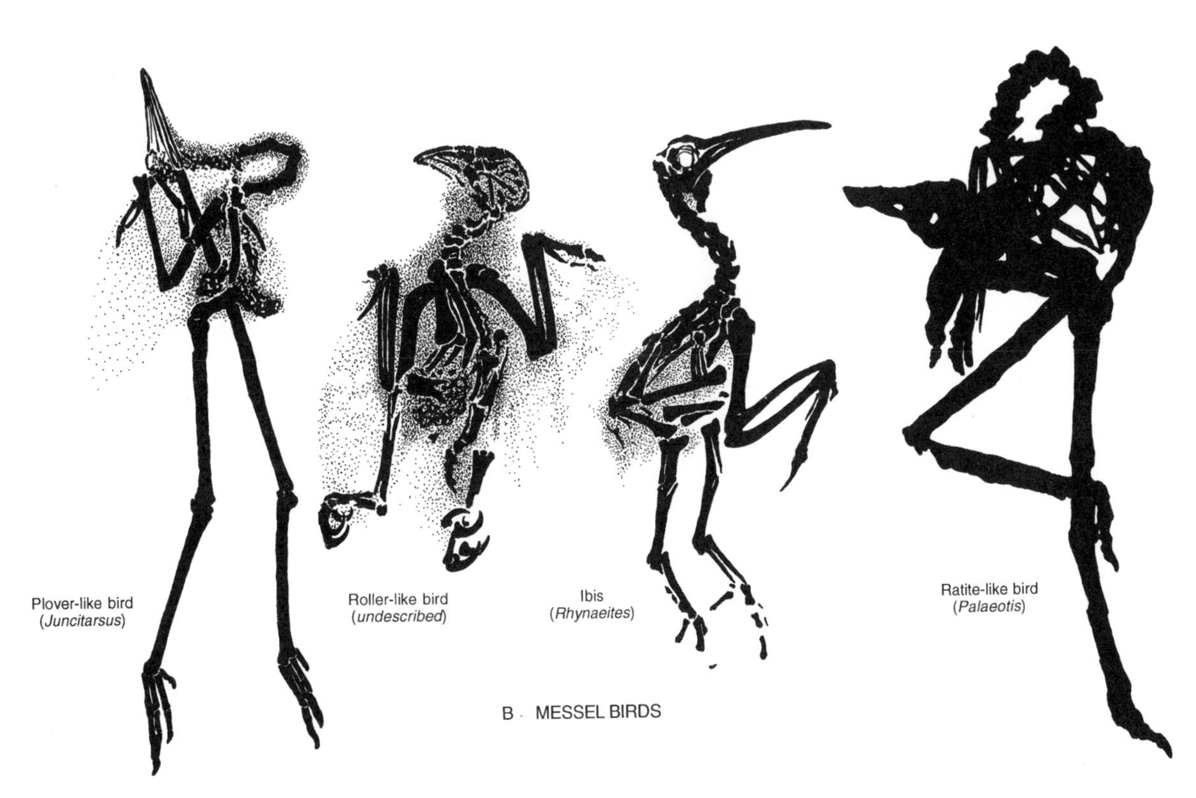

FIGURE 12.4

Messel bird assemblages of the Eocene epoch. *A,* Central Europe in the Eocene epoch showing the Messel fossil locality, near Frankfurt, Germany. *B,* some Messel birds—the plover-like *Juncitarsus,* an unnamed roller, the ibis *Rhynaeites,* and the ratite-like *Palaeotis*—preserved in the Messel shale (source: simplified from photographs in Peters 1992).

evolved and become established. The Pleistocene was also a period of expansion of many flightless species, including emus, cassowaries, moas, and elephant birds in the southern landmasses. The climate throughout this time continued its deterioration, culminating in the ice ages of the Pleistocene. The connection between North and South America was established, allowing a great interchange of fauna. The flightless, carnivorous ground bird *Titanis* moved from South to North America. In the Northern Hemisphere, the Pleistocene icecaps covered much of the continents.

FIGURE 12.5

The largest known flying bird, *Argentavis*, with a wing span of 7 or more meters, lived in Argentina during the Miocene period. The sheer scale of this giant condor is demonstrated by comparing its silhouette with that of the bald eagle (*Haliaeetus leucocephalus*) of today (source: Campbell and Marcus 1990).

The most remarkable Ice Age avifauna is known from the Rancho La Brea tar pits of California, where more than 135 species of birds were preserved between 4,000 and 40,000 years ago. Many of these birds represent predatory species, such as falcons, eagles, condors, and vultures, which were probably also carrion-feeders and may have been accidentally trapped in the asphalt while attempting to feed on trapped mammals (see fig. 13.3). The largest bird from these deposits, *Teratornis,* which is related to storks and New World vultures, had a wingspan of over 3 meters. Teratorns were ground-stalking birds as well as opportunist scavengers.

During the Pleistocene glaciation, a large number of avian species suffered extinction. The rapid rise of the Himalayan-Alpine range formed an orographic barrier for bird migration and dramatically altered the climate of Eurasia. Bird life as we know it today was essentially established some 20-50 thousand years ago. The dominant order today is the Passeriformes, which outnumbers all other orders combined.

The Phylogeny and Classification of Neornithine Birds

There are 9,672 recognized species of living birds (Sibley and Monroe 1990), and they come in all shapes and sizes. The smallest living form, the bee hummingbird of Cuba, is 6.3 cm long and weighs less than 3 g. The ostrich, the largest living bird, may stand 2.5 meters tall and weigh 135 kg. Some extinct birds, such as moas and elephant birds, were even larger and may have reached over 3 meters in height. A Pliocene condor-like bird, *Argentavis,* had an estimated wingspan of 8 meters and was by far the largest known flying bird (fig. 12.5).

Out of this stunning diversity, systematists have tried to classify birds in a meaningful order. All modern birds can be placed in a monophyletic clade, Neornithes. Avian taxonomy at the species level is well known, but classification at higher levels is in an unsatisfactory state. Our knowledge of the phylogenetic relationships among orders, suborders, and families of birds is inferior to that of mammals and reptiles. Birds are often classified on external characters like size, color, and pattern and on habitat. However, many such groupings are the result of convergent evolution from different parental stocks. The taxonomic characters that in other vertebrates would be used for lower-level identification are accorded ordinal rank in the classification of birds. The systematic positions of several bird groups remain open to question. Currently living birds are grouped

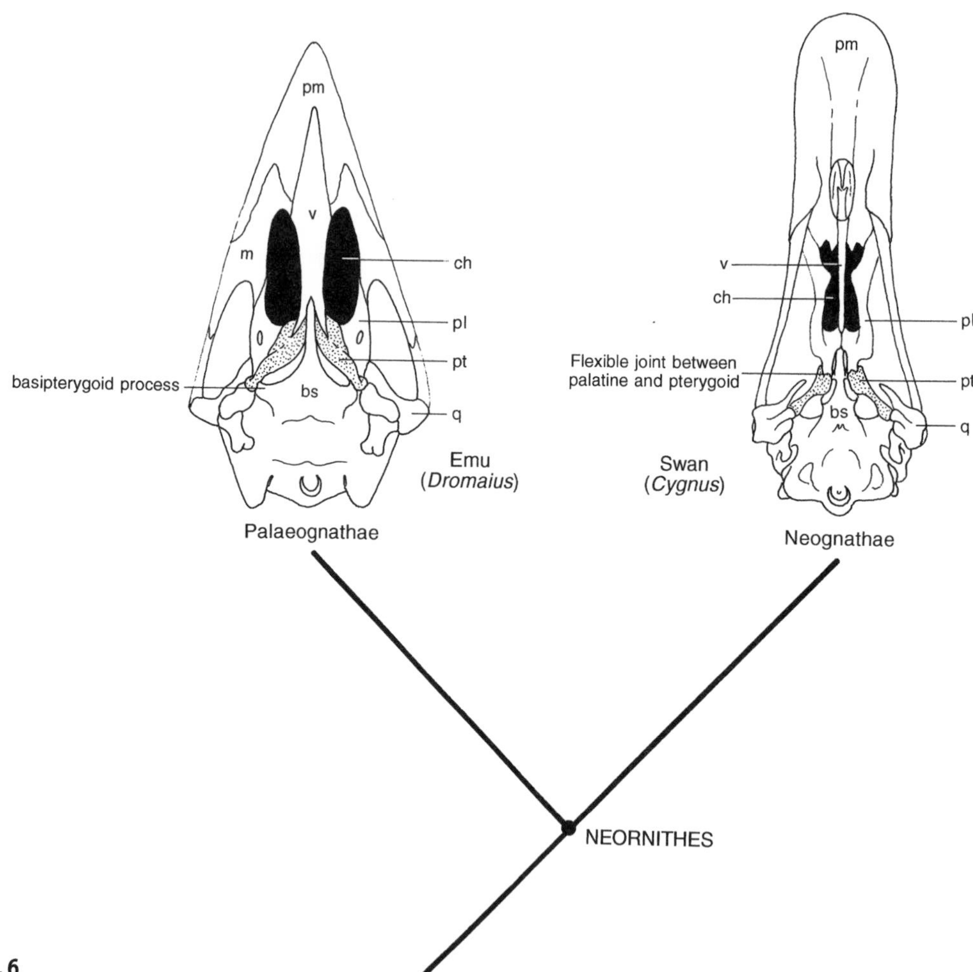

FIGURE 12.6

Cladogram showing the basal dichotomy of the neornithine birds into Palaeognathae and Neognathae on the basis of palatal structure. Reduction of the vomer and basipterygoid process and development of the flexible joint between the pterygoid and palatine are novelties for the neognaths. *bs,* basisphenoid; *ch,* choana; *m,* maxilla; *pl,* palatine; *pm,* premaxilla; *pt,* pterygoid; *q,* quadrate; *v,* vomer.

into twenty-seven orders on the basis of evolutionary systematics, but the hierarchical arrangements of taxa within orders and their interrelationships are poorly understood. Recently, there has been an attempt to present a phylogenetic classification of birds that reflects their genealogy and natural hierarchy, based on morphology (Cracraft 1974, 1981, 1988) and biochemical analysis (Sibley and Ahlquist 1990; Sibley and Monroe 1990).

Morphological Systematics

The most comprehensive classification of birds was proposed over a century ago by the German anatomist Hans Friedrich Gadow (1893) on the basis of morphological characters. Gadow used forty characters to develop his classification. He used several synapomorphies—palatal conformation, the exact formulation of the leg muscles, the configuration of the internal and external nares, the shape of the toe bones, the presence or absence of a fifth secondary feather, and the exact arrangement of scutes and scales on the legs—to group species into different hierarchical ranks. Although Gadow's classification has been modified and refined by several ornithologists over the years, it still forms the foundation for the morphological classification of birds.

Traditionally, neornithine birds are divided into two sister taxa, or supergroups—Palaognathae (ratites and tinamous) and Neognathae (all other modern birds)—on the basis of palatal configuration (Huxley 1867). Both cladistic analysis (Cracraft 1974) and molecular data (Sibley and Ahlquist 1990) support this basal dichotomy of neornithine birds. The different palatal structures of modern birds are linked to their different styles of feeding (fig. 12.6). It is widely believed that the avian palate can be derived from the basic archosaurian pattern with backward migration of the choanae, loss of the ectopterygoid, and development of palatal kinesis (Witmer and Martin 1987). The palaeognathous palate is the diagnostic feature of ratites and tinamous. In this type, the palatine and pterygoids are sutured in the primitive archosaur fashion; the vomer extends far back and articulates with the rostral ends of pterygoids; the elongate basipteryoid processes develop a flexible joint with the pterygoid.

The neognathous palate is characteristic of most flying birds. Here the basipterygoid processes are lost, a flexible joint is developed between the palatine and the pterygoid, and the vomers are reduced or lost. The movable pterygoid-palatine joint in the neognathous palate is an evolutionary novelty. Balouet (1982) considered virtually all other features of the neognathous palate to be consequences of pterygoid segmentation. The embryonic pterygoid in many neognaths splits into two portions: the rostral part becomes detached and fuses with the palatine, and the caudal part remains free and forms the adult pterygoid. Thus, a movable intrapterygoid joint is established between the rostral and caudal segments of an initially single bone but seems to lie between the pterygoid and the palatine in the adult. The adult palatine, in reality, is a compound bone, where the original palatine and the rostral portion of the pterygoid join to form a large element. The adult pterygoid is reduced and is represented by a stout bar connecting the parasphenoid rostrum with the condyle of the quadrate.

PALAEOGNATHOUS BIRDS

The origin and evolution of palaeognathous birds have been debated incessantly. The question is whether they are a natural group or merely an assemblage of unrelated forms that have followed a parallel line of evolution. Traditionally, six orders of palaeognathous birds are recognized: Tinamiformes (tinamous), Aepyornithiformes (elephant birds), Dinorthiformes (moas and kiwis), Casuariiformes (cassowaries, emus), Struthioniformes (ostriches), and Rheiformes (rheas). Of these, only tinamous can fly; the other groups are flightless.

The flightless groups are generally called *ratites.* The living ratites have a flat sternum, reduced wings, and a lack of ossification between the ilium and ischium around the ilioischiadic fenestra. It is generally believed that ratites descended from flying ancestors that lost their powers of flight as they evolved into medium-sized, grazing animals. This idea is supported by the discovery of *Lithornis*, a volant palaeognath from the Paleocene and Eocene of Wyoming (Houde and Olson 1981). There are ten living species of ratites, all restricted to southern continents: two species of rheas (*Rhea*) in South America, three species of cassowaries (*Casuaris*) in New Guinea, the emu (*Dromaius*) of Australia, three species of kiwis (*Apteryx*) in New Zealand, and the ostrich (*Struthio*) of Africa.

Some of the most spectacular ratites are now extinct: thirteen species of moas (Dinorthidae) of New Zealand, nine species of elephant birds (Aepyornithidae) of Madagascar, and eight species of mihirung birds (Dromornithidae) of Australia. These gigantic grazers were able to develop so successfully because of their isolation from significant competitors or predators. It is likely that their ancestors flew to these islands and then became flightless. Their fossils and eggs have been recovered from Pleistocene and Holocene localities.

The fossil record of ratites is spotty; the oldest record, *Remiornis*, extends back to the Paleocene of France (Martin 1992). Another flightless form, *Palaeotis* from the Eocene Messel Oil Shale, is regarded as a close relative of the ostrich (Peters 1992, 1995). It shows weak development of the sternum, pectoral girdle, and wings. African ratites are known from Eocene and Oligocene deposits. The discovery of North American, European, and Mongolian ancestral ratites has zoogeographic implication; it argues against the popular notion that ratites evolved in Gondwana, a thesis based on their present distribution.

Recent cladistic (Cracraft 1974) and biochemical (Sibley and Ahlquist 1990) analyses suggest that there are two basal lineages within the palaeognathous birds—tinamous and ratites. Cracraft (1974) recognized five successive clades in palaeognathous birds: Palaeognathae, Ratiti, Struthiones, Struthionoidea, and Struthionidae (fig. 12.7). Cracraft's cladogram is followed in our discussion of the relationships of palaeognathous birds.

Tinamidae
(tinamous)

NODE 1: PALAEOGNATHAE (TINAMIDAE + RATITI) The most primitive palaeognaths are the Tinamidae. The tinamous are partridge-like birds that range from southern Mexico to the tip of South America. They have weak bills and small tails. Although tinamous have the keeled sternum essential for flight, they are clumsy flyers. They are actually most closely related to moas and kiwis and are considered to be an outgroup of the ratites. Stratigraphic range: Upper Pliocene–Holocene; living species: 47 species; size: 15–50 cm.

Apteryges
(moas, kiwis)

NODE 2: RATITI (APTERYGES + STRUTHIONES) The moas are an extinct group of large, flightless, grazing birds that once flourished in New Zealand. Excessive hunting by Maori natives caused the extermination of these birds. There are twenty known species of moas; *Dinornis* stood over 3 meters tall and probably weighed about 250 kg. The kiwis of New Zealand are the smallest and most anomalous of the ratites. They are wingless, probing birds, with their nostrils at the tip of the flexible bill. They mainly feed on insects, worms, and berries. Stratigraphic range: Upper Miocene–Holocene; living species: 3; length 30-80 cm.

Aepyornithidae
(elephant birds)

NODE 3: STRUTHIONES (AEPYORNITHIDAE + STRUTHIONOIDEA) Elephant birds are another extinct group of large flightless birds; they lived on the island of Madagascar but were exterminated by human activity. They had long and powerful legs but retained vestiges of wings. The largest bird ever to exist, *Aepyornis maximus*, attained a height over 3 meters. Stratigraphic range: Upper Eocene–Holocene.

Casuariidae
(cassowaries, emus)

NODE 4: STRUTHIONOIDEA (CASUARIIDAE + STRUTHIONIDAE) Cassowaries spend much of the day hidden in the dense jungles of New Guinea and adjacent islands. They have a bony crest, or casque, on the top of the head, like a helmet, which may help to turn over loose soil or sand when they search for food. Their diet consists mainly of fruit, seeds, and berries. Emus, the world's second largest living birds, live in Australia. Both cassowaries and emus are large, flightless, cursorial birds with diminutive wings. Stratigraphic range: Pleistocene–Holocene; living species: 4; length: 130–190 cm.

NODE 5: STRUTHIONIDAE (STRUTHIONINAE + RHEINAE) The ostriches may date back to the Eocene and were widely distributed during the Neogene subperiod. Their fossils have been found in Europe, Africa, and Mongolia. Today only one species, *Struthio camelus,* remains. It is the largest living bird and inhabits the savanna or brushland of Africa. It can run faster than any other two-legged animal, and it lays the largest eggs of any living creature. The last ostriches in Arabia were killed during World War II. The ostriches have two toes (the 3rd

Struthioninae (ostriches)

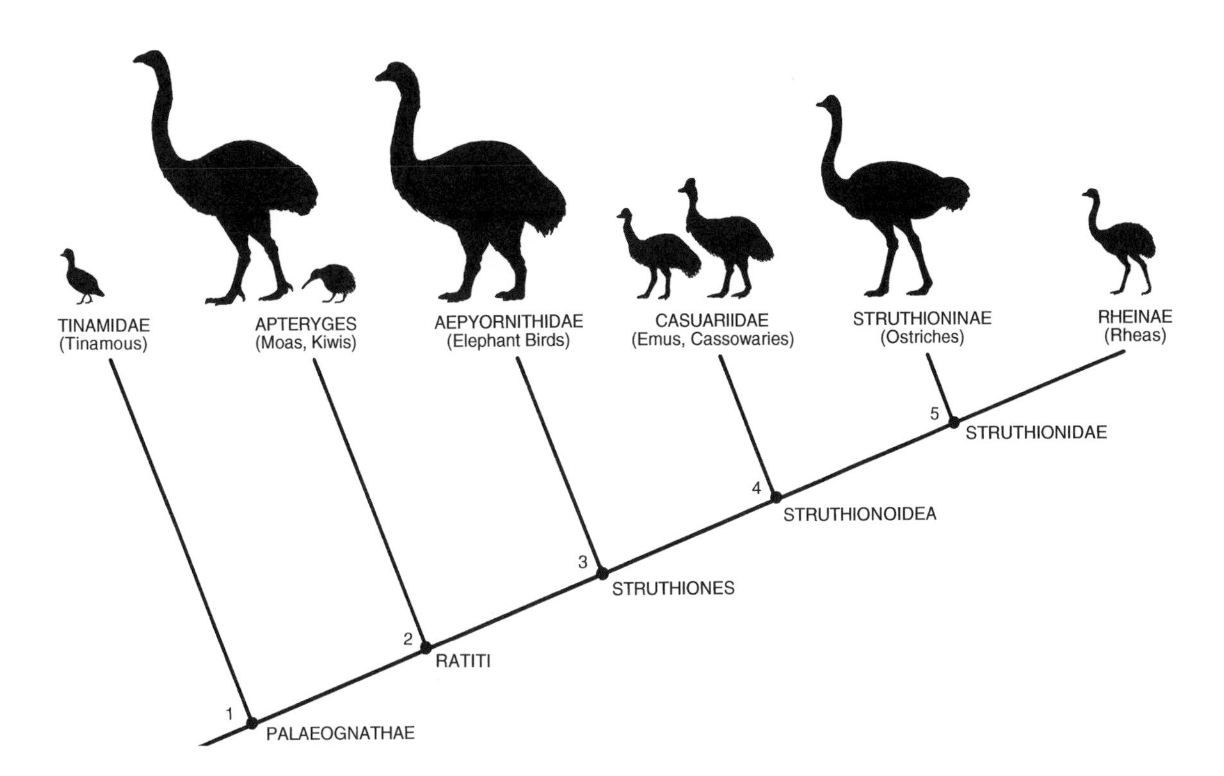

FIGURE 12.7

Phylogeny of palaeognathous birds (source: based on data in Cracraft 1974).

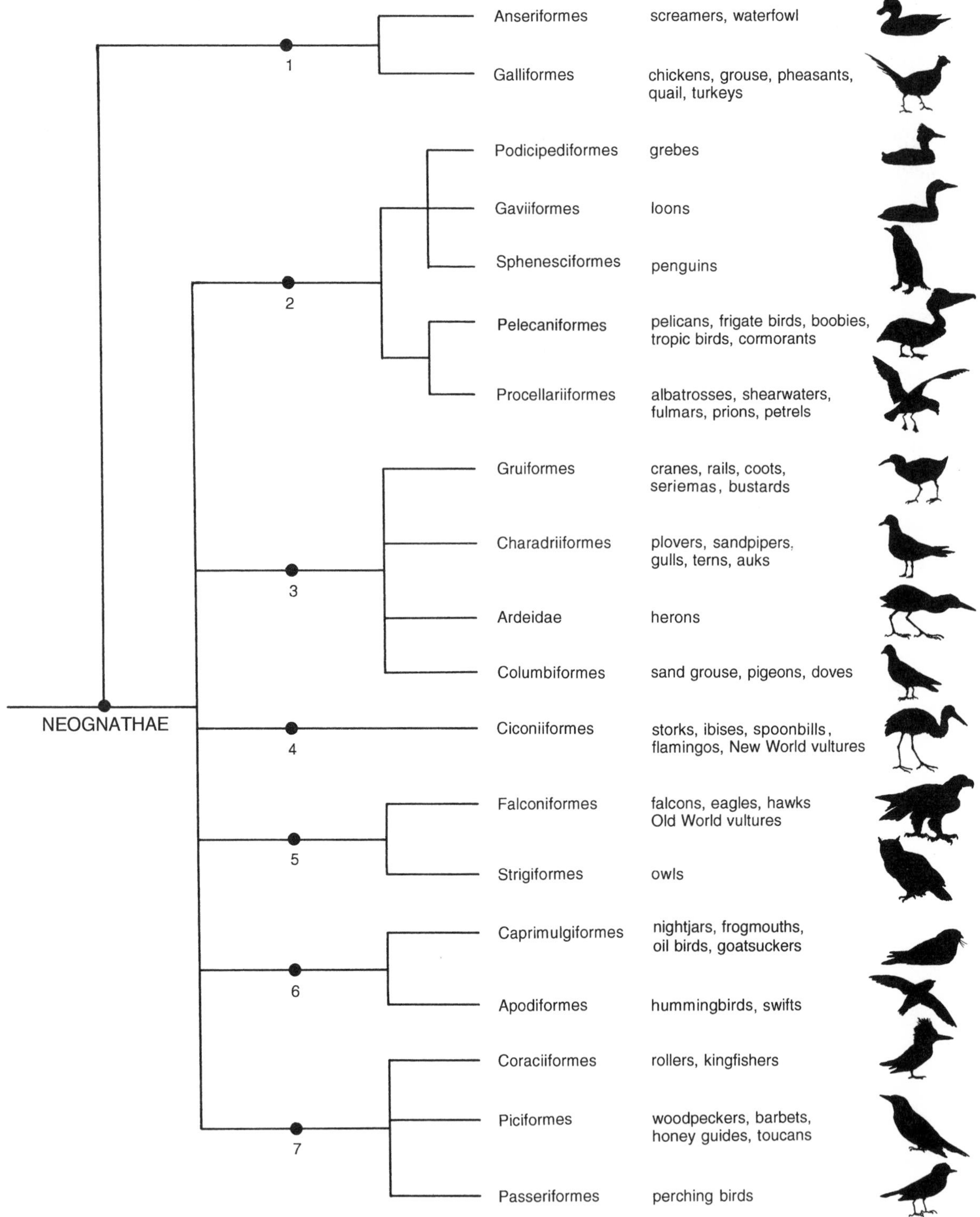

FIGURE 12.8

Morphological classification of neognathous birds (source: based on data in Cracraft 1986).

and 4th), and the lateral one seems to be in the process of vanishing. They are large, flightless, cursorial birds with small wings and may grow to be 2.6 meters tall. Stratigraphic range: Eocene–Holocene; living species: 1; length: 180 cm.

Rheinae (rheas)

Rheas have been restricted to South America throughout the Cenozoic. The two living species differ in size; the common rhea (approximately 1.4 meters) is taller than Darwin's rhea (approximately 90 cm). These flightless birds are capable of running long distances in the pampas. Rheas have small tails and only three toes; they lack an aftershaft in their loose, soft feathers. Stratigraphic range: Lower Eocene–Holocene; living species: 2; length: 90–130 cm.

NEOGNATHOUS BIRDS

The interrelationships of neognathous birds are highly controversial. Joel Cracraft (1988) provided a phylogenetic hypothesis covering the nineteen living orders of neognaths, which will be followed here for discussion (fig. 12.8).

Anseriformes (screamers, waterfowl)

NODE 1: ANSERIFORMES, GALLIFORMES Anseriforms are water birds with webbed or long, unwebbed toes. All waterfowl, such as ducks, geese, and swans, share webbed feet and broad bills that contain fine plates or lamellae to aid in food handling and in straining food organisms from water. The screamers are large-footed marsh birds of South America with chicken-like bills, and they bear little resemblance to other waterfowl. Stratigraphic range: Middle Eocene–Holocene; living species: 161; length: 29–160 cm.

Galliformes (chickens, grouse, pheasants, quail, turkeys)

Galliforms are chicken-like, vegetarian ground birds with short, stout beaks. They have strong, usually spurred legs that are well-adapted for running and walking. They possess short rounded wings and elaborate tail feathers and can fly for short distances. Stratigraphic range: Middle Eocene–Holocene; living species: 250; length: 13–198 cm.

Podicipediformes (grebes)

NODE 2: PODICIPEDIFORMES, GAVIIFORMES, SPHENESCIFORMES, PELECANIFORMES, AND PROCELLARIFORMES The grebes are highly specialized, foot-propelled diving birds with lobed toes and soft and silky plumage. The legs are short and located fairly far back on the body. They have many loonlike characteristics, such as long necks, pointed beaks, highly enlarged cnemial crests, and minute tails. Both grebes and loons can fly. The cnemial crest in grebes is formed by the patella and the tibia. Stratigraphic range: Lower Miocene–Holocene; living species: 17; length: 21–66 cm.

Gaviiformes (loons)

The loons are also foot-propelled diving birds with three webbed toes. They differ externally from the grebes in being larger and bulkier, lacking paddle-like lobes on the toes, and having rough plumage. The legs are short and located far back on the heavy body. The cnemial crest is formed entirely by the tibia. The loons are ancient birds known from the Cretaceous of Antarctica. Stratigraphic range: Late Cretaceous–Holocene; living species: 4; length: 66–95 cm.

Sphenesciformes (penguins)

The penguins are specialized seabirds that have completely lost the power of flight. They retain a deep keel and flipper-like wings for underwater propulsion. The wings lack flight feathers and cannot be folded. The legs are set well back along the body and bear webbed feet and strong nails that allow an erect posture. Penguins have a rich fossil record from Antarctica. Stratigraphic range: Eocene–Holocene; living species: 15; length 40–120 cm.

Pelecaniformes (pelicans, frigate birds, boobies, tropic birds, cormorants)

Pelecaniforms are large, fish-eating water birds with all four toes webbed for swimming. They are strong flyers but poor walkers. All have some form of flexible throat pouch that allows them to hold large fishes. They have long beaks that may be hooked, pointed, or straight. Their salt glands are located within the orbit. Among the living families, the frigate birds show the earliest record. Stratigraphic range: Eocene–Holocene; living species: 50; length: 50–180 cm.

Procellariiforms are large oceanic birds with webbed feet, prolonged tubular nostrils, and a characteristic musky smell. The nasal tubes entering into the bill may serve as outlets for excess salts taken into the system. The rhamphotheca of the bill is divided into plates. Members of this group spend most of the time soaring above the oceans and only come to land for nesting. Stratigraphic range: Middle Eocene–Holocene; living species: 81; length: 14–135 cm.

Procellariiformes (albatrosses, shearwaters, fulmars, prions, petrels)

NODE 3: GRUIFORMES, CHARADRIIFORMES, ARDEIDAE, AND COLUMBIFORMES Most gruiforms inhabit a marsh environment and are ground-nesting, but a few live on dry land and some even on deserts. Most species rarely fly, but some have managed to colonize oceanic islands. The Gruiformes are an ancient order, which includes large flightless forms such as phorusrhacoids and diatrymas. Stratigraphic range: Paleocene–Holocene; living species: 185; length: 11–152 cm.

Gruiformes (cranes, rails, coots, seriemas, bustards)

Charadriiforms are diverse, water-feeding shore birds. Some, such as plovers and sandpipers, are the familiar shorebirds with long legs, which usually feed on small animals in mud or water. The Lari include gulls and terns, which have webbed feet and are noted for their long migrations. They feed by plunging into water for fish, robbing other birds, or scavenging. The third group, the Alcae, are wing-propelled divers, rather like penguins, which feed on fish or invertebrates. Stratigraphic range: Late Cretaceous–Holocene; living species: 293; length: 13–76 cm.

Charadriiformes (plovers, sandpipers, gulls, terns, auks)

The herons are generally large wading birds in the family Ardeidae. Cracraft removes the family from ciconiiforms and places it closer to charadriiforms. Stratigraphic range: Eocene–Holocene, living species: 13; length: 91 cm.

Ardeidae (herons)

Columbiforms are fast-flying birds with long, pointed wings, a well-developed sternum, thick plumage, and weak bills. They feed mainly on seeds and fruits. Flightless forms, such as dodos and the solitaires of the Mascarene islands, became extinct dur-

Columbiformes (sand grouse, pigeons, doves)

ing the eighteenth century. Stratigraphic range: Eocene–Holocene; living species: 307; length: 15–84 cm.

Ciconiiformes (storks, ibises, spoonbills, flamingos, New World vultures)

NODE 4: CICONIIFORMES The systematic relationships of the taxa placed in this order are highly controversial. Ciconiiforms are large, fish-eating, long-legged wading birds with long bills, long necks, and broad wings. Stratigraphic range: Oligocene–Holocene; living species: 119; length: 28–152 cm.

Falconiformes (falcons, eagles, hawks, Old World vultures)

NODE 5: FALCONIFORMES, STRIGIFORMES Falconiforms are diurnal birds of prey, famous for their flying prowess. They have sharp, hooked beaks with nostrils set into the fleshy cere and powerful feet with long, sharp talons and an opposable hind toe. Stratigraphic range: Upper Paleocene–Holocene; living species: 287; length: 15–150 cm.

Strigiformes (owls)

Owls are raptorial birds, often with nocturnal habits. They possess hooked beaks, strong talons, and soft plumage. The eyes are frontally placed, with a large component of binocular vision. Stratigraphic range: Paleocene–Holocene; living species: 145; length: 13–69 cm.

Caprimulgiformes (nightjars, frogmouths, oil birds, goatsuckers)

NODE 6: CAPRIMULGIFORMES, APODIFORMES Caprimulgiforms are mainly nocturnal birds, with long, pointed wings, weak feet, small bills, and large mouths for catching insects. Stratigraphic range: Eocene–Holocene; living species: 95; length: 19–53 cm.

Apodiformes (hummingbirds, swifts)

Apodiforms are fast, acrobatic birds that feed on the wing, eating insects and nectars. Flying from flower to flower, the hummingbird pollinates the plants on which it feeds. Its long beak is ideal for feeding on nectar. Hummingbirds and swifts have powerful flight muscles and short feet with sharp claws that enable them to cling on perches. Stratigraphic range: Upper Eocene–Holocene; living species: 422; length: 6.3–23 cm.

NODE 7: CORACIIFORMES, PICIFORMES, PASSERIFORMES Coraciiforms nest in the hollows of trees. The birds in this diverse group share pointed bills, plumed crests, rounded wings, a long tail, and syndactyl feet. Stratigraphic range: Eocene–Holocene; living species: 194; length: 9–150 cm.

Coraciiformes (rollers, kingfishers)

Piciforms are tree-dwelling birds that nest in holes and have specialized bills, brilliantly colored feathers, and zygodactyl feet. Stratigraphic range: Oligocene–Holocene; living species: 387; length: 9–61 cm.

Piciformes (woodpeckers, barbets, honey guides, toucans)

The Passeriformes are the dominant group of modern birds, containing over half of the living species of birds. They include all song birds, such as robins, finches, thrushes, sparrows, and jays, and have a complex syrinx system. Their feet are adapted to cling to branches automatically. They are essentially land birds and show great variation in size and color. Their fossil record is sketchy; they became an important component of the avifauna in the Neogene subperiod. Stratigraphic range: Oligocene–Holocene; living species: 5,712; length: 9–61 cm.

Passeriformes (perching birds)

Molecular Systematics

Sibley and Ahlquist (1990) used a different method, the DNA-DNA hybridization technique, to propose a molecular classification of birds. In this classification, the deoxyribonucleic acid (DNA) of one bird species is compared against that of other birds on the basis of sequence data or genetic distance, and the birds are then grouped accordingly. This classification rests on a simple assumption: once two species separate in evolution, the DNA in the two lines accumulates changes or mutations. By measuring the genetic difference, the length of time since the divergence of two species can be estimated and the relationships between them can be established. In some cases, molecular classification is congruent with the traditional morphological approach. In other cases, however, there is serious discordance between the two methods. One of the unusual results of biochemical classification is the new composition of the Order Ciconiiformes, which now incorporates these former orders—Sphenesciformes, Gaviiformes, Podicipediformes, Procellariformes, Charadriiformes, Falconiformes, and Pelecaniformes (fig. 12.9). One of the reasons for the conflict between morphological and molecular classifications may be the failure to recognize convergences. Since birds have had a single evolutionary history or phylogeny, a compromise between the morphological and biochemical classifications is likely to emerge in the future as data are obtained and critically examined.

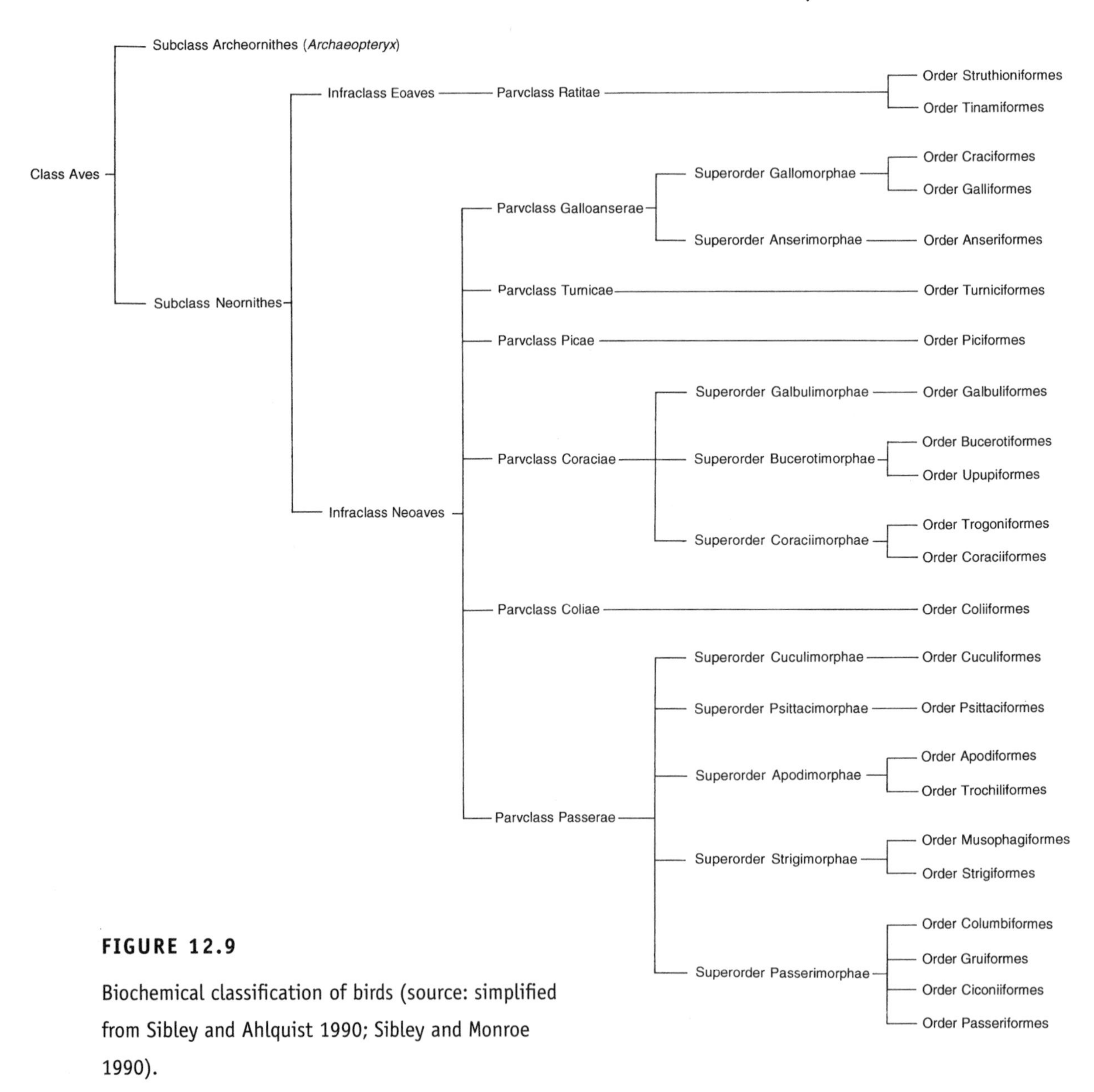

FIGURE 12.9

Biochemical classification of birds (source: simplified from Sibley and Ahlquist 1990; Sibley and Monroe 1990).

Birds and Humans

Hear the prayer of an earth that is stricken with pain:
In the green woods, O may the birds
Sing supreme again.

—Rabindranath Tagore, "Flying Man," 1940

CHAPTER 13

Since the beginning of humanity, we have worshipped birds as deities, used them as symbols of power and royalty, portrayed them in art and folklore, kept them as pets, hunted them for food and feathers, and watched them for aesthetic and recreational pleasure. Thousands of years ago our ancestors immortalized birds in numerous Stone Age cave paintings in France, Spain, Africa, India, and elsewhere. For millennia birds have enchanted us with their bright colors, striking behaviors, and melodious songs. Their awesome power of flight inspired us to conquer the air. During our long association with birds, however, we have also become their adoring assailants, causing countless species to disappear.

The Quaternary Extinction of Birds: Human Impact

Birds endured two mass extinctions, one at the end of the Triassic and the second at the end of the Cretaceous, but they rebounded from these crises. Since then, the number and variety of species has progressively increased to the present peak of diversity. However, birds are now at risk because of the effects of human activity. People are directly responsible for the extinction of many wild birds. Some species have been hunted to oblivion. Others have been killed off as people destroyed natural habitats to make room for their own settlements. Human beings have also polluted the environment, which has wreaked havoc on bird diversity. These assaults started a long time ago when early humans began migrating to new frontiers and confronting pristine habitats. Extinctions of birds were more severe on islands than on continents. Edward O. Wilson (1992), the famous American naturalist, named these human-induced catastrophes *centinelan extinctions* after the Ecuadorian ridge Centinela at the Andean foothills of Ecuador. Centinela witnessed the collapse of biodiversity and wept silently because of human interference. Today, humans seem to be creating a sixth major mass extinction event.

EXTINCTION ON ISLANDS

Before human colonization, many islands, such as Hawaii, New Zealand, Madagascar, and Mascarene, supported a rich and varied endemic biota. The tropical climate and abundant rainfall gave rise to dense rain forests. Because of isolation and lack of predation, many birds on these islands became secondarily flightless. Biotic crises began thousands of years ago when early human explorers began to colonize these islands and exploit their resources. They hunted the easy prey, destroyed natural habitats, and pushed many avian species to the brink of extinction. From archaeological sites, sinkholes, lava tubes, sand dunes, and flooded caverns, Storrs Olson, Helen James, and David Steadman have pieced together the telltale evidence of human-caused avicides on these Pacific islands. The statistics of destruction are staggering: about 2,000 species of birds have

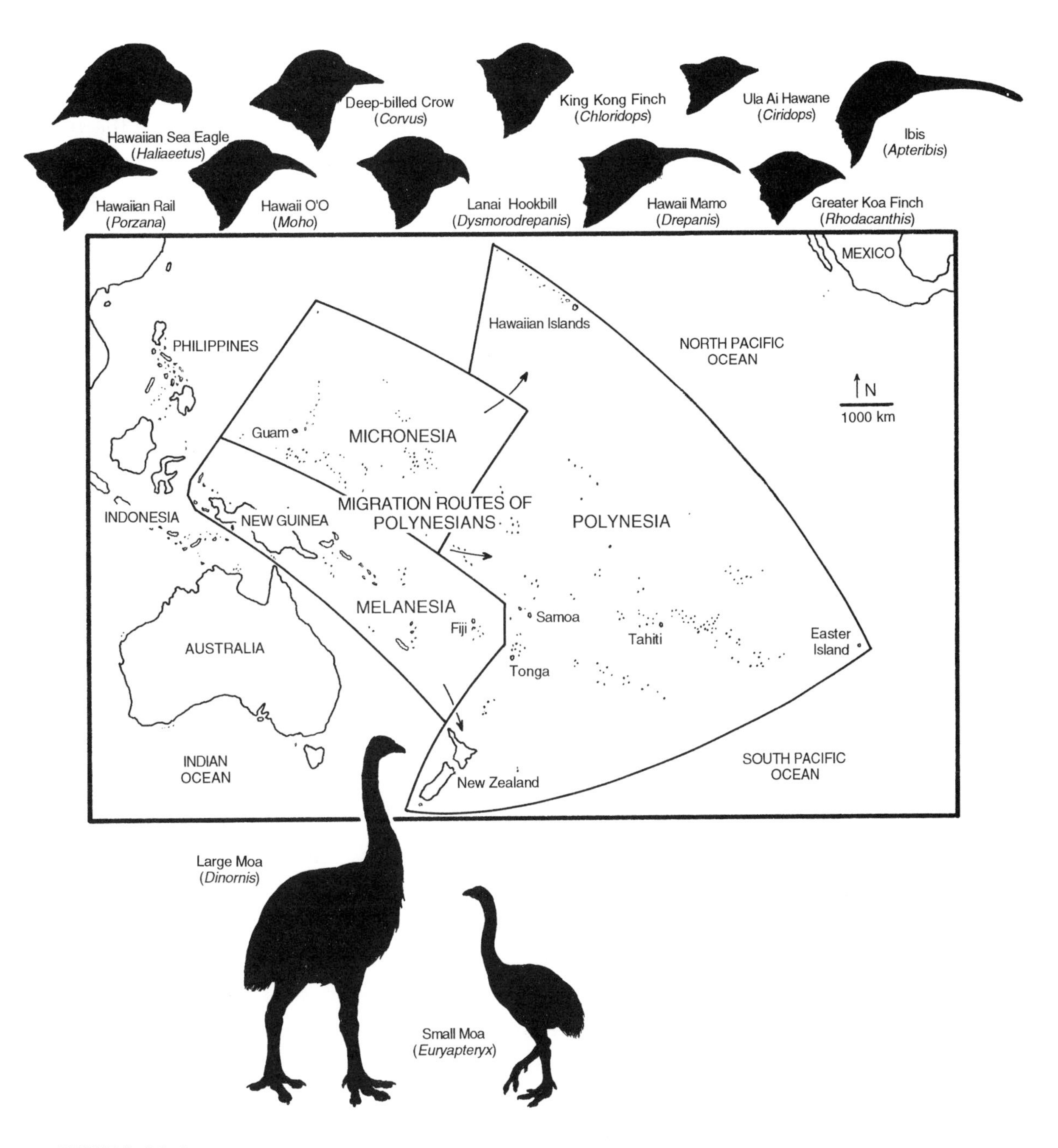

FIGURE 13.1

Human-induced extinctions of flightless birds in the Pacific islands; *top,* silhouettes of bird victims in Hawaii (source: Royte 1995); *center,* map of various Pacific islands colonized by the Polynesians; *bottom,* silhouettes of the giant and small moas of New Zealand.

been extinguished in the past 2,000 years, principally following the human occupation of these islands (Olson and James 1984, 1991; Steadman 1995).

The Hawaiian Islands (fig. 13.1) stretch in a great arc, some 2,500 km long, in the North Pacific Ocean, about midway between California and Japan. The first settlers in Hawaii were Polynesians. They reached these remote Pacific islands about 1,500 years ago by means of rafts and dugout canoes from western archipelagoes of Melanesia and Micronesia. They cut and burned the forests to plant taro and other crops and brought dogs, pigs, and rats into the virgin ecosystem. They encountered only two species of mammals—a bat and a seal—but abundant bird life. They ate fish, turtles, and a wide variety of flightless birds, which, being slow-footed, were easy to catch. Among the first to be extinguished were some geese, such as *Thambetochen;* the tortoise-jawed moa nalo (*Chelychelynechen*); several varieties of rails (*Porzana*), ibises (*Apteribis*), owls (*Grallistrix*), and finches; an eagle (*Haliaeetus*); a hawk; and a petrel. Clearing of forests by early settlers must have contributed heavily to the destruction of many species of passerine birds. Other crises may have been precipitated by the exotic predators that harmed endemic species through predation, competition, and disease. It is estimated that about sixty endemic species were lost because of the interference of Polynesian colonists.

A second wave of extinctions followed the arrival of Europeans in the eighteenth century, with more forest clearing and more introduced mammals, such as cattle and goats. Most of the natural habitat was lost to cultivation, browsing, and fire, eliminating another twenty to twenty-five species. Among the victims were an eagle, a flightless ibis, a strange owl, honeyeaters, honeycreepers, finches, and a dozen other land birds.

The human-induced extinctions on Hawaii were truly enormous, extinguishing 70 percent of the species that had made up the total avifauna. Centinelan extinctions on the same grand scale also occurred on other Polynesian islands, such as Henderson, Marquesas, Society, Cooks, Samoa, Tonga, and Polynesian outliers of Melanesia. Today, the avian fauna on Hawaii and other Pacific islands is highly impoverished because of human influence.

Another arena of destruction is New Zealand, the southwestern outpost of the Polynesian migrations. Lying about 1,600 km southeast of Australia, it has been separated from other Gondwana continents since the Cretaceous and lacked native mammals (except for bats) before human settlement. In this splendid isolation, giant flightless birds like moas thrived, especially in the South Island temperate forest. The first people to live in New Zealand were the Maoris, a Polynesian tribe that arrived about 900 A.D. Maoris lived mainly by fishing and hunting moas. When Maoris settled in New Zealand, they encountered fifteen species of moas ranging in size from that of large turkeys to giants weighing more than 250 kg. The moas were grazing animals and were defenseless. The Maoris aggressively hunted moas in large numbers and robbed their nests. Hundreds of archaeological sites testify to the indiscriminate slaughtering of moas all over the islands, with bones and eggs preserved in ancient kitchen middens. It took the Maoris about 600 years to hunt the moas to extinction. They also wiped out twenty other land birds, including nine additional flightless species. Humans had a devastating effect on the biota of New Zealand, which has lost forty-four endemic species of land birds during the past millennium.

Centinelan extinctions were also rampant on the islands of the Indian Ocean (fig. 13.2). Madagascar is the fourth largest island in the world. A rift separated it from Africa and India during the Cretaceous, and it moved southward and then, over millions of years, evolved an exotic biota that included several species of majestic elephant birds. These robust ratites were important grazers and browsers in this tropical forest, and the largest species attained heights of over 3 meters. They resembled heavyweight ostriches with massive legs and weighed about 450 kg. Immigrants from Indonesia moved to this island in successive waves between 1,500 and 2,000 years ago, followed by African settlers. It is likely that these Malagasy settlers hunted not only the elephant birds but also their huge eggs. The eggs, about 34 cm long and 11 liters in capacity, were used for both food and containers. The beaches of Madagascar are still littered with eggshells of elephant birds. These defenseless birds

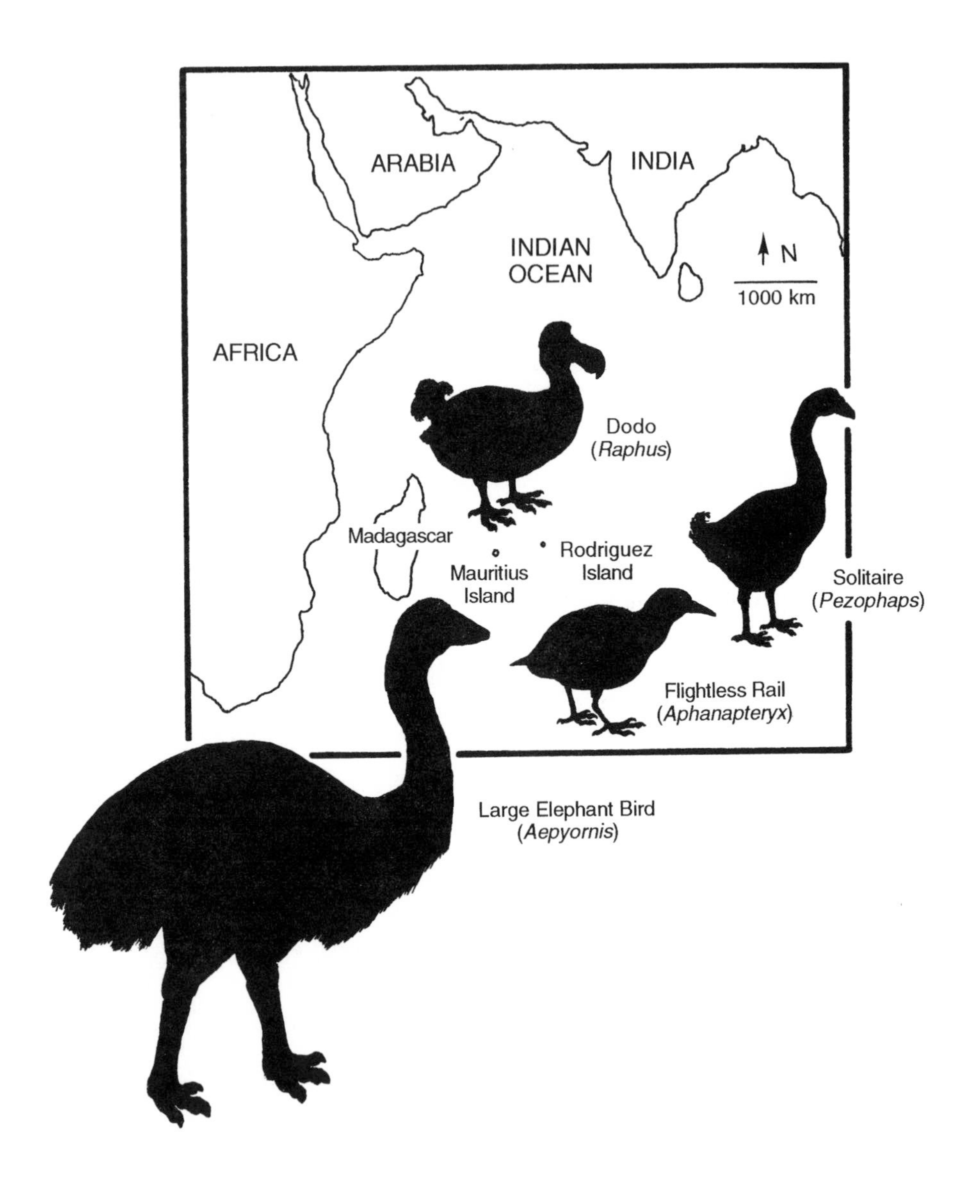

FIGURE 13.2

Human-induced extinctions of flightless birds in the islands of the Indian Ocean: the recently extinct dodo of Mauritius, the solitaire and flightless rail of Rodriguez Island, and the large elephant bird of Madagascar.

became easy prey for the arriving humans. Larger animals generally had smaller populations and lower reproductive rates, making them vulnerable to overexploitation. Continued hunting and egg collecting over centuries, along with habitat destruction, led to the extinction of the largest birds that ever lived. The last living elephant bird *Aepyornis* was reported in 1658.

A similar sad story of avian extinction was repeated in the Mascarene Islands with the arrival of Europeans during the fifteenth century. The Mascarene Islands lie east of Madagascar and constitute several land masses formed by the Deccan hotspot trails, including Réunion, Mauritius, and Rodriguez, which were isolated for millions of years during the northward drift of India. The total number of human-induced bird extinctions in the Mascarenes is close to thirty species, including several species of rails, waterfowl, and land birds. The most famous casualties are the dodo (*Raphus*) of Mauritius and the solitaire (*Pezophaps*) of Réunion and Rodriguez islands. These large, flightless pigeons had great body size (approximately 20 kg) and substantially reduced pectoral limbs. They were well adapted to these islands and evolved into many strange and colorful species before the arrival of their formidable enemy. The dodos and solitaires were mercilessly hunted by European sailors for sport and provisions. The newly introduced monkeys and pigs also relished their eggs. Their extinction was so rapid and so complete that the vague descriptions given of them by early navigators were long regarded as mythical. By 1680, dodos were gone forever. The extinction of solitaires probably happened a century later.

In the pattern of extinction among island birds, large flightless birds were the main victims because of overkill. These birds, having never before seen human beings, were unafraid of people and were easily approached and killed. Flightless rails were also severely threatened on these islands. Many of the extinct species come from six groups: ratites, rails, waterfowl, pigeons, crows, and birds of prey. The clearing and destruction of

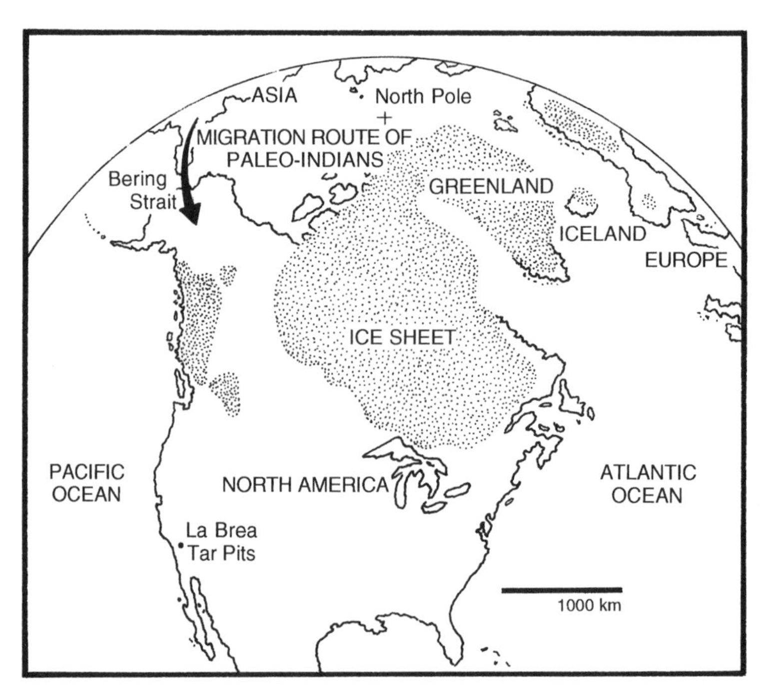

FIGURE 13.3

Top, silhouettes of various extinct species from the La Brea tar pits; *bottom,* paleogeography of North America during the last ice age. The Paleo-Indians arrived about 12,000 years ago; note the location of La Brea tar pits in Los Angeles (source: Harris and Jefferson 1985).

forests for agriculture fragmented the habitat of native species, contracted their range, reduced their resources, and caused their populations to dwindle. There was nowhere left to feed, hide, or escape. The surviving island birds are, for the most part, mere relics of a rich avifauna that vanished not long ago.

EXTINCTION ON CONTINENTS

On most continents where humans and birds evolved together for millions of years, the effect of humans on the diversity of birds was minimal—until recently. In North America, the major extinction of birds coincides with the arrival of Paleo-Indians 12,000 years ago, by way of the Bering land bridge. These Paleo-Indians were such successful big-game hunters that their populations expanded and spread rapidly—perhaps by about 16 km per year. By 10,000 years ago, humans occupied most parts of both North and South America. During this time most of the large mammals became extinct—mastodons, saber-tooth cats, camels, giant beavers, and lemurs—as did many species of birds that had fed on them (fig. 13.3). Whether early Indians armed with sophisticated stone tools and projectiles destroyed the mammalian megafauna remains controversial. There has been a continuing debate over whether human hunting or climatic change was to blame for the extinction of large mammals in the Americas. The close chronological coincidence between the extinctions of large mammals and the arrival of big-game hunters in North America is striking.

If, within a few thousand years of their arrival, the Paleo-Indians threatened the populations of large mammals of North America by overkill or blitzkrieg, as maintained by Paul Martin (1984), then human interference had indirect repercussions on the elimination of the late Pleistocene avifauna. This avian crisis is very unusual because, of nineteen extinct species, most were large predatory birds—teratorns, condors, vultures, carrion storks, hawks, and eagles—that scavenged on large mammals. None of the birds was suitable for human predation.

However, the extinction of large mammals surely had a ripple effect on the scavengers that fed on them. The short supply and eventual lack of large carrion disrupted the food chain. In most cases the avian extinctions in North America cannot be attributed directly to human interference, except for a flamingo and one anatid, *Chendytes* (Steadman and Martin 1984). In recent times, the passenger pigeon, Carolina parakeet, Labrador duck, great auk, and heath hen have been victims of humanity.

The Recent Crisis

The effects of humans on avian diversity and that of other biota have become increasingly disastrous. Since the time of Christ, 20 percent of the bird species worldwide have become extinct, 11 percent of the surviving species are seriously endangered, and more than 50 percent of the world's birds are declining in numbers (Wilson 1992). The parrot family is the most threatened group, with more than seventy species at risk. This is an alarming signal. Recent statistics are not encouraging. Since 1860, approximately eighty species of birds have been lost. Overkilling, habitat destruction, and the introduction of exotic species have contributed to the past demise of avian species. In recent times, the crisis has been compounded by pollution, pesticides, nuclear waste, and human overpopulation. The most alarming human assault upon the environment is the contamination of air, earth, rivers, and sea with dangerous pollutants.

We are stealing from future generations the basic necessities of life. We are not only polluting the environment, but also increasing our territory and range. It has been estimated that the human population will exceed 8 billion some time in the next fifty years, making ever greater demands on Earth's resources. In our thirst for water, our hunger for land, and our appetite for natural resources, we have altered large areas of habitat for human settlement. Many species in shrinking habitats will find it increasingly difficult to live in a world dominated by *Homo sapiens,* the most universally predatory species that has ever existed. With our technology and explosive growth, we are rapidly threatening many other species, ones that have been around millions of years longer than we.

It seems that our planet will increasingly become the home of a single aggressive species. This will have a cascade effect on the destruction of habitat and biodiversity. We are now victims of our own success and have not adequately assessed the dark side of our achievement. The pumping of greenhouse gases into the atmosphere and the depletion of stratospheric ozone have serious consequences on the global climate and ecology. Will the rain forests survive? Will there be any wilderness left for wildlife? Will the lakes and the rivers be contaminated with pesticides and pollutants, killing all their inhabitants? Will desertification claim more and more arable land? Will Nature take her revenge with recurrent crop failures, volcanic eruptions, earthquakes, floods, droughts, diseases, and pestilence to bring us in line with the rest of life? Are we precipitating another biotic catastrophe, the sixth great mass extinction? Many believe so. This time the agent of destruction will not be a bolide from outer space, but home-grown human beings. It is not clear whether the survivors will include our species after the recovery.

Against this gloomy background we must decide the future of biodiversity. The world's rare birds need our help if they are to survive. Birds are sensitive indicators for detecting environmental disturbance—they are closely linked to all other wildlife on the planet, and problems for birds serve as warning signals for the survival of plants, other animals, and people. As Wilson (1992) asserted, we have to develop an environmental ethic for long-range plans to save this fragile planet. The conservation of biodiversity is a global responsibility. Wildlife does not recognize political boundaries. Each county, each state, each nation, and each international agency has a necessary role to play in finding new ways to manage biological resources. The Nature Conservancy has done an admirable job in preserving plants and natural communities that represent the diversity of life by protecting the land and waters they need to survive. But many of these preserves are small, and we must take additional steps to conserve biodiversity before we are overtaken by another catastrophe. We must educate the public about the potential dangers created by ourselves. It is in our own best interests to safeguard the biosphere. This is the only home in the entire solar system that we can share together.

Bibliography

Alberch, P., S. J. Gould, G. F. Oster, and D. B. Wake. 1979. Size and shape in ontogeny and phylogeny. *Paleobiology* 5:296–317.

Alonso, R. N., and R. A. Marquillas. 1986. Nueva localidad con nuellas de dinosaurios y primer hallazo do huellas de Aves en la Formacion Yacoraite (Maastrichtian) del Norte Argentino. *Act. IV Congr. Argent. Paleontol. Biostratigraph. Mendoza* 2:33–41.

Alvarenga, H. M. V., and J. F. Bonaparte. 1992. A new flightless landbird from the Cretaceous of Patagonia. In Campbell, *Papers in Avian Paleontology*, 51–64.

Alvarez, L. W., W. Alvarez, F. Asaro, and H. V. Michel. 1980. Extraterrestrial cause for the Cretaceous-Tertiary extinction. *Science* 208:1095–1108.

Alvarez, W. 1986. Toward a theory of impact crises. *EOS Trans. Am. Geophys. Res.* 63 (35): 649–658.

Alvarez, W., and F. Asaro. 1990. An extraterrestrial impact. *Sci. Am.* 263 (4): 78–84.

Andors, A. V. 1995. *Diatryma* among the dinosaurs. *Nat. Hist.* 104 (6): 68–72.

Ash, S. R. 1972. Upper Triassic Dockum flora of eastern New Mexico and Texas. In *New Mexico Geological Society Guide Book 23rd Field Contr.* 124–128.

Bakker, R. T. 1975. Dinosaur renaissance. *Sci. Am.* 232 (4): 58–78.

Bakker, R. T. 1986. *The Dinosaur Heresies.* New York: William Morrow.

Balda, R. P., G. Caple, and R. R. Willis. 1985. Comparison of the gliding to flapping sequence with the flapping to gliding sequence. In Hecht et al., *The Beginnings of Birds,* 267–277.

Balouet, J.-C. 1982. Les paleognaths (Aves) sont-ils primitifs? *J. Soc. Zool. Fr.* 82:648–653.

Barsbold, R. 1983. Carnivorous dinosaurs from the Cretaceous of Mongolia (in Russian). *Trans. Joint Soviet-Mongolian Paleontol. Expeditions* 19:5–120.

Basu, A. R., P. Renne, D. K. Dasgupta, F. Teichmann, and R. J. Foreda. 1993. Early and late alkali igneous pulses and high $^{-3}$He plume origin of the Deccan Flood Basalts. *Science* 261:902–906.

Bock, W. J. 1964. Kinetics of the avian skull. *J. Morphol.* 114:1–41.

Bock, W. J. 1965. The role of adaptive mechanisms in the origin of higher levels of organisms. *Syst. Zool.* 14:272–287.

Bock, W. J. 1969. The origin and early radiation of birds. *Ann. N.Y. Acad. Sci.* 167:147–155.

Bock, W. J. 1983. On extended wings. *Sciences N.Y.* 23:16–20.

Bock, W. J. 1985. The arboreal theory for the origin of birds. In Hecht et al., *The Beginnings of Birds,* 199–207.

Bock, W. J. 1986. The arboreal origin of avian flight. In Padian, *Origin of Birds and Evolution of Flight,* 57–72.

Bock, W. J., and P. Bühler. 1995. Origin of birds: feathers, flight and homoiothermy. *Archaeopteryx* 13:5–13.

Bohor, B. F., P. J. Modreski, and E. E. Foord. 1987. Shocked quartz in the Cretaceous-Tertiary boundary clay: evidence for a global distribution. *Science* 224:705–709.

Brown, R. H. J. 1951. Flapping flight. *Ibis* 93:333–359.

Brown, R. H. J. 1963. The flight of birds. *Biol. Rev.* 38:460–489.

Browne, M. W. 1996. Feather fossil hints dinosaur-bird link. *New York Times,* 19 October, pp. 1, 11.

Buffetaut, E., J. Le Loeff, P. Mechin, and A. Mechin-Salessy. 1995. A large French Cretaceous bird. *Nature* 377:110.

Bühler, P. 1985. On the morphology of the skull of *Archaeopteryx.* In Hecht et al., *The Beginnings of Birds,* 135–140.

Burton, R. 1990. *Bird Flight.* New York: Facts on File.

Camp, C. S. 1936. A new type of small bipedal dinosaur from the Navajo Sandstone of Arizona. *Univ. Calif. Publ. Geol. Sci.* 24:39–53.

Campbell, K. E., Jr., ed. 1992. *Papers in Avian Paleontology: Honoring Pierce Brodkorb.* Science Series No. 36. Los Angeles: Natural History Museum of Los Angeles County.

Campbell, K. E., Jr., and L. Marcus. 1990. How big was it? Determining the size of the ancient birds. *Terra* 28 (4): 33–43.

Campbell, K. E., Jr., and E. Tonni. 1983. Size and locomotion in teratorns (Aves: Teratornithidae). *Auk* 100:390–403.

Caple, G., R. P. Balda, and W. R. Willis. 1983. The physics of leaping animals and the evolution of pre-flight. *Am. Naturalist* 121: 455–476.

Cartmill, M. 1985. Climbing. In *Functional Vertebrate Morphology,* 73–88. Edited by M. Hildebrand, D. M. Bramble, K. F. Lien, and D. B. Wake. Cambridge: Harvard University Press.

Chatterjee, S. 1983. An ictidosaur fossil from North America. *Science* 220:1151–1153.

Chatterjee, S. 1984. A new ornithischian dinosaur from the Triassic of North America. *Naturwissenschaften* 71:63–631.

Chatterjee, S. 1985. *Postosuchus,* a new thecodontian reptile from the Triassic of Texas and the origin of tyrannosaurs. *Philos. Trans. R. Soc. Lond. [Biol.]* 309:395–460.

Chatterjee, S. 1986. The Late Triassic Dockum vertebrates: their biostratigraphic and paleobiogeographic significance. In *The Beginning of the Age of Dinosaurs,* 139–150. Edited by K. Padian. New York: Cambridge University Press.

Chatterjee, S. 1987a. Skull of *Protoavis* and early evolution of birds. *Abs. J. Vert. Paleontol.* 7 (3): 16A.

Chatterjee, S. 1987b. A new theropod dinosaur from India with remarks on the Gondwana-Laurasia connection in the Late Triassic. In *Gondwana Six: Stratigraphy, Sedimentology and Paleontology,* 183–189. Geophysical Monograph 41. Washington, D.C.: American Geophysical Union.

Chatterjee, S. 1988. Functional significance of the semilunate carpal in archosaurs and birds. *Abs. J. Vert. Paleontol.* 8 (3): 11A.

Chatterjee, S. 1989. The oldest Antarctic bird. *Abs. J. Vert. Paleontol.* 9 (3): 16A.

Chatterjee, S. 1991. Cranial anatomy and relationships of a new Triassic bird from Texas. *Philos. Trans. R. Soc. Lond. [Biol.]* 332:277–342.

Chatterjee, S. 1992a. The dawn of the age of dinosaurs. In *Ultimate Dinosaurs,* 2–9. Edited by B. Preiss and R. Silverberg. New York: Bantam Press.

Chatterjee, S. 1992b. A kinematic model for the evolution of the Indian plate since the Late Jurassic. In *New Concepts in Global Tectonics,* 33–62. Edited by S. Chatterjee and N. Hotton. Lubbock: Texas Tech University Press.

Chatterjee, S. 1993. *Shuvosaurus,* a new theropod. *Natl. Geogr. Res. Expl.* 9:476–491.

Chatterjee, S. 1994. *Protoavis* from the Triassic of Texas: the oldest bird. *J. Ornithol.* 135:330.

Chatterjee, S. 1995. The Triassic bird *Protoavis. Archaeopteryx* 13:15–31.

Chatterjee, S. 1996. The last dinosaurs of India. In *The Dinosaur Society Report* 12–17.

Chatterjee, S. n.d. *Protoavis* and the early evolution of birds. *Palaeontographica.* In press.

Chatterjee, S., E. N. Kurochkin, and K. E. Mikhailov. n.d. A new embryonic bird from the Cretaceous of Mongolia. *Russian Academy of Sciences.* In press.

Chatterjee, S., and D. K. Rudra. 1996. KT events in India: impact, rifting, volcanism, and dinosaur extinction. In *Proceedings of the Gondwanan Dinosaur Symposium,* 39:489–532. Edited by F. A. Novas and R. E. Molnar. Queensland, Australia: Memoirs of the Queensland Museum.

Chiappe, L. M. 1991. Cretaceous avian remains from Patagonia shed new light on the early radiation of birds. *Alcheringa* 15:333–338.

Chiappe, L. M. 1992. Enantiornithine (Aves) tarsometatarsi and the avian affinities of the Late Cretaceous Avisauridae. *J. Vert. Paleontol.* 12: 344–350.

Chiappe, L. M. 1995a. The first 85 million years of avian evolution. *Nature* 378:349–354.

Chiappe, L. M. 1995b. The phylogenetic position of the Cretaceous birds of Argentina: Enantiornithes and *Patagopteryx deferrariisi. Courier Forsch. Inst. Senckenberg* 181:55–63.

Chiappe, L. M. 1996. Late Cretaceous birds of southern South America: anatomy and systematics of Enantiornithes and *Patagopteryx deferrariisi. Münchner Geowiss. Abh.* A30:203–244.

Chiappe, L. M., and J. O. Calvo. 1994. *Nenquernornis volans,* a Late Cretaceous bird (Enantiornithes: Avisauridae) from Argentina. *J. Vert. Paleontol.* 14:230–246.

Cope, E. D. 1893. A preliminary report on the vertebrate paleontology of the Llano Estacado. *Tex. Geol. Surv. Ann. Rep.* 4:11–87.

Courtillot, V. 1990. A volcanic eruption. *Sci. Am.* 263 (4): 85–92.

Cracraft, J. 1974. Phylogeny and evolution of the ratite birds. *Ibis* 116: 494–521.

Cracraft, J. 1981. Toward a phylogenetic classification of the recent birds of the world (class Aves). *Auk* 98:681–714.

Cracraft, J. 1986. The origin and early diversification of birds. *Paleobiology* 12:383–399.

Cracraft, J. 1988. The major clades of birds. In *The Phylogeny and Classification of the Tetrapods,* 1:339–362. Edited by M. J. Benton. Oxford, England: Clarendon Press.

Cummins, W. F. 1890. The Permian of Texas and its overlying beds. *Tex. Geol. Surv. Ann. Rep.* 1:183–197.

Currie, P. J. 1981. Bird footprints from the Gething Formation (Aptian, Lower Cretaceous) of northeastern British Columbia, Canada. *J. Vert. Paleontol.* 1:257–264.

Darwin, C. 1859. *On the Origin of Species.* London: John Murray.

Davis, P. G., and D. E. G. Briggs. 1995. Fossilization of feathers. *Geology* 23:783–786.

de Beer, G. 1956. The evolution of ratites. *Bull. Br. Mus. Nat. Hist. Zool.* 4:59–70.

de Beer, G. 1958. *Embryos and Ancestors.* Oxford, England: Clarendon Press.

Duncan, R. A., and D. G. Pyle. 1988. Rapid evolution of the Deccan flood basalts at the Cretaceous/Tertiary boundary. *Nature* 333:841–843.

Ellenberger, P. 1972. Contribution à la classification des pistes de vertebres du trias: les types du Stormberg d'Afrique du Sud. *Palaeovert. Mem. Ext. Mont.* 1:1–104.

Elzanowski, A. 1981. Results of the Polish-Mongolian paleontological expeditions—part IX. Embryonic bird skeletons from the Late Cretaceous of Mongolia. *Paleontol. Polon.* 42:147–179.

Elzanowski, A., and P. Wellnhofer. 1996. Cranial morphology of *Archaeopteryx:* evidence from the seventh skeleton. *J. Vert. Paleontol.* 16:81–94.

Feduccia, A. 1980. *The Age of Birds.* Cambridge: Harvard University Press.

Feduccia, A. 1985. On why the dinosaur lacked feathers. In Hecht et al., *The Beginnings of Birds,* 75–79.

Feduccia, A. 1993. Evidence from claw geometry indicating arboreal habits of *Archaeopteryx. Science* 259:790–793.

Feduccia, A. 1995a. The aerodynamic model for the evolution of feathers and feather misinterpretation. *Courier Forsch. Inst. Senckenberg* 181:65–78.

Feduccia, A. 1995b. Explosive evolution in Tertiary birds and mammals. *Science* 267:637–638.

Feduccia, A. 1996. *The Origin and Evolution of Birds.* New Haven: Yale University Press.

Feduccia, A., and H. B. Tordoff. 1979. Feathers of *Archaeopteryx:* asymmetric vanes indicate aerodynamic function. *Science* 203:1021–1022.

Ferrer-Condal, L. F. 1954. Notice préliminaire concernaut la présence d'une plume d'olsean dans le Jurassique superior du Montesch (Province de Lerida, Espagne). In *Acta II Congress International Ornithology*, 268–269.

Forster, C. A., L. M. Chiappe, D. W. Krause, and S. D. Sampson. 1996. The first Cretaceous bird from Madagascar. *Nature* 382:532–534.

Gadow, H. 1893. Vögel. II. Systematischer Theil. In *Bronn's Klassen Ordnungen des Thier-Reichs*, 6 (4): 1–303. Leipzig: C. F. Winter.

Galton, P. M. 1970. Ornithischian dinosaurs and the origin of birds. *Evolution* 24:448–462.

Gatesy, S. M. 1990. Caudofemoral musculature and the evolution of theropod locomotion. *Paleobiology* 16:170–186.

Gauthier, J. 1986. Saurischian monophyly and the origin of birds. In Padian, *Origin of Birds and Evolution of Flight*, 1–55.

Gauthier, J., and K. Padian. 1985. Phylogenetic, functional and aerodynamic analyses of the origin of birds and their flight. In Hecht et al., *The Beginnings of Birds*, 185–197.

Gill, F. B. 1990. *Ornithology.* New York: Freeman.

Gould, S. J. 1977. *Ontogeny and Phylogeny.* Cambridge: Harvard University Press.

Gould, S. J., and E. S. Vrba. 1982. Exaptation—a missing term in the science of form. *Paleobiology* 8:4–15.

Grande, L. 1980. Paleontology of the Green River Formation, with a review of the fish fauna. *Geol. Surv. Wyo. Bull.* 63:1–333.

Grimaldi, D., and G. R. Case. 1995. A feather in amber from the Upper Cretaceous of New Jersey. *Am. Mus. Novitates* 3126:1–5.

Haeckel, E. 1866. *Generelle Morphologie der Organismen: Allgemeine Grundzüge der organischen Formen-Wissenschaft, mechanisch begründet durch die von Charles Darwin reformirte Descendenz-Theorie.* Berlin: Reiner.

Hallam, A. 1987. End-Cretaceous mass extinction agent: argument for terrestrial causation. *Science* 238:1237–1247.

Harris, J. M., and G. T. Jefferson. 1985. *Rancho La Brea: Treasures of the Tar Pits.* Los Angeles: Natural History Museum of Los Angeles County.

Hecht, M. K., J. H. Ostrom, G. Viohl, and P. Wellnhofer, eds. *The Beginnings of Birds.* Eichstätt: Freunde des Jura-Museums.

Heilmann, G. 1926. *The Origin of Birds.* London: Witherby.

Hennig, W. 1966. *Phylogenetic Systematics.* Chicago: University of Illinois Press.

Hildebrand, A., G. T. Penfield, D. A. Kina, M. Pilkington, A. Z. Camaro, S. B. Jacobson, and W. B. Boynton. 1991. Chicxulub crater: a possible

Cretaceous/Tertiary boundary impact crater on the Yucatan Peninsula of Mexico. *Geology* 19:867–871.

Hildebrand, M. 1982. *Analysis of Vertebrate Structure.* New York: John Wiley.

Holtz, T. R., Jr. 1994. The phylogenetic position of the Tyrannosauridae: implication for theropod systematics. *J. Paleontol.* 68:1100–1117.

Hopson, J. A. 1980. Relative brain size in dinosaurs: implications for dinosaurian endothermy. In *A Cold Look at the Warm-blooded Dinosaurs,* 287–310. Edited by R. D. K. Thomas and E. C. Olson. Boulder, Colo.: West View.

Hou, L., and Z. Liu. 1984. A new fossil from Lower Cretaceous of Gansu and early evolution of birds. *Sci. Sin. B* 27:1296–1302.

Hou, L., L. D. Martin, Z. Zhou, and A. Feduccia. 1996. Early adaptive radiation of birds: evidence from fossils from northeastern China. *Science* 274:1164–1167.

Hou, L., Z. Zhou, L. D. Martin, and A. Feduccia. 1995. A beaked bird from the Jurassic of China. *Nature* 377:616–618.

Houde, P., and S. L. Olson. 1981. Paleognathous carinate birds from the Early Tertiary of North America. *Science* 214:1236–1237.

Hoyle, F., and C. Wickramasinghe. 1986. *Archaeopteryx, the Primordial Bird: A Case of Fossil Forgery.* London: Christopher Davis.

Huxley, T. H. 1867. On the classification of birds and on the modification of certain of the cranial bones observable in that class. *Proc. Zool. Soc. Lond.* 1867:415–472.

Huxley, T. H. 1868a. Remarks upon *Archaeopteryx lithographica. Proc. R. Soc. Lond.* 16:243–248.

Huxley, T. H. 1868b. On the animals which are most nearly intermediate between the birds and reptiles. *Ann. Mag. Nat. Hist.* 2:66–75.

Huxley, T. H. 1870. Further evidence of the affinity between dinosaurian reptiles and birds. *Proc. Geol. Soc. Lond.* 26 (1): 12–31.

James, H. F., and S. L. Olson. 1983. Flightless birds. *Nat. Hist.* 92:30–40.

Jenkins, F. A., Jr., K. P. Dial, and G. E. Goslow, Jr. 1988. A cinematographic analysis of bird flight: the wishbone in starling is a spring. *Science* 241:1495–1498.

Jerison, H. J. 1973. *Evolution of the Brain and Intelligence.* New York: Academic Press.

Kaufmann, J. 1970. *Birds in Flight.* New York: William Morrow.

Kellner, A. W. A., J. G. Maisey, and D. A. Campos. 1994. Fossil down feather from the Lower Cretaceous of Brazil. *Paleontology* 37 (3): 489–492.

Kurochkin, E. N. 1985. A true carinate bird from Lower Cretaceous deposits in Mongolia and other evidence of early Cretaceous birds in Asia. *Cretaceous Res.* 6:271–278.

Kurochkin, E. N. 1995. Synopsis of Mesozoic birds and early evolution of class Aves. *Archaeopteryx* 13:47–66.

Kurzanov, S. M. 1985. The skull structure of the dinosaur *Avimimus. Paleontol. J.* 4:81–89.

Kurzanov, S. M. 1987. Avimimidae and the problem of the origin of birds (in Russian). *Trans. Joint Soviet-Mongolian Paleontol. Expeditions* 31:1–87.

Kyte, F. T., J. A. Bostwick, and L. Zhou. 1994. The KT boundary in the Pacific plate. *Lunar Planet. Inst. Contr.* 825:64–65.

Lasca-Ruiz, A. 1989. Nuevo genero de Ave fosil del Yacimiento Neocomiense del Montsec (Provincia de Lerida, Espana). *Estidios Geol.* 45:417–425.

Lehman, T. M. 1994. The saga of the Dockum Group and the case of the Texas/New Mexico boundary fault. *N.M. Bur. Mines Min. Bull.* 150: 37–51.

Lehman, T. M., S. Chatterjee, and J. Schnable. 1992. The Cooper Canyon Formation (Late Triassic) of western Texas. *Tex. J. Sci.* 44:349–355.

Lilienthal, O. 1889. *Der Vögelflug als Grundlage der Fleigekurst.* Berlin: Gaertners.

Linnaes, C. von. 1758. Systema naturae per regna tria naturae, secundum classes, ordines, genera, species, cum characteribus, differentiis, synonimis, locis. Ed. 10, Tom 1–2. Holmiae, impensis direct. Laurentii Salvii. 1758–1759.

Livezey, B. C. 1995. Heterochrony and the evolution of avian flightlessness. In *Evolutionary Change in Heterochrony,* 169–193. Edited by K. J. McNamara. New York: John Wiley.

Lockley, M. G., S. Y. Yang, M. Matsukawa, and F. Fleming. 1992. The track record of Mesozoic birds: evidence and implications. *Philos. Trans. R. Soc. Lond. [Biol.]* 336:113–134.

Lull, R. S. 1953. Triassic life of the Connecticut Valley (revised). *Conn. State Geol. Nat. Hist. Bull.* 81:1–336.

Maderson, F. P. A. 1972. How an archosaurian scale might have given rise to an avian feather. *Am. Naturalist* 146:424–428.

Marsh, O. C. 1877. Introduction and succession of vertebrate life in America. *Am. J. Sci.* 14:337–378.

Marsh, O. C. 1880. Odontornithes: a monograph on the extinct toothed birds of North America. *Rep. Geol. Expl. Fortieth Parallel* 7:1–201.

Marshall, L. G. 1994. The terror birds of South America. *Sci. Am.* 270 (2): 90–95.

Martill, D. M., and J. B. M. Filgueira. 1994. A new feather from the Lower Cretaceous of Brazil. *Palaeontology* 37 (3): 483–487.

Martin, L. D. 1983a. The origin of birds and of avian flight. *Curr. Ornithol.* 1:105–129.

Martin, L. D. 1983b. The origin and early radiation of birds. In *Perspectives in Ornithology*, 291–338. Edited by A. H. Brush and A. Clark, Jr. Cambridge: Cambridge University Press.

Martin, L. D. 1984. A new hesperornithid and the relationships of the Mesozoic birds. *Trans. Kans. Acad. Sci.* 87:141–150.

Martin, L. D. 1985. The relationships of *Archaeopteryx* to other birds. In Hecht et al., *The Beginnings of Birds*, 177–183.

Martin, L. D. 1987. The beginnings of modern avian radiation. In *L'évolution des oiseux d'après le temorguae des fossiles*, 9–19. Edited by C. Mourer-Chauviré. Lyon: Documents des Laboratoires de Geologie de la Faculté des Sciences de Lyon.

Martin, L. D. 1991. Mesozoic birds and the origin of birds. In *Origins of the Higher Groups of Tetrapods*, 485–540. Edited by H.-P. Schultze and L. Trueb. Ithaca: Cornell University Press.

Martin, L. D. 1992. The status of the Late Paleocene birds *Gastornis* and *Remiornis*. In Campbell, *Papers in Avian Paleontology*, 97–108.

Martin, L. D. 1995a. A new skeletal model of *Archaeopteryx*. *Archaeopteryx* 13:33–40.

Martin, L. D. 1995b. The Enantiornithes: terrestrial birds of the Cretaceous. *Courier Forsch. Inst. Senckenberg* 181:23–36.

Martin, P. S. 1984. Prehistoric overkill: the global model. In *Quaternary Extinctions*, 354–403. Edited by P. S. Martin and R. G. Klein. Tucson: University of Arizona Press.

Maynard-Smith, J. 1952. The importance of the nervous system in the evolution of animal flight. *Evolution* 6:127–129.

McKinney, M. L., and K. J. McNamara. 1991. *Heterochrony*. New York: Plenum Press.

Meyer, H. von. 1861a. Vögel-Federn und Palpipes pricus von Solnhofen. *Neues Jahrb. Min. Geol. Paläontol.* 1861:561.

Meyer, H. von. 1861b. *Archaeopteryx lithographica* (Vögel-Feder) und *Pterodactylus* von Solnhofen. *Neues Jahrb. Min. Geol. Paläontol.* 1861: 678–679.

Mikhailov, K. E. 1992. The microstructure of avian and dinosaurian eggshell: phylogenetic implications. In Campbell, *Papers in Avian Paleontology*, 361–374.

Mourer-Chauviré, C. 1992. The Galliformes (Aves) from the Phosphorite du Quercy (France): systematics and biostratigraphy. In Campbell, *Papers in Avian Paleontology*, 67–95.

Nopsca, F. 1907. Ideas on the origin of flight. *Proc. Zool. Soc. Lond.* 1907:223–236.

Nopsca, F. 1923. On the origin of flight in birds. *Proc. Zool. Soc. Lond.* 1923:463–477.

Norberg, U. M. 1985. Evolution of vertebrate flight: an aerodynamic model for the transition from gliding to active flight. *Am. Naturalist* 126:303–327.

Norberg, U. M. 1990. *Vertebrate Flight.* Berlin: Springer-Verlag.

Officer, C. B., A. Hallam, C. L. Drake, J. D. Devine, and A. A. Meyerhoff. 1987. Late Cretaceous and paroxysmal Cretaceous/Tertiary extinctions. *Nature* 236:143–149.

Olmez, I., D. L. Finnegan, and W. H. Zoller. 1986. Iridium emissions from Kilauea volcano. *J. Geophys. Res.* 91:653–664.

Olson, S. L. 1977. A Lower Eocene frigatebird from the Green River Formation of Wyoming. *Smithson. Contr. Paleobiol.* 35:1–33.

Olson, S. L. 1992. *Neogaeornis wetzeli* Lambrecht, a Cretaceous loon from Chile (Aves: Gavidae). *J. Vert. Paleontol.* 12:122–124.

Olson, S. L., and A. Feduccia. 1979. Flight capability and the pectoral girdle of *Archaeopteryx. Nature* 278:247–248.

Olson, S. L., and A. Feduccia. 1980. *Presbyornis* and the origin of Anseriformes. *Smithson. Contr. Zool.* 323:1–24.

Olson, S. L., and H. F. James. 1984. The role of Polynesians in the extinction of the avifauna of the Hawaiian Islands. In *Quaternary Extinctions*, 768–780. Edited by P. S. Martin and R. C. Klein. Tucson: University of Arizona Press.

Olson, S. L., and H. F. James. 1991. Description of thirty-two new species of birds from the Hawaiian Islands, Part 1: Non-Passeriformes. *Ornithol. Monogr.* 45:1–88.

Olson, S. L., and D. C. Parris. 1987. The Cretaceous birds of New Jersey. *Smithson. Contr. Paleobiol.* 63:1–22.

Ostrom, J. H. 1969. Osteology of *Deinonychus antirrhopus*, an unusual theropod from the Lower Cretaceous of Montana. *Bull. Peabody Mus. Nat. Hist.* 30:1–165.

Ostrom, J. H. 1973. The ancestry of birds. *Nature* 242:136.

Ostrom, J. H. 1974. *Archaeopteryx* and the origin of flight. *Q. Rev. Biol.* 49:27–47.

Ostrom, J. H. 1976a. *Archaeopteryx* and the origin of birds. *Biol. J. Linn. Soc.* 8:91–182.

Ostrom, J. H. 1976b. Some hypothetical anatomical stages in the evolution of avian flight. *Smithson. Contr. Paleobiol.* 27:1–21.

Ostrom, J. H. 1976c. On a new specimen of the Lower Cretaceous theropod dinosaur *Deinonychus antirrhopus. Breviora* 439:1–21.

Ostrom, J. H. 1979. Bird flight: how did it begin? *Am. Sci.* 67:46–56.

Ostrom, J. H. 1985a. The meaning of *Archaeopteryx.* In Hecht et al., *The Beginnings of Birds,* 161–176.

Ostrom, J. H. 1985b. The Yale *Archaeopteryx:* the one that flew the coop. In Hecht et al., *The Beginnings of Birds,* 359–369.

Ostrom, J. H. 1986. The cursorial origin of avian flight. In Padian, *Origin of Birds and Evolution of Flight,* 73–81.

Ostrom, J. H. 1990. Dromaeosauridae. In *The Dinosauria,* 269–279. Edited by D. B. Weishampel, P. Dodson, and H. Osmólska. Berkeley: University of California.

Ostrom, J. H. 1991. The question of the origin of birds. In *Origins of the Higher Groups of Tetrapods,* 467–484. Edited by H.-P. Schultze and L. Trueb. Ithaca: Cornell University Press.

Owen, R. 1863. On the *Archaeopteryx* of von Meyer, with a description of the fossil remains of a long-tailed species from the lithographic stone of Solnhofen. *Philos. Trans. R. Soc. Lond.* 153:33–47.

Padian, K. 1982. Macroevolution and the origin of major adaptations: vertebrate flight as a paradigm for the analysis of patterns. *Proceedings of the Third North American Paleontological Convention* 2:387–392.

Padian, K. 1985. The origins and aerodynamics of flight in extinct vertebrates. *Palaeontology* 28:413–433.

Padian, K., ed. 1986. *The Origin of Birds and the Evolution of Flight.* San Francisco: California Academy of Sciences.

Padian, K. 1987. A comparative phylogenetic and functional approach to the origin of vertebrate flight. In *Recent Advances in the Study of Bats,* 3–22. Edited by M. B. Fenton, R. Racey, and J. M. V. Rayner. Cambridge: Cambridge University Press.

Padian, K. 1992. A proposal to standardize the tetrapod phalangeal formula designation. *J. Vert. Paleontol.* 12:260–262.

Parkes, K. C. 1966. Speculations on the origin of feathers. *Living Bird* 5:77–86.

Paul, G. S. 1988. *Predatory Dinosaurs of the World.* New York: Simon and Schuster.

Penfield, G. T., and Z. A. Camagro. 1982. Definition of a major zone in the central Yucatan platform with aeromagnetics and gravity. *Geophysics* 47:448–449.

Pennycuick, C. J. 1972. *Animal Flight.* London: Arnold.

Pennycuick, C. J. 1986. Mechanical constraints on the evolution of flight. In Padian, *Origin of Birds and Evolution of Flight*, 83–98.

Perle, A., M. A. Norell, L. M. Chiappe, and J. M. Clark. 1993. Flightless bird from the Cretaceous of Mongolia. *Nature* 362:623–626.

Peters, D. S. 1992. Messel birds: a land-based assemblage. In *Messel: An Insight into the History of Life and of the Earth*, 137–151. Edited by S. Schaal and W. Ziegler. Oxford, England: Clarendon Press.

Peters, D. S. 1994. Die Enstehung der Vögel verändern die jüngstern Fossilfunde das Modell? *Senckenberg-Buch* 70:403–424.

Peters, D. S. 1995. *Idiornis tuberculata* n. spec., ein weiterer ungewöhnlicher Vögel aus der Grube Messel (Aves: Gruiformes: Cariamidae: Idiornithinae). *Courier Forsch. Inst. Senckenberg* 181:107–120.

Pettigrew, J. D. 1979. Binocular visual processing in the owl's telencephalon. *Proc. R. Soc. Lond. [Biol.]* 204:435–454.

Pettigrew, J. D., and B. J. Frost. 1985. A tactile fovea in the Scolopacidae? *Brain Behav. Evol.* 26:185–195.

Raath, M. A. 1977. The anatomy of the Triassic theropod *Syntarsus rhodesiensis* (Saurischia: Podokesauridae) and a consideration of its biology. Ph.D. diss. Rhodes University, Grahamstown, South Africa.

Raath, M. A. 1985. The theropod *Syntarsus* and its bearing on the origin of birds. In Hecht et al., *The Beginnings of Birds*, 219–227.

Rahn, H., A. Ar, and C. Paganelli. 1979. How bird eggs breathe. *Sci. Am.* 242 (2): 46–55.

Rahn, H., C. V. Paganelli, and A. Ar. 1975. Relation of avian egg weight to body weight. *Auk* 92:750–765.

Rawles, M. E. 1963. Tissue interactions in scale and feather development as studied in dermal-epidermal recombinations. *J. Emb. Exp. Morphol.* 11:765–789.

Rayner, J. M. V. 1979. A new approach to animal flight mechanics. *J. Exp. Biol.* 80:17–54.

Rayner, J. M. V. 1981. Flight adaptations in vertebrates. In *Vertebrate Locomotion*, 137–172. Edited by M. H. Day. New York: Academic Press.

Rayner, J. M. V. 1985. Mechanical and ecological constraints on flight evolution. In Hecht et al., *The Beginnings of Birds*, 279–288.

Rayner, J. M. V. 1988. The evolution of vertebrate flight. *Biol. J. Linn. Soc.* 34:269–287.

Rayner, J. M. V. 1989. Mechanics and physiology of flight in fossil vertebrates. *Trans. R. Soc. Edinb.* 80:311–320.

Rayner, J. M. V. 1991. Avian flight evolution and the problem of *Archaeopteryx*. In *Biomechanics and Evolution*, 183–212. Edited by J. M. V. Rayner and R. J. Wootton. Cambridge: Cambridge University Press.

Regal, P. J. 1975. The evolutionary origin of feathers. *Q. Rev. Biol.* 50:35–66.

Riggs, N. R., T. M. Lehman, G. E. Gehrels, and W. R. Dickinson. 1996. Detrital zircon link between headwaters and terminus of the Upper Triassic Chinle-Dockum paleoriver system. *Science* 273:97–100.

Robin, E., J. Garyrand, L. Froget, and R. Rocchia. 1994. On the origin of the regional variation in spinel compositions at the KT boundary. *Lunar Planet. Inst. Contr.* 825:96–97.

Rosowki, J. J., and J. C. Saunders. 1980. Sound transmission through the avian interaural pathway. *J. Comp. Physiol.* 136:183–190.

Royte, E. 1995. Hawaii's vanishing species. *Natl. Geogr.* 188 (3): 2–37.

Ruben, J. 1991. Reptilian physiology and the flight capacity of *Archaeopteryx. Evolution* 45:1–17.

Rüppell, G. 1977. *Bird Flight.* New York: Van Nostrand.

Sanz, J. L., and J. F. Bonaparte. 1992. A new order of birds (class Aves) from the Lower Cretaceous of Spain. In Campbell, *Papers in Avian Biology*, 39–49.

Sanz, J. L., and A. D. Buscalioni. 1992. A new bird from the Early Cretaceous of Las Hoyas, Spain, and the early radiation of birds. *Palaeontology* 35:829–845.

Sanz, J. L., L. M. Chiappe, B. P. Pérez-Moreno, A. D. Buscalioni, J. J. Moratalla, F. Ortega, and F. J. Poyato-Aroza. 1996. An Early Cretaceous bird from Spain and its implications for the evolution of avian flight. *Nature* 382:442–445.

Saville, D. B. O. 1962. Gliding and flight in the vertebrates. *Am. Zool.* 2:161–166.

Schlee, D. 1973. Harzkonservierte fossile Vögelfedern aus der untersten Kreide. *J. Ornithol.* 114:207–219.

Schultz, P. H., and D. E. Gault. 1990. Prolonged global catastrophes from oblique impacts. *Geol. Soc. Am. Spec. Pap.* 247:239–261.

Seeley, H. G. 1876. On the British fossil Cretaceous birds. *Q. J. Geol. Soc. Lond.* 32:496–512.

Sereno, P. C. 1991. Basal archosaurs: phylogenetic relationships and functional implications. *J. Vert. Paleontol.* 11 (4): 1–52.

Sereno, P. C., and A. B. Arcucci. 1993. Dinosaur precursor from the Middle Triassic of Argentina. *Marasuchus lilloensis,* gen nov. *J. Vert. Paleontol.* 14:53–73.

Sereno, P. C., C. A. Forster, R. R. Rogers, and A. M. Monetta. 1993. Primitive dinosaur skeleton from Argentina and the early evolution of Dinosauria. *Nature* 361:64–66.

Sereno, P. C., and C. Rao. 1992. Early evolution of avian flight and perching: new evidence from the Lower Cretaceous of China. *Science* 255:845–848.

Sharov, A. G. 1970. An unusual reptile from the Lower Triassic of Fergana. *Paleontol. J.* 1:112–116.

Sharpton, V. L., and P. D. Ward, eds. 1990. Global catastrophes in earth history. *Geol. Soc. Am. Spec. Pap.* 247:1–631.

Shoemaker, E. M., and C. S. Shoemaker. 1994. The crash of P/Shoemaker-Levy 9 into Jupiter and its implications for comet bombardment on Earth. *Lunar Planet. Inst. Contr.* 825:113–114.

Sibley, C. G., and J. E. Ahlquist. 1990. *Phylogeny and Classification of Birds: A Study in Molecular Evolution.* New Haven: Yale University Press.

Sibley, C. G., and B. L. Monroe. 1990. *Distribution and Taxonomy of Birds of the World.* New Haven: Yale University Press.

Signor, P. W., III, and J. M. Lipps. 1982. Sampling bias, gradual extinction patterns, and catastrophes in the fossil record. *Geol. Soc. Am. Spec. Pap.* 190:291–296.

Silver, L. T., and P. H. Schultz, eds. 1982. Geological implications of impacts of large asteroids and comets on the earth. *Geol. Soc. Am. Spec. Pap.* 190:1–528.

Stanley, S. M. 1987. *Extinction.* New York: Freeman.

Starck, J. M. 1993. Evolution of avian ontogenies. *Curr. Ornithol.* 10:275–366.

Steadman, D. W. 1995. Prehistoric extinctions of Pacific Island birds: biodiversity meets zooarchaeology. *Science* 267:1123–1131.

Steadman, D. W., and P. S. Martin. 1984. Extinction of birds in the Late Pleistocene of North America. In *Quaternary Extinctions,* 466–477. Edited by P. S. Martin and R. G. Klein. Tucson: University of Arizona Press.

Stolpe, M. 1932. Physiologisch-anatomische Untersuchungen über die hintere Extremität der Vögel. *J. Ornithol.* 80:161–247.

Storer, R. W. 1956. The fossil loon *Columboides minutus. Condor* 58: 413–426.

Tarsitano, S. F., and M. Hecht. 1980. A reconsideration of the reptilian relationships of *Archaeopteryx. Zool. J. Linn. Soc.* 69:149–182.

Tennekes, H. 1996. *The Simple Science of Flight: From Insects to Jumbo Jets.* Cambridge: MIT Press.

Thulborn, R. A. 1985. Birds as neotenous dinosaurs. *Rec. N.Z. Geol. Surv.* 9:90–92.

Thulborn, R. A., and T. L. Hamley. 1985. A new palaeoecological role for *Archaeopteryx.* In Hecht et al., *The Beginnings of Birds,* 81–89.

Unwin, D. M. 1988. Extinction and survival in birds. In *Extinction and*

Survival in the Fossil Record, 295–318. Edited by G. P. Larwood. Oxford, England: Clarendon Press.

Vandamme, D., and V. Courtillot. 1992. Paleomagnetic constraints on the structure of the Deccan Traps. *Phys. Earth Planet. Inst.* 74:241–261.

Vasquez, R. J. 1992. Functional osteology of the avian wrist and the evolution of flapping flight. *J. Morphol.* 211:259–268.

Vasquez, R. J. 1994. The automating skeletal and muscular mechanisms of the avian wing. *Zoomorphology* 114:59–71.

Viohl, G. 1985. Geology of the Solnhofen Lithographic Limestones and the habitat of *Archaeopteryx.* In Hecht et al., *The Beginnings of Birds,* 31–44.

Wagner, J. A. 1862. On a new fossil reptile supposed to be furnished with feathers. Translated by W. S. Dallas. *Ann. Mag. Nat. Hist. 3rd Ser.* 9: 261–267.

Waldman, M. 1970. A third specimen of a Lower Cretaceous feather from Victoria, Australia. *Condor* 72:377.

Walker, A. D. 1972. New light on the origin of birds and crocodiles. *Nature* 237:257–263.

Walker, A. D. 1974. Evolution organic. In *McGraw-Hill Year Book of Science and Technology, 1974,* 177–179. New York: McGraw-Hill.

Walker, A. D. 1977. Evolution of the pelvis in birds and dinosaurs. In *Problems in Vertebrate Evolution,* 319–357. Edited by S. M. Andrews, R. S. Miles, and A. D. Walker. Linnaes Society Symposium Series No. 4. London: Linnaes Society.

Walker, A. D. 1985. The braincase of *Archaeopteryx.* In Hecht et al., *The Beginnings of Birds,* 123–134.

Walker, C. A. 1981. New subclass of birds from the Cretaceous of South America. *Nature* 92:51–53.

Walls, G. L. 1963. *The Vertebrate Eye and Its Adaptive Radiation.* London: Hufner.

Weems, R. E., and P. G. Kimmel. 1993. Upper Triassic reptile footprints and coelacanth fish scales from the Culpepper Basin, Virginia. *Proc. Biol. Soc. Wash.* 106 (2): 390–401.

Weishampel, D. B., P. Dodson, and H. Osmólska, eds. 1990. *The Dinosauria.* Berkeley: University of California Press.

Wellnhofer, P. 1974. Das funfte Skelettexemplar von *Archaeopteryx. Palaeontographica A* 147:169–216.

Wellnhofer, P. 1990. *Archaeopteryx. Sci. Am.* 262(5):70–77.

Wellnhofer, P. 1993. Das siebte Exemplar von *Archaeopteryx* aus den Solnhofener Schichlen. *Archaeopteryx* 11:1–47.

Williston, S. W. 1879. Are birds derived from dinosaurs? *Kans. City Rev. Sci.* 3:457–460.

Wilson, E. O. 1992. *The Diversity of Life.* New York: Norton.

Witmer, L. M. 1990. The craniofacial air sac system of Mesozoic birds (Aves). *Zool. J. Linn. Soc.* 100:327–378.

Witmer, L. M. 1991. Perspectives on avian origins. In *Origins of the Higher Groups of Tetrapods,* 427–466. Edited by H. P. Schultze and L. Trueb. Ithaca: Cornell University Press.

Witmer, L. M., and L. D. Martin. 1987. The primitive features of the avian palate with special reference to Mesozoic birds. In *L'évolution des oiseaux d'après le temorguae des fossiles,* 9–19. Edited by C. Mourer-Chauviré. Lyon: Documents des Laboratoires de Geologie de la Faculté des Sciences de Lyon.

Wolbach, W. S., I. Gilmore, E. Angers, C. J. Orth, and R. R. Book. 1988. Global fire at the Cretaceous-Tertiary boundary. *Nature* 324:665–669.

Yalden, D. W. 1985. Forelimb function in *Archaeopteryx.* In Hecht et al., *The Beginnings of Birds,* 91–97.

Zhou, Z. 1995a. The discovery of Early Cretaceous birds in China. *Courier Forsch. Inst. Senckenberg* 181:9–22.

Zhou, Z. 1995b. Is *Mononykus* a bird? *Auk* 112:958–963.

Zhou, Z., F. Jin, and J. Y. Zhang. 1992. Preliminary report on a Mesozoic bird from Liaoning, China. *Chinese Sci. Bull.* 37:1365–1368.

Zou, H., and L. Niswander. 1996. Requirement for BMP signaling in interdigital apoptosis and scale formation. *Science* 272:738–741.

Index